Advanced Research on Structure-Function Relationships of Membrane Proteins, 2nd Edition

Advanced Research on Structure-Function Relationships of Membrane Proteins, 2nd Edition

Guest Editors

Akira Naito

Izuru Kawamura

MDPI

Basel • Beijing • Wuhan • Barcelona • Belgrade • Novi Sad • Cluj • Manchester

Guest Editors
Akira Naito
Graduate School of Engineering
Yokohama National Universuty
Yokohama
Japan

Izuru Kawamura
Graduate School of Engineering
Yokohama National University
Yokohama
Japan

Editorial Office
MDPI AG
Grosspeteranlage 5
4052 Basel, Switzerland

This is a reprint of the Special Issue, published open access by the journal *Membranes* (ISSN 2077-0375), freely accessible at: https://www.mdpi.com/journal/membranes/special_issues/Membrane_Proteins_Volume_II.

For citation purposes, cite each article independently as indicated on the article page online and as indicated below:

Lastname, A.A.; Lastname, B.B. Article Title. *Journal Name* **Year**, *Volume Number*, Page Range.

ISBN 978-3-7258-8025-6 (Hbk)
ISBN 978-3-7258-8026-3 (PDF)
https://doi.org/10.3390/books978-3-7258-8026-3

Contents

About the Editors

Akira Naito

Akira Naito is a Professor emeritus at the Graduate School of Engineering of Yokohama National University. He is studying the structure–function relationship of biologically active molecules using mainly high-resolution solid-state NMR spectroscopy. In the first, the three-dimensional structure of [^{13}C, ^{15}N]-labeled Leu-enkephaline (Tyr-Gly-Gly-Phe-Leu) dehydrate on the basis of six sets of accurately determined ^{13}C-^{15}N interatomic distances by rotational echo double resonance (REDOR) and some additional constraints obtained from ^{13}C chemical shifts. In the second, using the magnetically oriented vesicle systems, dynamic structures of bee venom melittin bound to the vesicles were investigated by analyzing the anisotropic and isotropic ^{13}C chemical shifts of selectively ^{13}C-labeled carbonyl carbons of melittin under the static and magic-angle spinning conditions. In the third, Amyloid fibril formation mechanism of human calcitonin (hCT) was investigated by high-resolution solid-state ^{13}C NMR spectroscopy. In the fourth, photo-intermediates of photoreceptor membrane proteins have been detected using a novel in situ photo-irradiation solid-state NMR. In the fifth, A novel two-dimensional NMR experiment, state–correlated two-dimensional NMR (SC-2D NMR) spectroscopy, was successfully developed and performed to obtain SC-2D NMR spectrum between the nematic phase and the isotropic phase of 4′-methoxybenzylidene-4-acetoxyaniline (APAPA) by using pulsed microwave heating.

Izuru Kawamura

Izuru Kawamura is a Professor at Yokohama National University.

He received his Ph.D. in Engineering from Yokohama National University in 2007 (Supervisor: Prof. Akira Naito). After serving as a researcher at Yokohama National University and as a visiting researcher at the University of Guelph, Canada (Supervisors: Prof. V. Ladizhansky and Prof. L.S. Brown), he was appointed Assistant Professor at Yokohama National University in 2012 and promoted to Associate Professor there in 2013. He has been a Professor at the Faculty of Engineering, Yokohama National University, since 2024. His research interests include biophysical chemistry and solid-state nuclear magnetic resonance spectroscopy of membrane proteins and membrane-bound peptides. In particular, he has been actively advancing NMR research on microbial rhodopsins and antimicrobial peptides containing D-amino acids. He is also engaged in the development of bio-materials based on polymer science through structural and physicochemical analyses of cellulose nanofibers derived from food and agricultural waste. He received the Progress Award from the Nuclear Magnetic Resonance Society of Japan in 2023.

Preface

This Special Issue focuses on advanced studies on structure–function relationships using advanced methods such as Solid-state NMR, Solution NMR, X-ray crystallography, and Cryo-electron microscopy to determine the high-resolution structure of membrane proteins. Within the scope of this Special Issue, it is shown that such advanced methods play important roles not only in determining the complete structure but also in local structure changes and dynamic properties of membrane proteins. This information gives us insight of the structure–function relationship of membrane proteins. In this Special Issue 2nd Edition, functions of large-sized membrane proteins are also extensively discussed for such as transporter and photosynthetic reaction center membrane proteins.

Akira Naito and Izuru Kawamura
Guest Editors

membranes

Review

Roles of a Glycolipid MPIase in Sec-Independent Membrane Protein Insertion

Kaoru Nomura [1,*], Shoko Mori [1] and Keiko Shimamoto [1,2]

1 Bioorganic Research Institute, Suntory Foundation for Life Sciences, 8-1-1 Seikadai, Seika-cho, Soraku-gun, Kyoto 619-0284, Japan; mori@sunbor.or.jp (S.M.); shimamot@sunbor.or.jp (K.S.)
2 Department of Chemistry, Graduate School of Science, Osaka University, 1-1 Machikaneyama, Toyonaka, Osaka 560-0043, Japan
* Correspondence: nomura@sunbor.or.jp

Abstract: Membrane protein integrase (MPIase), an endogenous glycolipid in *Escherichia coli* (*E. coli*) membranes, is essential for membrane protein insertion in *E. coli*. We have examined Sec-independent membrane protein insertion mechanisms facilitated by MPIase using physicochemical analytical techniques, namely solid-state nuclear magnetic resonance, fluorescence measurements, and surface plasmon resonance. In this review, we outline the physicochemical characteristics of membranes that may affect membrane insertion of proteins. Subsequently, we introduce our results verifying the effects of membrane lipids on insertion and estimate the impact of MPIase. Although MPIase is a minor component of *E. coli* membranes, it regulates insertion by altering the physicochemical properties of the membrane. In addition, MPIase promotes insertion by interacting with substrate proteins. We propose comprehensive mechanisms for the membrane insertion of proteins involving MPIase, which provide a physicochemical basis for understanding the roles of glycolipids in protein translocation.

Keywords: glycolipids; membrane protein; Sec-independent insertion; membrane lipids; physicochemical property of membrane

1. Introduction

Efficient functioning of membrane proteins relies on their accurate insertion into the membrane. The mechanisms of membrane insertion have been the subject of considerable research. Membrane insertion depends on the size of the protein to be inserted. Large membrane proteins, which are characterized by multiple transmembrane domains, require Sec translocons, which are membrane protein complexes that form pores for proteins to permeate the membrane (Sec-dependent membrane insertion process) [1,2]. Since highly homologous Sec translocons are present in all organisms, the mechanism of Sec-dependent insertion is considered to be basically conserved. In contrast, Sec-independent processes (Figure 1a) have been shown to occur in a subset of small membrane proteins and tail-anchored (TA) membrane proteins that cannot use translocons [3,4]. In eukaryotes, chaperones for membrane targeting of TA proteins have been identified [5,6]. However, because such chaperones have not been found in *Escherichia coli* (*E. coli*), the Sec-independent process in *E. coli* was previously considered to be a spontaneous process. Then, it became evident that spontaneous insertion into liposomes composed only of commercially available *E. coli* phospholipids (EPLs) does not reflect the correct biological phenomena. For example, mannitol permease (MtlA), a Sec-dependent protein, spontaneously inserted into EPL liposomes even without the Sec translocon, whereas such improper insertion did not occur in the inner membrane vesicles [7]. These results suggested that a blocking factor for spontaneous insertion was present in the inner membrane of *E. coli*. Diacylglycerol (DAG) was subsequently identified as a factor that blocks unregulated spontaneous insertion into EPL membranes. DAG is a simple molecule with two fatty acid chains covalently bonded to a glycerol moiety via ester bonds. Nishiyama et al. reported that spontaneous membrane insertion of proteins was significantly inhibited in EPL liposomes containing a physiological concentration of DAG (2–3 *w*/*w*% of EPLs) [7,8]. Furthermore, in a later study, they discovered that a minor component of *E. coli* membranes, membrane protein integrase (MPIase), restored Sec-independent insertions that were suppressed by DAG [7,9,10]. MPIase is also involved in Sec-dependent integration and translocation [7,11]. Despite its name, structural analysis revealed that MPIase is a novel glycolipid composed of a DAG anchor moiety, a long sugar chain consisting of approximately ten trisaccharide units, and a pyrophosphate linker (Figure 1b) [10]. MPIase dose-dependently enhances Sec-independent insertion into DAG-containing liposomes [9].

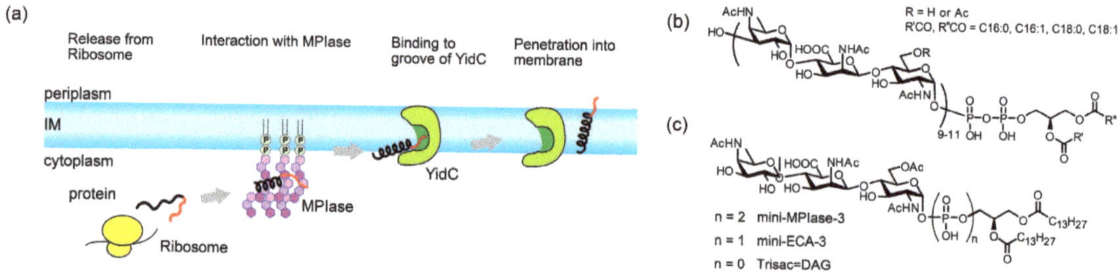

Figure 1. (**a**) Schematic model of the Sec translocon-independent membrane protein integration pathway into the inner membrane (IM) of *E. coli*. (black: hydrophobic region, red: hydrophilic region). MPIase is presumed to function prior to an insertase YidC. (**b**) Molecular structure of MPIase. Approximately one-third of the 6 position on GlcNAc residues is *O*-acetylated. The number of repeating trisaccharide units ranges from 7 to 14, but most are from 9 to 11. (**c**) Molecular structures of synthesized MPIase analogs.

Moreover, in the Sec-independent insertion process, substrate proteins require the help of a proteinaceous factor called YidC as the hydrophilic proportion of the substrate proteins increases [12–14]. YidC transiently captures the hydrophilic regions of a substrate protein in its hydrophilic groove in the membrane (Figure 1a) [15]. In addition, the short transmembrane domain of YidC locally thins the lipid bilayer, facilitating the crossing of the hydrophilic region of the protein through the bilayer [12,16,17]. Thus, YidC is recognized as an insertase; however, most previous studies have been conducted in the absence of DAG and MPIase. Nishiyama et al. reported that substrate proteins were not inserted into DAG-containing liposomes without MPIase, even in the presence of YidC, whereas insertion was dependent on MPIase without YidC. In contrast, when YidC coexists with MPIase in DAG-containing liposomes, it accelerates insertion [18,19]. Therefore, we consider that MPIase functions at the initial step of protein integration into the membrane, whereas YidC completes insertion later.

To date, several MPIase analogs have been synthesized to elucidate insertion mechanisms [20,21]. A minimal structural unit of MPIase, mini-MPIase-3 (Figure 1c), consisting of only one trisaccharide unit, displayed significant activity, although it was less efficient than natural MPIase [21]. Trisac-DAG, a mini-MPIase-3 analog lacking pyrophosphate, and mini-ECA-3, which has a monophosphate linker instead of a pyrophosphate linker (Figure 1c), exhibited lower integration activity than mini-MPIase-3. Structure–activity relationship studies have enabled us to analyze the insertion process at the atomic level using natural MPIase and its synthetic analogs [20,21]. We examined the Sec-independent membrane protein insertion mechanisms facilitated by MPIase based on physicochemical analytical techniques such as solid-state nuclear magnetic resonance (NMR), fluorescence measurements, and surface plasmon resonance (SPR) [22–24].

In this review, we describe the mechanism of the initial step of Sec-independent membrane insertion induced by MPIase. Before discussing the role of MPIase, we outline the physicochemical characteristics of membranes that may affect protein membrane insertion. Subsequently, we introduce our results to verify the effects of membrane properties on insertion using physicochemical analytical methods and describe how MPIases support membrane insertion into DAG-containing membranes. Next, we discuss the interactions between MPIase and membrane proteins, which affect the efficiency prior to insertion. Finally, the proposed insertion mechanism facilitated by MPIase is described in detail.

2. Factors Influencing Membrane Protein Insertion

Membrane insertion that does not require Sec translocons is affected by the physicochemical characteristics of the membranes. Historically, the membrane insertion of peptides, especially antimicrobial peptides (AMPs), has been extensively studied in various membranes. In this section, we describe the general effects of membrane lipids on alterations in membrane properties.

2.1. Charges of Membrane Lipids

Membrane insertion can be affected by the charges of membrane lipids. The polar head of phospholipids commonly has a negatively charged monovalent phosphate diester group. The net charge of the lipids depends on the attached functional group; phosphatidylcholine (PC) and phosphatidylethanolamine (PE), which have positively charged choline and ethanolamine groups, respectively, are neutral, whereas phosphatidylglycerol (PG), which has an uncharged glycerol group, and phosphatidylserine (PS), which has a twitterionic serine residue, are acidic with negative net charges. The lipid composition varies considerably among organisms; eukaryotic cell membranes contain large amounts of PC and cholesterol, whereas bacterial cell membranes are rich in PE and PG. In general, positively charged peptides exhibit a higher insertion ability into negatively charged membranes than into electrically neutral membranes. For example, many positively charged AMPs demonstrate greater binding and insertion into PG-rich membranes than into PC-only membranes [25,26]. This difference is thought to be one of the reasons why antimicrobial peptides selectively disrupt bacterial membranes.

2.2. Acyl Chain Ordering

The acyl chain ordering of membrane lipids also influences membrane insertion. In cholesterol-containing membranes, the acyl chains of the membrane lipids are ordered. Since the polar head of cholesterol is a hydroxyl group, which is much smaller than that of phospholipids, cholesterol tends to fit between lipids with large head groups, such as PC. Under the "umbrella" of the phospholipid head groups, the acyl chains of the phospholipid and sterol ring of cholesterol become tightly packed, making insertion difficult [27]. Indeed, the presence of cholesterol has been shown to significantly inhibit peptide membrane binding and insertion [25]. Cholesterol is abundant in mammalian membranes but absent in bacterial membranes. As mentioned later, in *E. coli* membranes, DAG functions in a manner similar to that of cholesterol.

2.3. Spontaneous Lipid Curvature

Non-cylindrical lipids in the bilayers tend to modify the physicochemical properties of membranes. When non-cylindrical lipids form a bilayer, their curvature differs from the spontaneous curvature of the lipid. To compensate for this difference, lateral pressure was induced in the membrane. For example, a membrane composed of cone-shaped lipids (lipids with a small head group and larger acyl chain area) increases the lateral pressure in the acyl chain region, whereas that of inverted cone-shaped lipids (lipids with a large head group and smaller acyl chain area) results in higher lateral pressure in the head group region. Alterations in the lateral pressure influence the ability to insert a protein into the membrane. The depth of peptide membrane insertion depends on the spontaneous curvature of the lipids constituting the membrane [28]. In membranes containing lipids with negative spontaneous curvature, peptides bind easily to the membrane surface because of the lower pressure in the region of the head group. However, deeper insertions of peptides become difficult owing to the higher lateral pressure inside the membrane. In contrast, in membranes containing lipids with a positive spontaneous curvature, peptides have difficulty binding to the membrane surface; however, once bound, the peptides can easily penetrate deeper into the bilayer.

In addition, binding and insertion into membranes containing cone-shaped PE depends on the type of peptide. The intervesicular transfer of the hydrophobic transmembrane helix peptide, $(AALALAA)_3$, decreased as the PE fraction increased [29]. The tightness of the PE membranes hampered their transfer. For the amphipathic α-helical peptide, Ac-18A-NH_2, however, there was an increased partitioning of the peptide to the membrane correlating with the rise in the PE mole fraction [30]. In this case, the positively charged side of the amphipathic peptide was in contact with the hydrophilic membrane surface. Although the small head group of PE is advantageous for accepting the peptide, the amphipathic peptide could not be further embedded in the hydrophobic core region and accumulated on the membrane surface.

2.4. Membrane Defects

The induction of positive curvature in the membrane causes structural defects. With structural defects, the hydrophobic core of the membrane is more exposed to the membrane surface, making it easier for the hydrophobic side chains of the peptides to interact with the membrane. For example, the amphiphilic helical peptide EpN-18, derived from the N-terminus of epsin-1, induces positive membrane curvature. Coexistence with EpN-18 enhances the plasma membrane permeability of the octa-arginine peptide, possibly because EpN-18 increases structural defects in the membrane, exposing its hydrophobic core to the membrane surface [31,32].

2.5. Conformational Changes of Proteins Associated with Membranes

Conformational changes are often observed when peptides bind into membranes. When an amphiphilic peptide is inserted into the membrane, favorable interactions between nonpolar residues and the lipid tail and between charged residues and the solvent molecules or charged sites of the lipid head group convert the peptide from a random coil to a helix. In some cases, amphiphilic peptides are further inserted into the hydrophobic core. Hydrophobic peptides are readily inserted into the hydrophobic core of the membrane and are simultaneously converted to the α-helix structure.

3. Effect of Membrane Properties on Membrane Insertion

In this section, the verification of the effect of membrane properties on membrane insertion using a Sec-independent substrate is described. Generally, membrane proteins have one or two hydrophobic transmembrane regions and basic amino acid residues at the cytoplasmic loop, which is known as the "positive-inside rule" [33,34]. As a model substrate for Sec-independent integration, we have used the Pf3 coat protein (hereafter referred to as Pf3 coat), a major coat protein of the bacteriophage Pf3, and its derivatives (Figure 2a). To fully penetrate the extramembrane region, the insertion of intact Pf3 coat into the lipid bilayer sometimes requires YidC or a membrane potential. Because we focused on the initial insertion process, truncated Pf3 coat (Pf3_24) corresponding to the transmembrane region was used as the simplest model for verification.

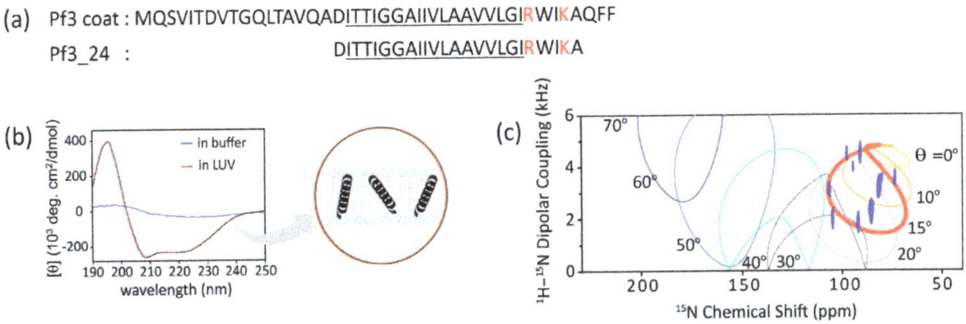

Figure 2. (**a**) Amino acid sequence of Pf3_24 and substructure of Pf3 coat, Pf3_24. Hydrophobic region is underlined, and basic residues are colored in red. (**b**) Circular dichroism (CD) spectra of Pf3_24 in 1,2-dimyristoyl-*sn*-glycero-3-phosphocholine (DMPC) liposomes (dark red) or in buffer (blue) [23]. (**c**) ^{1}H–^{15}N SAMPI4 spectrum of Pf3_24 in DMPC/1,2-diheptanoyl-*sn*-glycero-3-phosphocholine (DHPC) bicelles [23]. The simulated PISA wheels with varied tilt angles θ (0–70°) were overlaid onto the spectrum.

3.1. Conformational Changes and Topology of Pf3_24 in Membranes

Typically, the transmembrane regions of membrane proteins are inclined to adopt α-helical or β-strand conformations. Circular dichroism (CD) spectra showed that Pf3_24 adopts an α-helical conformation when inserted into the lipid bilayer (Figure 2b, dark red), although it is difficult to determine the secondary structure of Pf3_24 in buffer solution due to its tendency to form aggregates and precipitates (Figure 2b, blue). Solid-state NMR is a useful technique for obtaining the accurate tilt angles of helical peptides or helical parts of proteins with respect to the membrane. The topology of Pf3_24 in bicelles was determined using ^{1}H–^{15}N SAMPI4 spectra (Figure 2c) [35]. A characteristic PISA wheel pattern of Pf3_24 (Figure 2c, blue) implies the formation of an α-helical structure in bicelles [36]. The insertion angle of Pf3_24 into the membrane from the bilayer normal was determined to be approximately 15° by fitting to account for the wobbling motion of the bicelles. Based on its angle and hydrophobicity, Pf3_24 directly accessed the deeper regions of the membrane. Simultaneously, conversion to the α-helix structure likely occurs.

3.2. Insertion of Pf3_24 into Membranes

We estimated insertion efficiency based on the strong correlation between ^{15}N NMR chemical shifts and protein secondary structure [37,38]. It was reported that valine residues take the averaged ^{15}N chemical shift values 123.27 ppm, 119.66 ppm, and 119.53 ppm for the β-strand, random coil, and α-helix, respectively [37]. As aforementioned, Pf3_24 has a complete α-helical structure in large unilamellar vesicles (LUVs) of 1,2-dimyristoyl-*sn*-glycero-3-phosphocholine (DMPC) (Figure 3a, top), which was verified by the chemical shift value for valine. Conversely, in the absence of membranes, Pf3_24 formed aggregates, yielding a stronger signal at 125 than at 120 ppm, which indicate that most of it adopted β-strand-rich structures, whereas some parts formed random coils and α-helical structures (Figure 3a, middle). To observe membrane insertion, as shown in Figure 3b, Pf3_24 solubilized in 1,2-diheptanoyl-*sn*-glycero-3-phosphocholine (DHPC) solution was mixed with LUVs composed of various lipid types. During incubation, some Pf3_24 was inserted into the LUVs, whereas the remainder formed aggregates outside the LUVs. The third spectrum was obtained from the experiment conducted in the presence of DMPC LUVs (Figure 3a, bottom). We considered that the third spectrum represents the weighted sum of the first and second spectra, with the coefficient of the first spectrum corresponding to the insertion efficiency. Thus, the relative peak intensities corresponding to the α-helix and the aggregates in the third spectrum provided an insertion efficiency of 46% for the DMPC membrane. Similarly, we calculated the insertion efficiency values for membranes with various lipid compositions (Figure 3c). However, this approach is applicable for each membrane system when the protein forms a complete α-helical conformation in the membrane and a β-sheet rich conformation outside the membrane with a constant ratio of conformations in the aggregates.

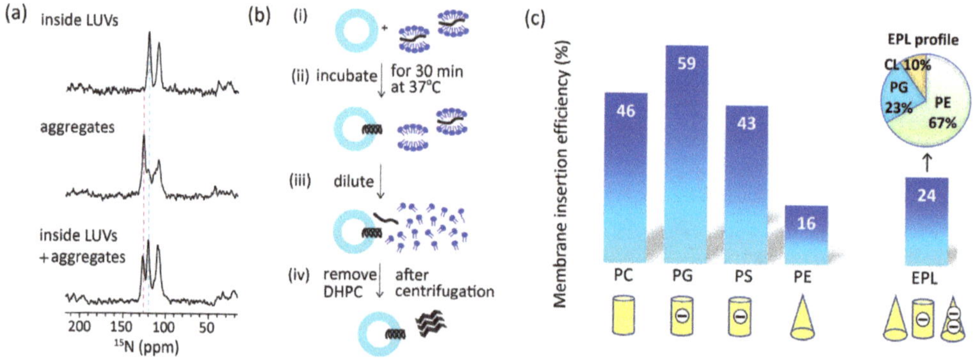

Figure 3. (**a**) ^{15}N cross polarization under magic angle spinning (CPMAS) NMR spectra of Pf3_24 [23]. (top) Pf3_24 was fully reconstituted into DMPC LUVs. (middle) Pf3_24 in aggregates. The sample was prepared without the use of LUVs in the procedure shown in (**b**). (bottom) The sample was prepared using the procedure shown in (**b**) for DMPC LUVs. The pink dashed line shows the chemical shift for the β-strand conformation, and the light blue dashed line shows that for α-helix and/or random-coil conformations. The signal observed at approximately 110 ppm originated from glycine. (**b**) Procedure for inserting Pf3_24 into the membranes. (i) Pf3_24 (dark gray) solubilized in DHPC solution was mixed with LUVs composed of various lipid types. (ii) The mixture was incubated for 30 min at 37 °C. (iii) Then, it was diluted with a buffer solution until the DHPC concentration reached below its critical micelle concentration. (iv) The supernatant containing DHPC was removed by centrifugation. Precipitates containing both LUVs and aggregates were subjected to ^{15}N CPMAS NMR. (**c**) Membrane insertion efficiency of Pf3_24, determined using the procedure shown in (**b**), into several types of membranes.

In comparing the efficiencies of the major membrane lipids, we noted significantly lower levels in PE and EPL membranes, which contained cone-shaped lipids. Thus, the membrane insertion efficiency depends more on the molecular shape than on the charge of the lipids.

3.3. *Acyl Chain Mobility and Membrane Surface Packing*

The mobility inside the membrane was analyzed by measuring the anisotropy of the rod-shaped fluorescent probe 1,6-diphenyl-1,3,5-hexatriene (DPH), which was incorporated near the bilayer core region (Figure 4a,b). DPH exhibits strong polarization characteristics in membranes due to its shape, and these polarization characteristics are influenced by the membrane environment. An increase in the fluorescence anisotropy values of DPH reflects an increase in the molecular order in the membrane core region [39]. The membrane mobility of PE was significantly lower than that of the other bulk lipids (Figure 4a). In addition, the membrane mobility decreased as the PE concentration increased (Figure 4b).

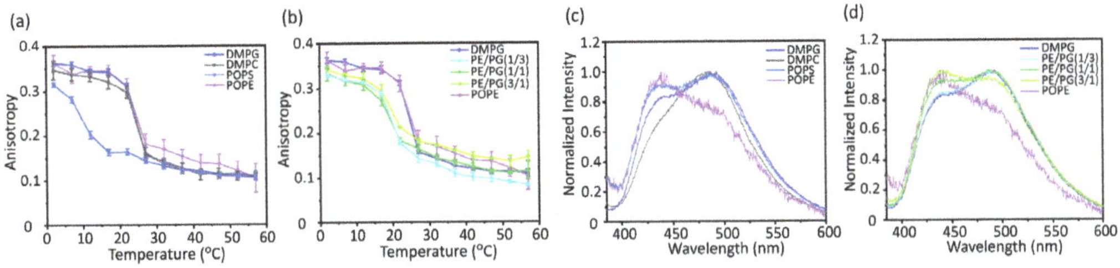

Figure 4. Fluorescence anisotropy of DPH (**a**,**b**) and fluorescence emission spectra of Laurdan at 37 °C (**c**,**d**) incorporated into various types of membranes [23]. DMPG, 1,2-dimyristoyl-*sn*-glycero-3-phospho-(1′-rac-glycerol); DMPC, 1,2-dimyristoyl-*sn*-glycero-3-phosphocholine; POPS, 1-palmitoyl-2-oleoyl-*sn*-glycero-3-phospho-L-serine; POPE, 1-palmitoyl-2-oleoyl-*sn*-glycero-3-phosphoethanolamine.

Membrane packing was also analyzed by measuring the emission spectra of Laurdan, which resides around the hydrophilic–hydrophobic interface of the lipid bilayers (Figure 4c,d) [40]. The emission spectrum of Laurdan is sensitive to the water content of the surrounding environment. The red shift of Laurdan's emission spectrum indicates more hydration by loosening the lipid packing, whereas the blue shift of the spectrum indicates membrane tightening. The generalized polarization (GP) values $[GP = (I_{440} - I_{490})/(I_{440} + I_{490})]$ express the degree of hydration. The membranes composed of the cone-shaped lipid PE showed the strongest tightening of membrane packing (Figure 4c). Furthermore, the membrane packing became remarkably tight as the PE concentration increased (Figure 4d).

3.4. Correlation between Membrane Physicochemical Properties and Membrane Insertion Efficiency of Pf3_24

Correlation plots of acyl chain mobility (DPH anisotropy) and membrane surface packing (Laudan GP) in various membranes against insertion efficiency (Figure 5a,b) indicate the contribution of physicochemical properties to membrane insertion. Both the anisotropy and GP values strongly correlated with the membrane insertion efficiency. Thus, in membranes composed only of bulk phospholipids, acyl chain mobility and membrane packing play direct roles in insertion, with regression lines serving as reliable guidelines. For example, proteins are readily inserted into loosely packed membranes formed by cylindrical lipids, such as PC, PG, and PS. However, as the concentration of PE increased, the acyl chain mobility decreased and the surface packing tightened, leading to the suppression of insertion. PE easily results in negative curvature stress in membranes owing to its small head group, which makes it difficult for the protein to interact with the core of the membranes.

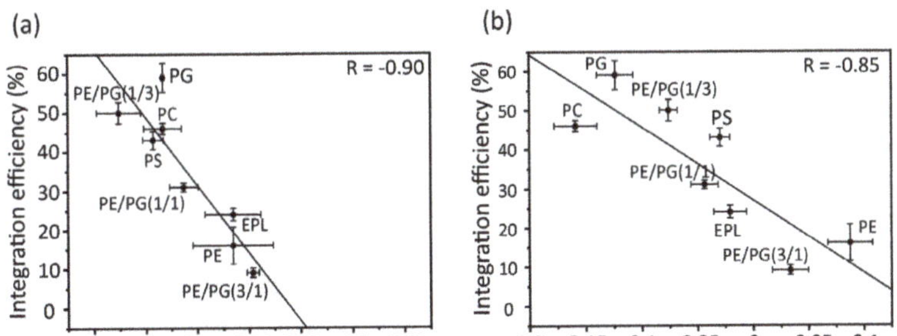

Figure 5. Correlation between membrane insertion efficiency values and fluorescence anisotropy values of DPH at 37 °C (**a**) and Laurdan GP values, $GP = (I_{440} - I_{490})/(I_{440} + I_{490})$, (**b**) in various types of membranes. Global linear fits were performed for all bulk phospholipids.

4. Role of MPIase on Membrane Protein Insertion

Thus far, we have evaluated the insertion of Pf3_24 into membranes composed only of bulk lipids. However, in actual *E. coli* inner membranes, the presence of minor endogenous components, DAG and MPIase, efficiently regulates membrane protein insertion. Spontaneous membrane insertion is blocked by DAG, whereas suppressed insertion is recovered by MPIase. In our study, at physiological concentrations, neither MPIase nor DAG destabilized the lipid bilayer structure or caused membrane fusion [24]. Thus, the membrane insertion of proteins supported by MPIase is not accompanied by membrane disruption, as is the case with antimicrobial peptides. DAG is a small cone-shaped lipid, whereas MPIase is an inverted cone-shaped glycolipid containing a DAG substructure and an additional head group consisting of a long sugar chain and pyrophosphate. Owing to their unique molecular structures compared to other membrane lipids, they exert distinct effects on membrane properties. In this section, we describe the mechanism by which MPIase supports membrane protein insertion into *E. coli* membranes in the presence of DAG [23].

4.1. Mobility of MPIase in Membranes

The relative molecular mobility of MPIase and bulk membrane lipids was examined by measuring the ^{13}C cross-polarization (CP) and direct polarization (DP) NMR spectra of EPL liposomes containing uniformly ^{13}C-labeled natural MPIase (5 wt%) using magic angle spinning (MAS) (Figure 6a). It is generally known that NMR signals of all components are observed using DP, whereas those of rigid ones are selected by the dipolar transfer in CP. A comparison of the relative signal intensity observed in the DP (Figure 6a, top) and CP (Figure 6a, bottom) spectra suggests a higher mobility of the MPIase sugar chain than its lipid part and the other lipids in the liposomes. Higher fluctuation of sugar chain head groups compared to its lipid moiety has also been observed in other glycolipids [41–43], and generally, longer sugar chains amplify this effect [44]. Moreover, in the ^{31}P T1 values of the head group of bulk lipids, fluctuation of the MPIase sugar chain increased their movement. DAG had little influence on the lipid head group mobility because its small head groups were not exposed to the membrane surface [24].

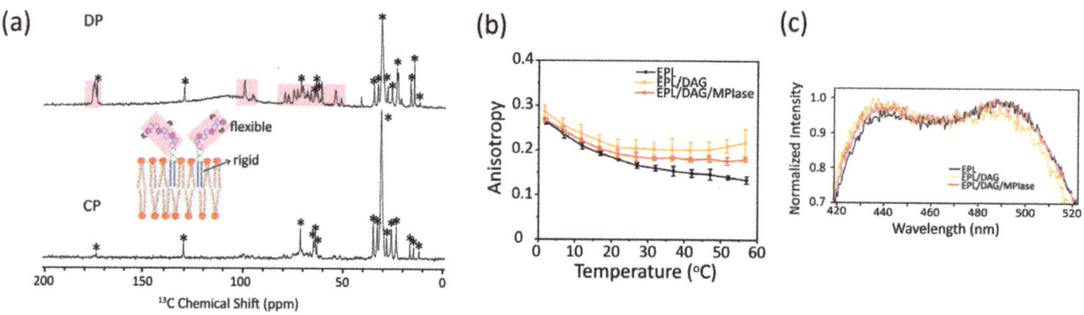

Figure 6. (**a**) Direct polarization (DP) (top) and cross-polarization (CP) (bottom) ^{13}C NMR spectra of EPL liposomes in the presence of uniformly ^{13}C-labeled natural MPIase (5 wt%) under MAS with a spinning speed of 5 kHz [24]. The sugar chain signals that appeared only in the DP spectrum are shown in salmon pink. The peaks from EPL are marked with *. (**b**,**c**) Fluorescence anisotropy of DPH (**b**) and fluorescence emission spectra of Laurdan at 37 °C (**c**) incorporated into EPL-based membranes [23].

4.2. Effects on Flip–Flop Motion of DAG

DAG has a very rapid flip–flop rate compared to phospholipids owing to their compact shape [45]. The rapid flip–flop motion of DAG likely disturbs membrane protein insertion. Once a transient defect is formed among phospholipids in the membrane, DAG can flip across the lipid bilayer, filling the space and reducing unfavorable exposure to the hydrophobic acyl chains. This weakens the contact of proteins with the inside of the membrane and blocks the insertion of membrane proteins. In contrast, the flip–flop is diminished in the presence of mini-MPIase-3 [24]. Locally, unfilled hydrophobic spaces are formed more frequently and are retained longer, allowing proteins to insert more easily into the membrane.

4.3. Effects of DAG and MPIase on Acyl Chain Mobility and Membrane Surface Packing

Acyl chain mobility in the EPL membrane core region was analyzed using the anisotropy of DPH (Figure 6b). The acyl chain mobility was significantly reduced by DAG. However, MPIase partially restored the reduced mobility of the DAG-containing membranes. Lipid acyl chain ordering was also studied by measuring the ^{2}H NMR spectra. Quadrupole splitting increased in the presence of DAG but returned to the original levels after the addition of natural MPIase (5 wt%) into the DAG-containing liposomes. Thus, DAG promotes the ordering of lipid acyl chains, whereas MPIase compensates for the membrane ordering induced by DAG. The increased packing of the hydrophobic segment by DAG further reduced lateral lipid diffusion. Similar to cholesterol, DAG fits under the polar head groups of lipids, enhancing the order of acyl chains. However, MPIase rebalances the head group and acyl chain volumes of lipids in the membranes and releases the packing pressure inside the membrane.

Membrane surface packing was analyzed by measuring the emission spectra of Laurdan (Figure 6c). DAG slightly tightened the membrane packing, whereas MPIase partially loosened the packing in DAG-containing membranes. Thus, MPIase counteracted the effects of DAG on membrane properties, both in the membrane core region and on the surface. However, the influence of DAG and MPIase on the membrane surface region was less than that on the membrane core region, suggesting that access to the membrane core region is more critical than initial contact with the surface region for successful protein insertion.

The correlation plots clearly demonstrate how alterations in the physicochemical properties of the lipid membranes containing DAG and MPIase contribute the insertion of proteins into EPL membranes (Figure 7). In the presence of DAG (5 mol% to EPLs), the mobility of the core region was dramatically reduced (Figure 7a); however, the mobility of the membrane surface region was not significantly reduced (Figure 7b). Therefore, we conclude that the drastic inhibition of insertion by DAG is attributable to decreased mobility of the core region. Moreover, we considered that the high insertion efficiency of MPIase was due, at least partly, to compensation for the effects in shape. Thus, we compared MPIase with lyso-PC, which has an inverse-cone shape similar to that of MPIase, because it is known that the presence of lyso-PC enhances the insertion of antimicrobial peptides in PC membranes [46]. The addition of lyso-PC (5 mol%) to the DAG-containing EPL membrane restored both the membrane mobility and the insertion efficiency of Pf3_24. The correlation is on the regression line, indicating that the effect of lyso-PC on insertion can be explained only by the alteration of membrane physicochemical properties. Conversely, after adding MPIase (1 mol%) to the DAG-containing EPL membrane, the insertion efficiency was much higher than what was expected from the correlation plot against anisotropy values. Therefore, factors other than membrane properties should also support the high insertion efficiency in the presence of MPIase. To identify other causes, we focused on the intermolecular interactions between MPIase and substrate proteins.

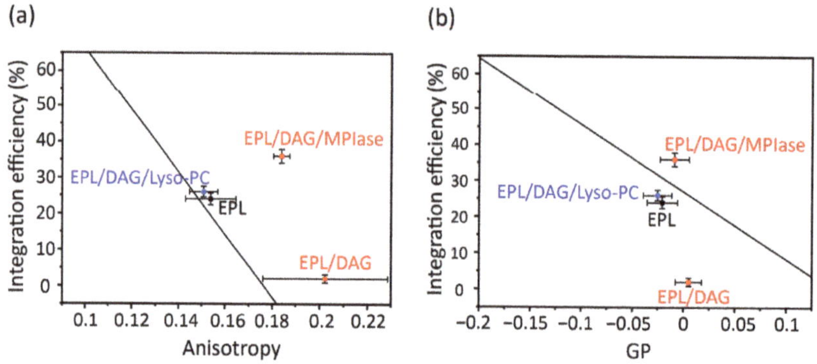

Figure 7. Correlation between membrane insertion efficiency values and fluorescence anisotropy values of DPH at 37 °C (Figure 6b) (**a**) and membrane insertion efficiency values and Laurdan GP values at 37 °C calculated from the emission spectra (Figure 6c) (**b**) in the EPL-based membranes. The black line shows the regression line obtained from the global linear fits performed on the bulk phospholipids in Figure 5.

4.4. Identification of the Contact Sites by Using Docking Simulations

The functional groups required for molecular interactions between MPIase and the protein were estimated by docking simulations using the MOE software, version 2019.0102 [47]. Docking simulations of pyrophosphorylated trisaccharide moieties, Trisac-PP, with Pf3 coat suggested the hydrophobic interactions between the acetyl groups in MPIase and the hydrophobic residues of Pf3 coat and electrostatic interactions between the pyrophosphate or carboxylate of MPIase and the basic amino acid residues of Pf3 coat (Figure 8). The reduction in contact in all combinations observed when using MPIase analogs lacking 6-*O*-acetyl of GlcNAc or pyrophosphate emphasizes the importance of these functional groups.

4.5. Binding Kinetics and Affinity of MPIase—Pf3 Coat Interactions

In an integration activity assay using in vitro translation [48], MPIase with a long sugar chain showed much higher activity than mini-MPIase-3 [21]. In addition, Trisac-DAG, a mini-MPIase-3 analog without a pyrophosphate group, exhibited much lower insertion than mini-MPIase-3 [20]. These differences in insertion activity can be understood by analyzing binding kinetics and affinity using SPR [49,50]. In the SPR measurements, the glycolipid analytes flowed over the flow cells on which Pf3 coat had been immobilized. While direct comparisons of response levels are unsuitable because of notable variations in molecular weight among ligands, the comparison of responses induced by MPIase, mini-MPIase-3, and Trisac-DAG highlights the involvement of pyrophosphate in the interaction of MPIase-Pf3 coat (Figure 9a) [22]. The on–off rate map, in which the association (k_a) and dissociation (k_d) rates determined by fitting analyses were plotted, clearly demonstrated a greater contribution of the long sugar chain of MPIase to rapid contact with Pf3 coat (Figure 9b). This is likely the reason for the higher integration activity of MPIase compared with that of mini-MPIase-3.

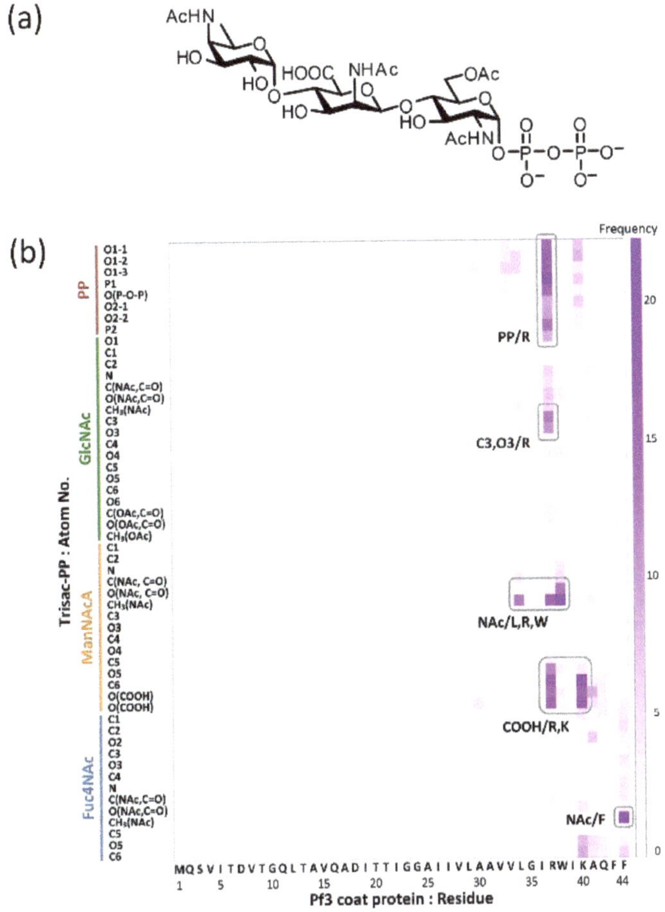

Figure 8. (**a**) Molecular structure of Trisac-PP. (**b**) Heat map of frequencies in the docking simulation. It illustrates the frequencies of interaction between each atom of Trisac-PP and each amino acid residue of Pf3 coat [22].

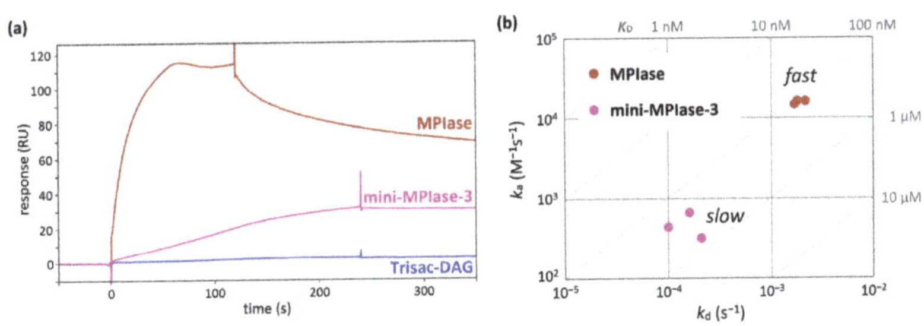

Figure 9. (**a**) SPR sensorgrams of the binding of natural MPIase, mini-MPIase-3, and Trisac-DAG to Pf3 coat [22]. (**b**) The on–off rate map showing the kinetic profiles of the interaction of MPIase and mini-MPIase-3 with Pf3 coat [22].

4.6. Interaction of the MPIase Pyrophosphate Group with Basic Residues in Pf3_27

As can be seen from the results of the docking simulations and SPR experiments, the MPIase pyrophosphate linker is one of the key structures involved in membrane insertion activity. The ^{1}H-^{15}N FSLG-HETCOR spectrum of Pf3_27, which has a transmembrane region followed by cytoplasmic residues (Figure 10a), exhibited a selective interaction between the pyrophosphate in MPIase and Arg residues (Figure 10b–d). The signal of the R20 H_η protons was shifted 0.53 ppm downfield in the presence of mini-MPIase-3 (Figure 10b), whereas the proton signal did not change with mini-ECA-3 that has a monophosphate linker (Figure 10c) [20,23]. Generally, basic residues in membrane proteins are abundant near the membrane surface on the cytoplasmic side; this is known as the positive-inside rule [33,34]. Therefore, the double negatively charged pyrophosphate group of MPIase attracts basic residues to the membrane surface (Figure 10d). The downfield shift value of 0.53 ppm due to hydrogen bonding of Pf3_27 with MPIase is smaller than that observed for the ligand-receptor complex (about 2–3 ppm) [23,51]. Therefore, the binding of the guanidino group of R20 to the pyrophosphate of MPIase is weaker than that in the ligand-receptor complex and more transient. Transient binding allows MPIase to detach the protein after insertion into the membrane.

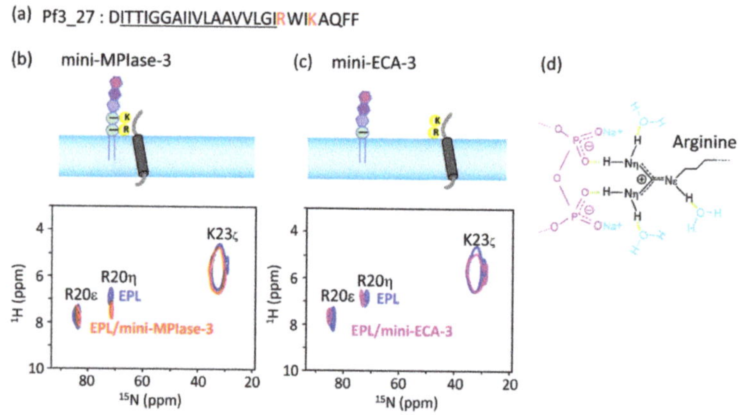

Figure 10. (**a**) Amino acid sequence of substructure of Pf3 coat, Pf3_27. Hydrophobic region is underlined, and basic residues are colored in red. (**b**,**c**) Arg and Lys side-chain signal regions of the 2D ^{1}H-^{15}N FSLG-HETCOR spectrum of Pf3_27 in EPL membranes in the presence of mini-MPIase-3 ((**a**), blue) [23] or mini-ECA-3 ((**b**), pink) and absence of them ((**a**,**b**), blue). (**d**) Schematic models of Arg side-chain (black) interaction with the pyrophosphate linker part of MPIase (purple). Light green dots show hydrogen bonding.

4.7. Inhibition of Protein Aggregation by MPIase

The interactions between MPIase and Pf3 coat induce changes in the secondary structure of Pf3 coat [22]. The CD spectra of Pf3 coat clearly exhibited that without MPIase, Pf3 coat did not form a definite secondary structure (Figure 11, orange), while it exhibited mostly an α-helical structure with MPIase (Figure 11, green). The long sugar chain of MPIase is likely to capture nascent proteins through multiple interactions, such as the acetyl groups of MPIase and the hydrophobic region of the proteins or the carboxyl groups of MPIase and the basic amino acid residues of the proteins, which prevent the protein from aggregating. Therefore, MPIase might act as a chaperone during membrane insertion. In this case, to effectively inhibit protein aggregation, MPIase might form a self-aggregated structure on the inner membrane of *E. coli*.

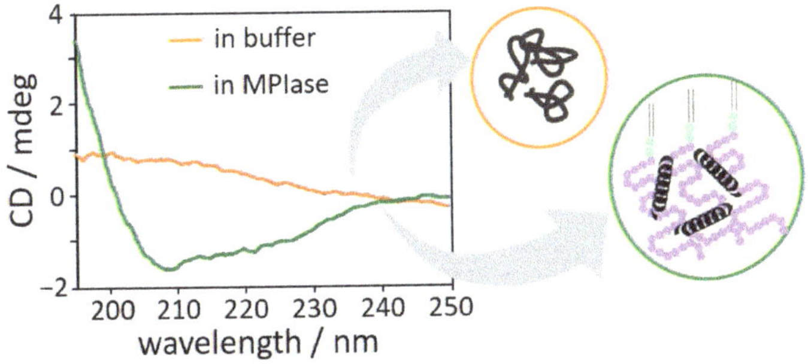

Figure 11. CD spectra of Pf3 coat in buffer in the presence (green) and absence (orange) of MPIase [22].

5. Mechanism of Sec-Independent Membrane Protein Insertion by MPIase

Finally, the insertion mechanism of small hydrophobic proteins which contain basic amino acids at their C-terminus, like Sec-independent membrane protein, into *E. coli* membranes under the regulation of MPIase and DAG is summarized in Figure 12.

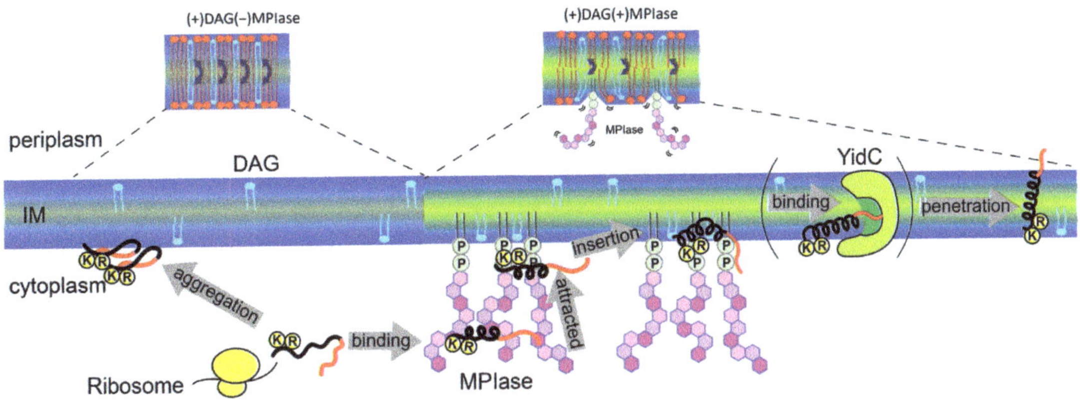

Figure 12. Schematic model of the membrane insertion of a small hydrophobic substrate protein (black: hydrophobic region, red: hydrophilic region) supported by MPIase in *E. coli* membranes.

1. DAG blocks spontaneous insertion by reducing acyl chain mobility in the membrane core. The rapid flip–flop of DAG also reduces lateral diffusion and fills membrane defects, exposing acyl chains and making it difficult for proteins to contact the inside of the membranes. However, the presence of MPIase cancels out the effect of DAG and restores mobility. The flexible bulky sugar chain of MPIase makes the membrane surface flexible, increases the mobility of the membrane core region, and reduces the flip–flop motion of DAG. Small hydrophobic membrane proteins preferentially associate with the membrane core region rather than with the aqueous environment outside the membrane.
2. After a protein is released from a ribosome, the long sugar chain of MPIase captures the protein and prevents its aggregation. The protein efficiently changes its conformation through rapid association with and dissociation from the flexible long sugar chains of MPIase.
3. The basic residues that generally exist in the cytoplasmic loops near the hydrophobic transmembrane regions of the protein are attracted towards the membrane surface through electrostatic interactions with the pyrophosphate of MPIase.
4. The protein easily gains access to the loosened membrane core through hydrophobic interactions and finally integrates into the membrane.

MPIase interacts with proteins via its pyrophosphate group. However, this binding is weak, and MPIase can easily detach from the protein. It is likely that the protein is then delivered to YidC in the inner membrane. The detached MPIase can assist another substrate that has not yet been inserted into the membrane. Thus, MPIase efficiently supports membrane insertion.

6. Conclusions

This study discusses the various membrane properties that affect Sec-independent membrane insertion and the contributions of MPIase and DAG to membrane properties. Both MPIase and DAG are minor components of *E. coli* membranes. However, by acting in opposite directions, they effectively alter membrane properties. Comprehensive analysis of the membrane insertion activity, physicochemical properties, and molecular interactions between MPIase and its substrates provides an understanding of the atomic-level molecular mechanisms by which MPIase prevents substrate protein aggregation, attracts substrates near the membrane surface, and promotes membrane insertion. These findings reveal novel biological roles of glycolipids. Since MPIase is also involved in other protein translocation pathways across membranes [7,11], further studies on MPIase clustering on the membrane and its distribution in cells will provide insights into the diverse functions of MPIase, including membrane transport.

Author Contributions: Conceptualization, K.N. and K.S.; methodology, K.N. and K.S.; formal analysis, K.N. and S.M.; investigation, K.N. and S.M.; resources, K.N., S.M. and K.S.; writing—original draft preparation, K.N.; writing—review and editing, K.N., S.M. and K.S.; supervision, K.S.; funding acquisition, K.N. and K.S. All authors have read and agreed to the published version of the manuscript.

Funding: This work was funded by JSPS KAKENHI, grant numbers JP18K06143 and JP22K05323 to K.N. and JP19H02843 and JP22H02213 to K.S.

Institutional Review Board Statement: Not applicable.

Data Availability Statement: No new data are reported in this review article.

Acknowledgments: We are grateful to Ken-ichi Nishiyama (Iwate University), Kohki Fujikawa, Tsukiho Osawa, Toshiyuki Yamaguchi (Suntory Foundation for Life Sciences), Tsuyoshi Shirai, Masafumi Shionyu, Takao Yoda (Nagahama Institute of Bio-Science and Technology), Taku Yoshiya, Hideki Nishio, Shun Masuda, Kumiko Yoshizawa-Kumagaye, Shugo Tsuda (Peptide Institute, Inc.), Kenichi Morigaki (Kobe University), and Toshinori Shimanouchi (Okayama University) for their contribution to the research on MPIase.

Conflicts of Interest: The authors declare no conflicts of interest.

Abbreviations

The following abbreviations are used in this manuscript:

membrane protein integrase	MPIase
Escherichia coli	*E. coli*
diacylglycerol	DAG
E. coli phospholipids	EPLs
phosphatidylethanolamine	PE
phosphatidylglycerol	PG
cardiolipin	CL
antimicrobial peptides	AMPs
phosphatidylcholine	PC
phosphatidylserine	PS
1,2-dimyristoyl-*sn*-glycero-3-phosphocholine	DMPC
1,2-diheptanoyl-*sn*-glycero-3-phosphocholine	DHPC
large unilamellar vesicles	LUV
cross-polarization under magic angle spinning	CPMAS
1,6-diphenyl-1,3,5-hexatriene	DPH
generalized polarization	GP

1,2-dimyristoyl-*sn*-glycero-3-phospho-(1′-rac-glycerol)	DMPG
1-palmitoyl-2-oleoyl-*sn*-glycero-3-phospho-L-serine	POPS
1-palmitoyl-2-oleoyl-*sn*-glycero-3-phosphoethanolamine	POPE
magic angle spinning	MAS
surface plasmon resonance	SPR
association rate	k_a
dissociation rate	k_d
circular dichroism	CD

References

1. Luirink, J.; von Heijne, G.; Houben, E.; de Gier, J.W. Biogenesis of inner membrane proteins in *Escherichia coli*. *Annu. Rev. Microbiol.* **2005**, *59*, 329–355. [CrossRef]
2. Wolfe, P.B.; Rice, M.; Wickner, W. Effects of two sec genes on protein assembly into the plasma membrane of *Escherichia coli*. *J. Biol. Chem.* **1985**, *260*, 1836–1841. [CrossRef] [PubMed]
3. Geller, B.L.; Wickner, W. M13 procoat inserts into liposomes in the absence of other membrane proteins. *J. Biol. Chem.* **1985**, *260*, 13281–13285. [CrossRef]
4. Kiefer, D.; Hu, X.; Dalbey, R.; Kuhn, A. Negatively charged amino acid residues play an active role in orienting the Sec-independent Pf3 coat protein in the *Escherichia coli* inner membrane. *EMBO J.* **1997**, *16*, 2197–2204. [CrossRef] [PubMed]
5. Mariappan, M.; Li, X.; Stefanovic, S.; Sharma, A.; Mateja, A.; Keenan, R.J.; Hegde, R.S. A ribosome-associating factor chaperones tail-anchored membrane proteins. *Nature* **2010**, *466*, 1120–1124. [CrossRef] [PubMed]
6. Mariappan, M.; Mateja, A.; Dobosz, M.; Bove, E.; Hegde, R.S.; Keenan, R.J. The mechanism of membrane-associated steps in tail-anchored protein insertion. *Nature* **2011**, *477*, 61–66. [CrossRef]
7. Nishiyama, K.; Ikegami, A.; Moser, M.; Schiltz, E.; Tokuda, H.; Muller, M. A derivative of lipid A is involved in signal recognition particle/SecYEG-dependent and -independent membrane integrations. *J. Biol. Chem.* **2006**, *281*, 35667–35676. [CrossRef]
8. Kawashima, Y.; Miyazaki, E.; Müller, M.; Tokuda, H.; Nishiyama, K.-I. Diacylglycerol Specifically Blocks Spontaneous Integration of Membrane Proteins and Allows Detection of a Factor-assisted Integration. *J. Biol. Chem.* **2008**, *283*, 24489–24496. [CrossRef]
9. Nishiyama, K.; Maeda, M.; Abe, M.; Kanamori, T.; Shimamoto, K.; Kusumoto, S.; Ueda, T.; Tokuda, H. A novel complete reconstitution system for membrane integration of the simplest membrane protein. *Biochem. Biophys. Res. Commun.* **2010**, *394*, 733–736. [CrossRef]
10. Nishiyama, K.-I.; Maeda, M.; Yanagisawa, K.; Nagase, R.; Komura, H.; Iwashita, T.; Yamagaki, T.; Kusumoto, S.; Tokuda, H.; Shimamoto, K. MPIase is a glycolipozyme essential for membrane protein integration. *Nat. Commun.* **2012**, *3*, 1260. [CrossRef]
11. Moser, M.; Nagamori, S.; Huber, M.; Tokuda, H.; Nishiyama, K. Glycolipozyme MPIase is essential for topology inversion of SecG during preprotein translocation. *Proc. Natl. Acad. Sci. USA* **2013**, *110*, 9734–9739. [CrossRef]
12. Chen, M.; Samuelson, J.C.; Jiang, F.; Muller, M.; Kuhn, A.; Dalbey, R.E. Direct Interaction of YidC with the Sec-independent Pf3 Coat Protein during Its Membrane Protein Insertion. *J. Biol. Chem.* **2002**, *277*, 7670–7675. [CrossRef] [PubMed]
13. Samuelson, J.C.; Chen, M.; Jiang, F.; Möller, I.; Wiedmann, M.; Kuhn, A.; Phillips, G.J.; Dalbey, R.E. YidC mediates membrane protein insertion in bacteria. *Nature* **2000**, *406*, 637–641. [CrossRef] [PubMed]
14. Serek, J.; Bauer-Manz, G.; Struhalla, G.; van den Berg, L.; Kiefer, D.; Dalbey, R.; Kuhn, A. *Escherichia coli* YidC is a membrane insertase for Sec-independent proteins. *EMBO J.* **2004**, *23*, 294–301. [CrossRef] [PubMed]
15. Kumazaki, K.; Chiba, S.; Takemoto, M.; Furukawa, A.; Nishiyama, K.; Sugano, Y.; Mori, T.; Dohmae, N.; Hirata, K.; Nakada-Nakura, Y.; et al. Structural basis of Sec-independent membrane protein insertion by YidC. *Nature* **2014**, *509*, 516–520. [CrossRef] [PubMed]
16. McDowell, M.A.; Heimes, M.; Sinning, I. Structural and molecular mechanisms for membrane protein biogenesis by the Oxa1 superfamily. *Nat. Struct. Mol. Biol.* **2021**, *28*, 234–239. [CrossRef] [PubMed]
17. Wickles, S.; Singharoy, A.; Andreani, J.; Seemayer, S.; Bischoff, L.; Berninghausen, O.; Soeding, J.; Schulten, K.; van der Sluis, E.O.; Beckmann, R. A structural model of the active ribosome-bound membrane protein insertase YidC. *Elife* **2014**, *3*, e03035. [CrossRef] [PubMed]
18. Endo, Y.; Shimizu, Y.; Nishikawa, H.; Sawasato, K.; Nishiyama, K.I. Interplay between MPIase, YidC, and PMF during Sec-independent insertion of membrane proteins. *Life Sci. Alliance* **2022**, *5*, e202101162. [CrossRef]
19. Sasaki, M.; Nishikawa, H.; Suzuki, S.; Moser, M.; Huber, M.; Sawasato, K.; Matsubayashi, H.T.; Kumazaki, K.; Tsukazaki, T.; Kuruma, Y.; et al. The bacterial protein YidC accelerates MPIase-dependent integration of membrane proteins. *J. Biol. Chem.* **2019**, *294*, 18898–18908. [CrossRef]
20. Fujikawa, K.; Han, Y.; Osawa, T.; Mori, S.; Nomura, K.; Muramoto, M.; Nishiyama, K.-I.; Shimamoto, K. Structural Requirements of a Glycolipid MPIase for Membrane Protein Integration. *Chem. Eur. J.* **2023**, *29*, e202300437. [CrossRef]
21. Fujikawa, K.; Suzuki, S.; Nagase, R.; Ikeda, S.; Mori, S.; Nomura, K.; Nishiyama, K.I.; Shimamoto, K. Syntheses and Activities of the Functional Structures of a Glycolipid Essential for Membrane Protein Integration. *ACS Chem. Biol.* **2018**, *13*, 2719–2727. [CrossRef] [PubMed]

22. Mori, S.; Nomura, K.; Fujikawa, K.; Osawa, T.; Shionyu, M.; Yoda, T.; Shirai, T.; Tsuda, S.; Yoshizawa-Kumagaye, K.; Masuda, S.; et al. Intermolecular Interactions between a Membrane Protein and a Glycolipid Essential for Membrane Protein Integration. *ACS Chem. Biol.* **2022**, *17*, 609–618. [CrossRef] [PubMed]
23. Nomura, K.; Mori, S.; Fujikawa, K.; Osawa, T.; Tsuda, S.; Yoshizawa-Kumagaye, K.; Masuda, S.; Nishio, H.; Yoshiya, T.; Yoda, T.; et al. Role of a bacterial glycolipid in Sec-independent membrane protein insertion. *Sci. Rep.* **2022**, *12*, 12231. [CrossRef] [PubMed]
24. Nomura, K.; Yamaguchi, T.; Mori, S.; Fujikawa, K.; Nishiyama, K.I.; Shimanouchi, T.; Tanimoto, Y.; Morigaki, K.; Shimamoto, K. Alteration of Membrane Physicochemical Properties by Two Factors for Membrane Protein Integration. *Biophys. J.* **2019**, *117*, 99–110. [CrossRef] [PubMed]
25. Ding, B.; Soblosky, L.; Nguyen, K.; Geng, J.; Yu, X.; Ramamoorthy, A.; Chen, Z. Physiologically-relevant modes of membrane interactions by the human antimicrobial peptide, LL-37, revealed by SFG experiments. *Sci. Rep.* **2013**, *3*, 1854. [CrossRef] [PubMed]
26. Lee, T.H.; Heng, C.; Separovic, F.; Aguilar, M.I. Comparison of reversible membrane destabilisation induced by antimicrobial peptides derived from Australian frogs. *Biochim. Biophys. Acta* **2014**, *1838*, 2205–2215. [CrossRef] [PubMed]
27. Huang, J.; Feigenson, G.W. A microscopic interaction model of maximum solubility of cholesterol in lipid bilayers. *Biophys. J.* **1999**, *76*, 2142–2157. [CrossRef] [PubMed]
28. Strandberg, E.; Ulrich, A.S. AMPs and OMPs: Is the folding and bilayer insertion of beta-stranded outer membrane proteins governed by the same biophysical principles as for alpha-helical antimicrobial peptides? *Biochim. Biophys. Acta* **2015**, *1848*, 1944–1954. [CrossRef]
29. Yano, Y.; Yamamoto, A.; Ogura, M.; Matsuzaki, K. Thermodynamics of Insertion and Self-Association of a Transmembrane Helix: A Lipophobic Interaction by Phosphatidylethanolamine. *Biochemistry* **2011**, *50*, 6806–6814. [CrossRef]
30. Shintou, K.; Nakano, M.; Kamo, T.; Kuroda, Y.; Handa, T. Interaction of an Amphipathic Peptide with Phosphatidycholine/Phosphatidylethanolamine Mixed Membranes. *Biophys. J.* **2007**, *93*, 3900–3906. [CrossRef]
31. Murayama, T.; Masuda, T.; Afonin, S.; Kawano, K.; Takatani-Nakase, T.; Ida, H.; Takahashi, Y.; Fukuma, T.; Ulrich, A.S.; Futaki, S. Loosening of Lipid Packing Promotes Oligoarginine Entry into Cells. *Angew. Chem. Int. Ed.* **2017**, *56*, 7644–7647. [CrossRef]
32. Pujals, S.; Miyamae, H.; Afonin, S.; Murayama, T.; Hirose, H.; Nakase, I.; Taniuchi, K.; Umeda, M.; Sakamoto, K.; Ulrich, A.S.; et al. Curvature engineering: Positive membrane curvature induced by epsin N-terminal peptide boosts internalization of octaarginine. *ACS Chem. Biol.* **2013**, *8*, 1894–1899. [CrossRef]
33. Bogdanov, M.; Xie, J.; Dowhan, W. Lipid-protein interactions drive membrane protein topogenesis in accordance with the positive inside rule. *J. Biol. Chem.* **2009**, *284*, 9637–9641. [CrossRef]
34. Von Heijne, G. Membrane-protein topology. *Nat. Rev. Mol. Cell. Biol.* **2006**, *7*, 909–918. [CrossRef] [PubMed]
35. Nevzorov, A.A.; Opella, S.J. Selective averaging for high-resolution solid-state NMR spectroscopy of aligned samples. *J. Magn. Reson.* **2007**, *185*, 59–70. [CrossRef] [PubMed]
36. Nevzorov, A.A.; Opella, S.J. A "magic sandwich" pulse sequence with reduced offset dependence for high-resolution separated local field spectroscopy. *J. Magn. Reson.* **2003**, *164*, 182–186. [CrossRef] [PubMed]
37. Wang, Y.; Jardetzky, O. Probability-based protein secondary structure identification using combined NMR chemical-shift data. *Protein Sci.* **2002**, *11*, 852–861. [CrossRef] [PubMed]
38. Wishart, D.S.; Sykes, B.D.; Richards, F.M. Relationship between nuclear magnetic resonance chemical shift and protein secondary structure. *J. Mol. Biol.* **1991**, *222*, 311–333. [CrossRef] [PubMed]
39. Lentz, B.R. Use of fluorescent probes to monitor molecular order and motions within liposome bilayers. *Chem. Phys. Lipids* **1993**, *64*, 99–116. [CrossRef] [PubMed]
40. Antollini, S.S.; Soto, M.A.; Bonini de Romanelli, I.; Gutiérrez-Merino, C.; Sotomayor, P.; Barrantes, F.J. Physical state of bulk and protein-associated lipid in nicotinic acetylcholine receptor-rich membrane studied by laurdan generalized polarization and fluorescence energy transfer. *Biophys. J.* **1996**, *70*, 1275–1284. [CrossRef]
41. Carrier, D.; Giziewicz, J.B.; Moir, D.; Smith, I.C.; Jarrell, H.C. Dynamics and orientation of glycolipid headgroups by 2H-NMR: Gentiobiose. *Biochim. Biophys. Acta* **1989**, *983*, 100–108. [CrossRef] [PubMed]
42. Jarrell, H.C.; Wand, A.J.; Giziewicz, J.B.; Smith, I.C. The dependence of glyceroglycolipid orientation and dynamics on head-group structure. *Biochim. Biophys. Acta* **1987**, *897*, 69–82. [CrossRef]
43. Renou, J.P.; Giziewicz, J.B.; Smith, I.C.; Jarrell, H.C. Glycolipid membrane surface structure: Orientation, conformation, and motion of a disaccharide headgroup. *Biochemistry* **1989**, *28*, 1804–1814. [CrossRef]
44. Yamaguchi, T.; Sakae, Y.; Zhang, Y.; Yamamoto, S.; Okamoto, Y.; Kato, K. Exploration of conformational spaces of high-mannose-type oligosaccharides by an NMR-validated simulation. *Angew. Chem. Int. Ed.* **2014**, *53*, 10941–10944. [CrossRef]
45. Bai, J.; Pagano, R.E. Measurement of spontaneous transfer and transbilayer movement of BODIPY-labeled lipids in lipid vesicles. *Biochemistry* **1997**, *36*, 8840–8848. [CrossRef] [PubMed]
46. Strandberg, E.; Tiltak, D.; Ehni, S.; Wadhwani, P.; Ulrich, A.S. Lipid shape is a key factor for membrane interactions of amphipathic helical peptides. *Biochim. Biophys. Acta* **2012**, *1818*, 1764–1776. [CrossRef]
47. Chemical Computing Group ULC. *MOE (Molecular Operating Environment)*; Version 2022.02; Chemical Computing Group ULC: Montreal, QC, Canada, 2022.
48. Nishikawa, H.; Sasaki, M.; Nishiyama, K.I. In vitro Assay for Bacterial Membrane Protein Integration into Proteoliposomes. *Bio-Protoc.* **2020**, *10*, e3626. [CrossRef] [PubMed]

49. Jason-Moller, L.; Murphy, M.; Bruno, J. Overview of Biacore systems and their applications. *Curr. Protoc. Protein Sci.* **2006**, *45*, 19.13.1–19.13.14. [CrossRef]
50. O'Shannessy, D.J.; Brigham-Burke, M.; Soneson, K.K.; Hensley, P.; Brooks, I. Determination of rate and equilibrium binding constants for macromolecular interactions by surface plasmon resonance. *Meth. Enzymol.* **1994**, *240*, 323–349.
51. Yamazaki, T.; Pascal, S.M.; Singer, A.U.; Forman-Kay, J.D.; Kay, L.E. NMR Pulse Schemes for the Sequence-Specific Assignment of Arginine Guanidino 15N and 1H Chemical Shifts in Proteins. *J. Am. Chem. Soc.* **1995**, *117*, 3556–3564. [CrossRef]

membranes

Review

There Are No Insurmountable Barriers: Passage of the *Helicobacter pylori* VacA Toxin from Bacterial Cytoplasm to Eukaryotic Cell Organelle

Miroslaw Jarzab and Joanna Skorko-Glonek *

Department of General and Medical Biochemistry, Faculty of Biology, University of Gdansk, Wita Stwosza 59, 80-308 Gdansk, Poland; miroslaw.jarzab@ug.edu.pl
* Correspondence: joanna.skorko-glonek@ug.edu.pl; Tel.: +48-58-523-60-56

Abstract: The Gram-negative bacterium *Helicobacter pylori* is a very successful pathogen, one of the most commonly identified causes of bacterial infections in humans worldwide. *H. pylori* produces several virulence factors that contribute to its persistence in the hostile host habitat and to its pathogenicity. The most extensively studied are cytotoxin-associated gene A (CagA) and vacuolating cytotoxin A (VacA). VacA is present in almost all *H. pylori* strains. As a secreted multifunctional toxin, it assists bacterial colonization, survival, and proliferation during long-lasting infections. To exert its effect on gastric epithelium and other cell types, VacA undergoes several modifications and crosses multiple membrane barriers. Once inside the gastric epithelial cell, VacA disrupts many cellular-signaling pathways and processes, leading mainly to changes in the efflux of various ions, the depolarization of membrane potential, and perturbations in endocytic trafficking and mitochondrial function. The most notable effect of VacA is the formation of vacuole-like structures, which may lead to apoptosis. This review focuses on the processes involved in VacA secretion, processing, and entry into host cells, with a particular emphasis on the interaction of the mature toxin with host membranes and the formation of transmembrane pores.

Keywords: VacA; *Helicobacter pylori*; autotransporter; membrane insertion; pore-forming toxin; anion-selective pore; endosome; vacuolation

1. Introduction

H. pylori is a prevalent Gram-negative, spiral-shaped human gastric pathogen whose estimated carrier rate is 50% worldwide. The *H. pylori* infections are usually asymptomatic. However, this bacterium is also known as a major contributing factor leading to chronic inflammation of the gastric mucosa, development of gastric ulcers, and cancer [1]. *H. pylori* produces several virulence factors of which VacA (vacuolating cytotoxin A) and CagA (cytotoxin-associated gene A) are most intensively studied.

The most plausible functions of VacA are to enhance the availability of nutrients (or other factors, such as metals) and interference with the functions of the immune system. Those actions of VacA assist bacterial colonization, survival, and proliferation during long-lasting infections [2]. VacA exerts various effects on mammalian cells by affecting functions and the integrity of the plasma membrane and membranes of other organelles [3]. The most notable effect of the VacA protein on the host cells is the formation of large cytoplasmic vacuole-like structures [4,5]. The mechanism of vacuole formation by VacA is dependent on its ability to form channels across membranes. As such, VacA represents the largest class of bacterial toxins, i.e., the class of pore-forming toxins (PFTs) [6]. The target of a PFT is the plasma membrane, also known as the cell membrane, whose role is to maintain cell integrity and separate the cell contents from the external environment while ensuring the exchange of chemical compounds, energy, and information [7,8]. A disruption of the selective permeability of the cell membrane, regardless of the cause, usually leads to the loss of valuable substances from the cell. However, it may also allow the entry of harmful substances, significantly affecting cellular homeostasis and the ability to detect and respond to external stimuli [9]. After internalization by the host cell, VacA creates ion channels in endosomal membranes and leads to the formation of vacuoles. Moreover, it penetrates the membranes of mitochondria and other organelles, causing changes in their functioning.

VacA is an excellent example illustrating the ability of a protein toxin to overcome various membrane barriers in order to reach its target location and induce appropriate effects. However, before the active toxin causes vacuolation in host cells, it must undergo a "journey," during which it is subject to many post-translational modifications and conformational changes. The protein is synthesized as a precursor containing a signal peptide at the N-terminus, directing it to the periplasm, and a translocator domain at the C-terminus, which is responsible for the secretion of the toxin out of the cell. Between these regions is the passenger domain that contains the actual toxin. The toxin is cut out from the precursor sequentially while crossing the individual membranes of the *H. pylori* cell, which include the inner membrane (IM) and the outer membrane (OM). First, during transport across the IM, the signal peptide is removed. Second, another cut occurs, which frees the passenger domain. This process occurs after the passenger domain passes through the OM and causes the release of the active toxin to the extracellular environment. Finally, the toxin is cleaved into two subunits, designated p33 and p55. It should be noted that the precursor and the cleaved toxin do not have a toxic effect on the bacterial cell, only on the host cells. This means that VacA recognizes target cells by binding to the appropriate receptor on their surface. The following chapters present a detailed description of the individual stages related to the maturation, secretion, and interaction of the toxin with host cell membranes. An overview of the fate of VacA in the bacterial cell and host cell, along with the division into stages, is presented in Figure 1.

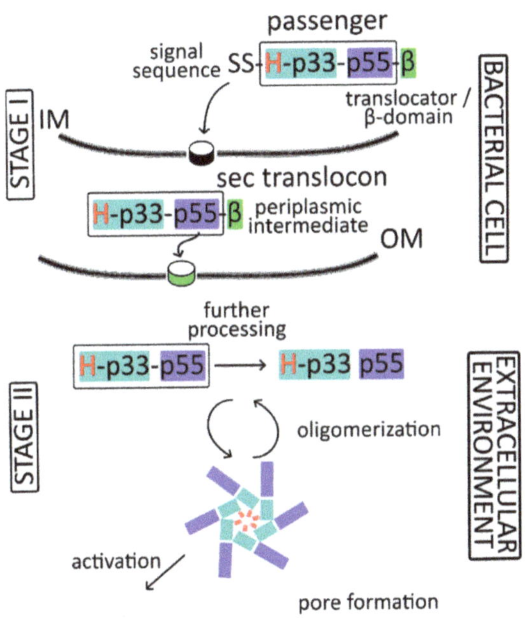

Figure 1. *Cont.*

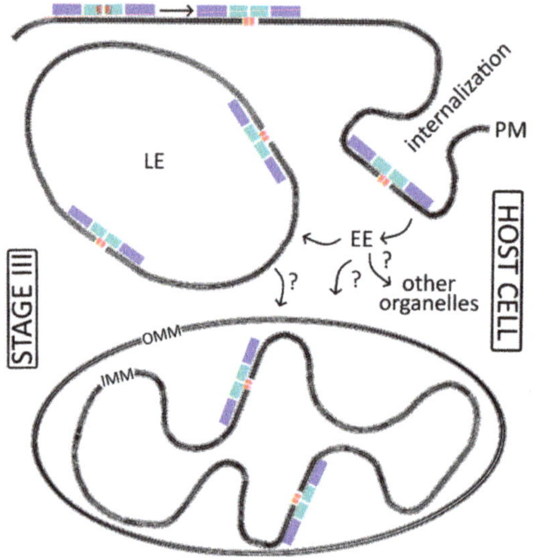

Figure 1. An overview of the fate of VacA in the bacterial and host cells. Stage I. The precursor of VacA is synthesized in the cytoplasm of the *H. pylori* cell, and it contains a signal sequence (SS) for the SEC-dependent translocation across the inner membrane (IM), the passenger domain with two subdomains of the toxin (p33 and p55), and the translocator β-domain (shown in green). The passenger domain crosses the outer membrane (OM) via a pore in the translocator β-domain. Stage II. The released passenger domain is further processed to form a mature toxin that undergoes a reversible oligomerization process. Stage III. The toxin interacts with the host cell plasma membrane (PM). The N-terminal part of the p33 domain (marked as red H) is incorporated into the lipid bilayer and participates in the formation of the pore. VacA becomes internalized in the process of endocytosis, and it can be found in late endosomes (LE), inner mitochondrial membranes (IMM), and the membranes of other organelles. Question marks denote the hypothetical mechanisms that still lack unequivocal experimental evidence. Based on [10–13].

2. Stage I—Events in the *Helicobacter pylori* Cell

2.1. The vacA Gene

The *vacA* gene is found in almost all isolated *H. pylori* strains [3,10,14–17]. In the case of the *H. pylori* 60190 strain, the transcription start site of *vacA* is located 119 bp upstream of the ATG translation start codon. The transcription of this gene depends on the extended −10 motif: tgaTAAAAG, compared to *E. coli* consensus sequences: TATAAT or TTGACA, and −35 motif sequence TTTATG compared to *E. coli* consensus sequence TTGACA [18]. The mRNA stem–loop-forming structure in the 5′ untranslated region (UTR) of the *vacA* transcript was also identified, and its disruption affects the half life and level of *vacA* mRNA and the VacA protein level [19]. The expression of *vacA* was reported to be influenced by environmental conditions to which the bacterium is exposed, such as acidic pH [20], iron limitation [21], contact with gastric epithelial cells [22], or NaCl [19,23,24]. Higher VacA levels were detected in the supernatants of high-salt-treated bacterial cultures, compared to those grown in conventional media [24], and at least a part of this increase could be attributed to increased *vacA* transcription [19,23,24].

The *vacA* gene varies in length (3.9 kb ± 35 bp) among strains, is highly polymorphic, and has at least six known heterogenic regions that may influence toxin activity and contribute to variable disease outcomes. These include variability in the signal sequence (s1a-d/s2) [16,25,26], as well as in intermediate (i1-3) [27,28] and middle (m1a-c/m2a-b/m3) regions [16,25,29–32]. Three new regions were also described: d1 and d2 subtypes [33] further classified to K-, Q-, or E-types [34], a tail region, which includes n1 and n2 [34,35], and c region with c1 and c2 subtypes [35] (Figure 2).

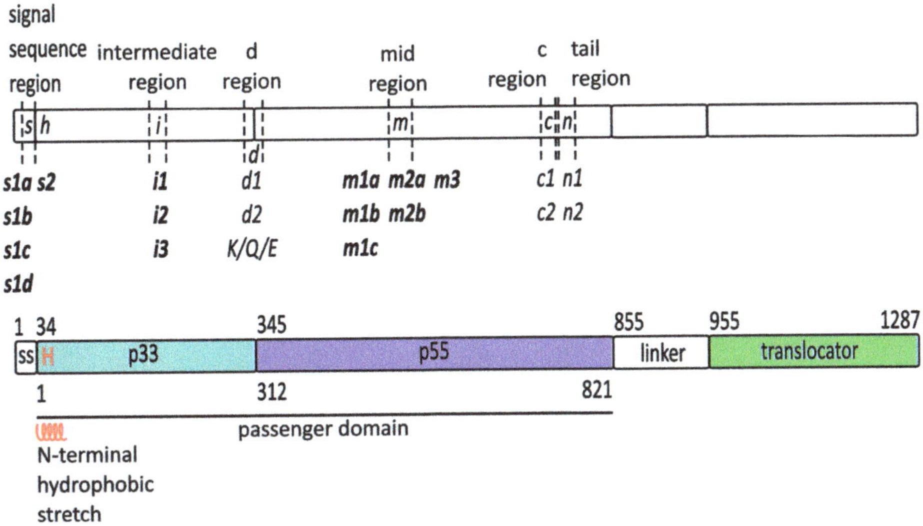

Figure 2. Allelic diversity in regions of the vacA gene and the VacA preprotoxin domain structure (based on VacA from *H. pylori* 60190 strain). Based on [36–38].

Due to the polymorphism, only about 50% of the *H. pylori* isolates can produce detectable amounts of this cytotoxin and exhibit vacuolating activity [4,10,39–41].

For clarity, the description of the structure and function of VacA in this work concerns the s1m1 variant, unless otherwise stated.

2.2. VacA Protein in the H. pylori Cell

The *H. pylori vacA* gene encodes a precursor protein (preprotoxin) of approximately 140 kDa [10,14,15], which is known to be secreted out of the cell and released as a mature toxin.

As all proteins are synthesized in the cytoplasm, the proteins destined for secretion must cross the cell envelope, which in Gram-negative bacteria is composed of the IM, the OM, and the periplasmic space with a peptidoglycan layer between them. To ensure efficient export, several secretion pathways with specialized protein machineries have evolved (Types I–IX secretion systems). These systems facilitate translocation across both membranes and, eventually, navigation through the periplasm. Two main strategies can be used for secretion. Proteins can be secreted directly (one step) from the cytoplasm bypassing the periplasm (Types I, III, IV, VI, and VII secretion systems), or via a two-step strategy. In the latter case, the secretory protein must first pass through the IM using the general secretory (SEC) pathway or twin arginine (TAT) translocation pathway. Then, it traverses the periplasm and becomes incorporated into the OM or exported outside the cell (Type II, V, VIII, or IX secretion systems) [42]. VacA is expected to use the Type V secretion system (T5SS). Although the exact mechanism of this process is not known, the detection of VacA in the periplasm supports a two-step secretion pathway [43].

2.2.1. Translocation of VacA across the Inner Membrane

The recognition and binding of the signal sequence (SS) is the first and necessary step leading to protein translocation through the IM. An analysis of the VacA precursor sequence revealed the presence of a 33 amino-terminal SS with typical features recognized by the general SEC translocon including positive N-terminal charges, a central hydrophobic stretch [10], and the signal peptidase cleavage site between Ala^{33}–Ala^{34} [10,15]. The amino-terminal amino acid of VacA purified from the *H. pylori* 60190 culture medium was determined as Ala^{34} [44] (Figure 3).

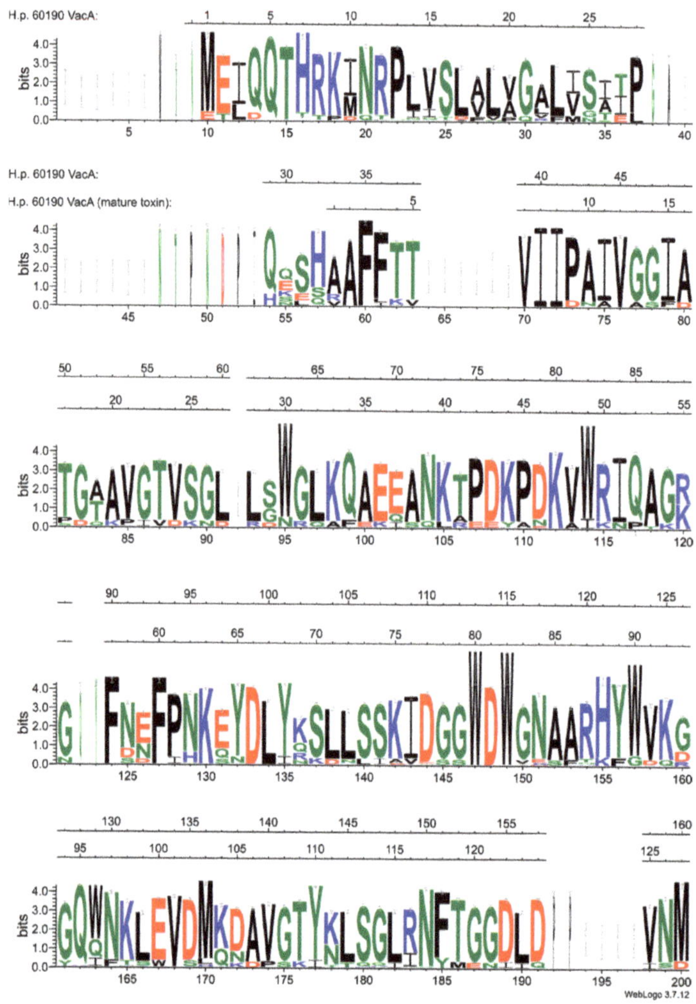

Figure 3. Sequence logo of signal sequence and hydrophobic motif of VacA. First 160 columns from the alignment of 3846 VacA sequences retrieved from UNIPROT using "vacA helicobacter" query subjected to Clustal X2 alignment. Logo generated using WebLogo 3.7.12 (https://weblogo.threeplusone.com, accessed on 14 September 2023). The amino acid residues are grouped and colored based on the R group of their side chain. Red denotes polar acidic amino acid residues (D, E); Blue denotes polar basic amino acid residues (K, R, H); Green denotes polar uncharged amino acid residues (C, G, N, Q, S, T, Y); Black denotes non-polar hydrophobic amino acid residues (A, F, I, L, M, P, V, W). Reference numbering for *H. pylori* 60190 VacA full sequence and mature toxin included above the sequence logo.

In *H. pylori*, the proteins responsible for the SS recognition, binding, and delivery to the SEC complex (core complex: SecYEG, SecY–HP1300, SecE–HP1203a, and SecG–HP1255 with ancillary subcomplex SecDF–YajC, SecD–HP1550, SecF–HP1549, YajC–HP1551, and YidC–HP1450) [45] have not been studied in detail. By analogy to the mechanism described for *E. coli*, it can be expected that the trigger factor homolog (HP0795) binds and protects the nascent proteins that exit the ribosome [46]. The homolog of the *E. coli* SecB chaperone, which delivers preproteins to SecA in the post-translational pathway, was not identified in *H. pylori*. An alternate, co-translational means of transport involving a signal recognition particle (SRP) homolog HP1152 and FtsY (HP0763) [47] is not likely, similar to other autotransporter proteins, but it cannot be excluded [48]. After translocation, the signal peptide will most likely be cut off by the protein HP0576 (a homolog of *E. coli* Type I signal peptidase). However, this process has not been studied in *H. pylori*.

2.2.2. Periplasmic Transit

Proteins pass across the IM via the SEC system in unfolded forms [49]. Therefore, once a polypeptide reaches the periplasm, components of the extracytoplasmic protein quality control system, including a set of periplasmic chaperones and folding helpers, stabilize nascent proteins to prevent misfolding or aggregation [50]. The periplasm of *H. pylori* contains several putative chaperones showing varying degrees of homology to the *E. coli* counterparts. One example is PPIase (peptidyl–prolyl cis–trans isomerase) HP0175, which resembles the *E. coli* SurA protein [47]. SurA plays a major role in (1) The binding of the proteins exiting the Sec channel [51]; (2) Trafficking them across the periplasm; and (3) Incorporating of the β-barrel outer membrane proteins into the OM [52]. The other putative chaperones/folding factors include HP0977 (a homolog of PpiC) and a SurA-like protein HP0659. However, it should be stressed that there are no experimental data on the involvement of these proteins in the transport and folding of exported proteins in *H. pylori*, and their function has only been proposed based on their similarity to their counterparts in other bacterial species. Nevertheless, autotransporters have been shown to interact with chaperones during the periplasmic transit (reviewed in [53]). Therefore, it can be assumed that VacA is also protected in this way before it is incorporated into the OM and exported outside the cell.

2.2.3. Translocator Domain and Translocation across the Outer Membrane

Based on the *vacA* gene structure, VacA has been classified in the autotransporter (AT) family of secreted proteins [10,15], i.e., Type V secretion system (T5SS). The T5SS is regarded as the simplest and one of the most common secretion systems in Gram-negative bacteria. Based on gene organization and protein structure, six sub-classes of the T5SS are distinguished (Va-f) [54]. The VacA protein is classified as Va type, i.e., classical AT [3]. Proteins of this type are composed of the N-terminal passenger domain and the C-terminal translocator (also termed autotransporter) domain, which adopts in the OM a β-barrel structure with a central transmembrane pore. Crystal structure of the translocator domain of VacA is not known, but the type Va ATs generally consist of a 12-stranded β-barrel domain that usually functions as an anchor in the outer membrane [55]. Indeed, the C-terminal fragment of VacA has a predicted β-barrel structure [56–58] and is required for secretion of the 88 kDa toxin [10,15,59].

Despite the name "autotransporter," the stage of insertion and folding of the translocator in the lipid bilayer is assisted by the β-barrel assembly machinery (BAM) or translocation and assembly module (TAM) machineries. The VacA β-barrel translocator domain becomes inserted into the OM in a process that is probably assisted by the BamA/TamA homolog protein HP0655. Once in the membrane, the translocator domain promotes passage of the passenger domain to the cell surface. There, the passenger can adopt the functional conformation.

2.2.4. Release of the Passenger Domain

Following translocation through the translocator channel, the passenger domain is cut off and released as the protoxin of approximately 88 kDa. However, varied masses ranging from 87 to 95 kDa were reported [10,14,15,44,60–65]. A cleavage site between amino acids 991 and 992 was determined by collision-induced dissociation mass spectrometry for *H. pylori* 26695 VacA [61], but the toxin may be processed further yielding of proteins consisting of 821 aa [60]. At present, there is considerable controversy regarding the process of VacA cleavage. Since no protease responsible for VacA processing has been identified, an autocatalytic cut was proposed. This is how the passenger domains of some

E. coli autotransporters, such as EspP [66] and Hbp [67], are processed. There are some indications that VacA may have autoproteolytic activity as the cytotoxic activity of VacA can be blocked by treatment of VacA with 3,4-dichloro-isocoumarin (a serine protease inhibitor). In addition, some features of the amino-terminal portion of VacA are related to serine proteases [68]. However, the *H. pylori* mutant strains with deletion within the p33 region of *vacA* (but encoding p55 and the translocator domains) expressed truncated VacA, which was C-terminally processed and secreted [69,70], indicating that an intact amino-terminal portion of VacA is not required for proteolytic processing of the protoxin. Nevertheless, no direct experimental evidence that VacA possesses proteolytic activity is available.

It is worth noting that although the VacA protoxin is generally secreted to the extracellular environment as a soluble protein [10,15,44,60,61], some of it may remain associated with the cell surface [63,71,72] (Figure 4). Alternatively, it may become part of OMV vesicles (Figure 4, Table 1).

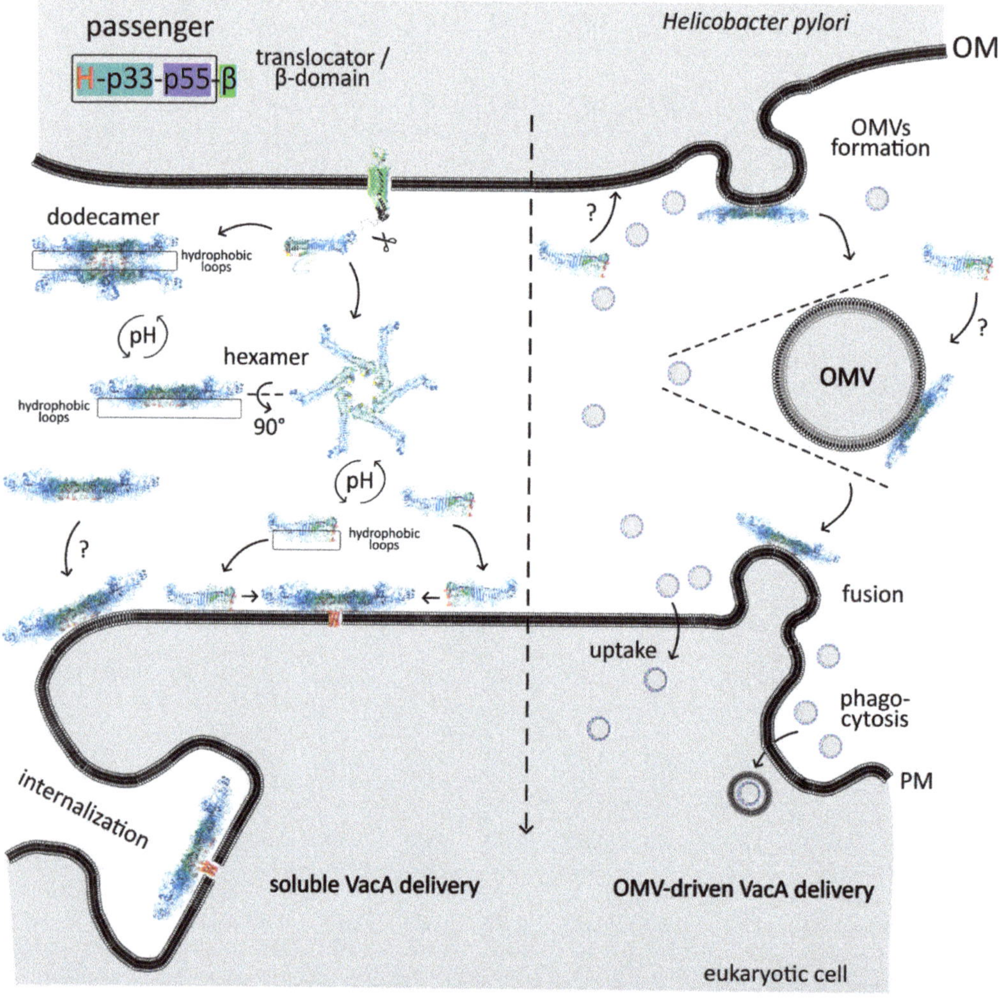

Figure 4. VacA in the extracellular space. The figure shows a fragment of the *H. pylori* cell (top), the extracellular space, and a fragment of the eukaryotic (host) cell (bottom). For translocation across the

outer membrane (OM), the C-terminal ~33 kDa translocator domain inserts into the outer membrane and translocates the passenger domain of VacA to the external side of the outer membrane. Proteolytic cleavage results in the release of the active 88 kDa cytotoxin molecule, which can then oligomerize to form mainly dodecamers. Acidic conditions can promote disassembly of VacA oligomers, which then expose hydrophobic loops, allowing for interaction with membrane lipids. VacA can probably bind to bacterial OM or directly to OMVs to form VacA-containing OMVs (blue/gray circles), which transfer VacA to epithelial cell (by fusion, uptake, or phagocytosis of OMVs). VacA molecules bind to the surface of epithelial cells where they oligomerize, become inserted into plasma membrane, and create a pore. Membrane-bound VacA can be subsequently internalized. Pore forming α-helices are shown in red, p33 domain—deep teal, p55 -marine, passenger-translocator linker—gray, translocator—green; VacA model structure used: PDBid: 6NYG, 1SEW and AF-P55981-F1. OMVs, VacA molecules, and cellular membranes are not drawn to scale. One vesicle with bound VacA is shown in enlargement. Question marks denote the hypothetical mechanisms, which lack unequivocal experimental evidence. Based on [13,73].

Table 1. Presence of VacA in OMVs.

Strain	Ref.
CCUG 17874	[74] *
60190, 84–183	[75] *
4767-C, 2074-Cd	[76] *
NCTC11637	[77] *
CCUG 17875	[78] *
60190, SS1, SS1^{s1i1}	[79] *
251 (and mutants: tolB, tolB+, pal, pal+, tolBpal, cagPAI)	[80]
HP99	[81] *
251 mutant cagPAI	[82]
B128_7.13	[83]
26695	[84–86]
NCTC11637, *H. pylori* 400 (CGMCC 15126)	[87]
TN2	[88]
60190	[89]

* Confirmed by specific antibodies.

3. Stage II—Events in the Extracellular Space

3.1. Processing of the Passenger Domain

The 88 kDa protein can be further processed to form separate p33 and p55 fragments, which are detected in the bacteria culture supernatant [14,60,90,91]. It is now accepted that proteolytic cleavage between p33 and p55 subunits occurs primarily between amino acids A311 and K312 (mature protein numbering) of secreted VacA toxin from *H. pylori* strain 60190 (and possibly several adjacent sites) [92–94]. Based on the MEROPS database and an analysis of the available data on the properties of the M03.006 subclass, PepF (HP0470) was proposed as a peptidase that cleaves the linker sequence between the p33 and p55 VacA domains (within aa 311 and 320) [95]. The p33 and p55 domains copurify, suggesting that these fragments remain associated after cleavage [14]. The experiments using various techniques, including yeast two-hybrid system [94], gel filtration [96], and co-immunoprecipitation from transiently transfected cells [97], further confirmed interaction between p33 and p55 domains.

3.2. Structure of the Mature VacA Toxin

Although the released toxin molecule has a predicted molecular mass of ~88 kDa, it is recovered from growth media as a large >600 kDa complex [44,96]. Using deep etch electron microscopy, native cytotoxin has been shown to form regular ~30 nm oligomers resembling hexagonal "flowers," each composed of a ~15 nm central ring surrounded by six ~6 nm globular "petals". The intact VacA oligomer was found to consist of 12 subunits of approximately 88 kDa that were assembled into two interlocked six-membered arrays, the overlapping of which gave a flower-like appearance [91,96]. Further analysis by cryo-negative staining of the VacA preparations showed multiple types of oligomeric VacA structures, including single-layered astral arrays, bilayered forms, and two-dimensional crystalline arrays [98]. Using cryo-negative staining electron microscopy, views of the different oligomeric structures in multiple orientations were also classified and analyzed, and three-dimensional models of the bilayered forms of VacA were constructed with a resolution of about 19 angstroms [99].

Summarizing the results obtained using various techniques, water-soluble VacA forms several oligomeric structures, such as hexamers [96,98–104] and heptamers [96,99,100,102–104]. However, they predominantly organize into double-layered oligomeric structures [102], mainly dodecamers [98,99,101,103,104] and tetradecamers [99,101,103,104].

A complete high-resolution structure of the active VacA toxin has not been obtained. Only the crystal structures of the isolated p55 subunit and the non-oligomerizing VacA variant containing the p33 subunit have been solved. Nevertheless, structural data obtained by other methods [103–106], including circular dichroism (CD) spectra of the purified VacA protein [107], confirm that the VacA toxin has predominantly a parallel β-strand structure, which is characteristic of autotransporter passenger domains.

4. Stage III—Events in the Host Cell

4.1. Interaction with the Host Cell Membranes

VacA was shown to bind to plasma membranes of different types of human/mammalian cells e.g., HeLa [90,108,109], AGS [90], G401 [110,111], AZ-521 [110–114], RK-13 [111], primary mouse glandular stomach epithelial cells [115], primary human T lymphocytes [116], and others. Most cellular alterations caused by VacA, like permeabilization, increased current, membrane depolarization, and ion conductivity, are attributed to membrane channel formation, either in the plasma membrane or in the membranes of endosomes, lysosomes, or mitochondria [2,101,117–128]. VacA is found in glycosylphosphatidyl inositol anchored protein (GPI-AP)-enriched early endosomal compartments (GEECs) within 10 min after internalization and within next 10ths of minutes in early endosomes (EEs), and then in late endosomes (LEs) [129]. This means that most of VacA is rapidly internalized upon contact with cells and can exert its activity on cell organelles. There is probably no release of the pore-forming moiety of the toxin into the cytosol [130]. It cannot be excluded that a fraction of VacA remains in the plasma membrane where it can also affect the cell physiology and contribute to *H. pylori* virulence.

Overall, the interaction with and internalization of VacA into target cell requires several steps: (1) Binding to the plasma membrane (via receptor); (2) Oligomerization in lipid rafts and formation of the toxin pore; and (3) Internalization by lipid raft-dependent, clathrin-independent endocytosis, or clathrin-independent carriers [130].

4.1.1. VacA Receptors

The first step in pore formation by PFTs is the binding of the toxin to a receptor on the surface of the target cell membrane. The most common types of receptors for PFTs are glycan receptors, protein receptors, and lipid receptors (cholesterol in particular). The binding of the toxin via receptors increases the local concentration of the toxin and promotes oligomerization [131]. Despite years of research, the type of molecule (i.e., lipid or protein) that serves as the VacA receptor on the host cell is still not clearly defined. The interaction of secreted VacA with the target eukaryotic cell via specific surface proteins was reported for numerous cell types, e.g., epidermal growth factor receptor, EGFR (HeLa cells) [108], receptor-type protein tyrosine phosphatase α, RPTPα (kidney cells) [110], (stomach cells) [111], RPTPβ (stomach cells) [111,112,114], (gastric epithelial cells) [115], (kidney cells) [132], integrin beta chain-2, CD18 (T cells) [116], low-density lipoprotein receptor-related protein-1, LRP1 (stomach cells) [113], and Multimerin-1 (platelet cells) [133]. Non-protein molecules were also found to be bound by VacA, e.g., heparan sulphate [113,134], sphingomyelin [135,136], glycosphingolipids [137], and phospholipids [138]. Of the many putative VacA-binding receptors identified in the cell membrane, only the presence or absence of sphingomyelin affected the degree of VacA binding and cell sensitivity to the toxin [135]. Interestingly, the length of the sphingomyelin acyl chain also determines the intracellular transport of VacA [136]. Sphingolipids and cholesterol facilitate the organization of relatively small, highly dynamic, and transient plasma membrane platforms. These platforms attract specific proteins, like lipidated proteins and glycosylphosphatidylinositol-anchored proteins (GPI-APs), and are known as lipid rafts [139]. Consequently, the binding of VacA to sphingomyelin probably accounts for the localization of the toxin to lipid rafts. Several lines of evidence support this hypothesis. First, VacA has an affinity to supported lipid bilayers of various compositions ([total brain lipids], [90% sphingomyelin (SM)-10% cholesterol], [45% dioleoylphosphatidylcholine (DOPC)-45% SM-10% cholesterol], and [50% DOPC-50% SM]) [140]. Furthermore, VacA binds to lipid rafts under conditions that preclude the recruitment of new proteins, suggesting that the VacA receptor resides in lipid rafts permanently [141]. VacA was also found to be enriched in lipid rafts isolated as detergent-resistant membranes (DRMs) [114,125,129,135,142,143], and VacA localization to lipid rafts was visualized using giant plasma membrane vesicles (GPMVs) from HeLa cells [13]. Acid activation has been shown to significantly increase the efficiency of VacA binding to lipid rafts [125].

Cholesterol, an important component of lipid rafts, also influences the binding of VacA [140], as the VacA-induced cell vacuolation was inhibited by the treatment of the cells with either the cholesterol-depleting agent methyl-β-cyclodextrin [125,142,144] or the cholesterol-binding agent nystatin [125,145].

Many membrane-embedded proteins acting as receptors for exogenous ligands are either permanently residing in or are permanently excluded from lipid rafts. However, some membrane receptor proteins that normally reside outside lipid rafts can be recruited into lipid rafts upon ligand binding and crosslinking of receptor–ligand complexes [146]. VacA, after binding to its receptor RPTPβ in non-lipid microdomains of cell surface rafts, also can relocate to lipid rafts [114]. VacA–receptor complex recruitment to lipid rafts can increase toxin local concentration favoring toxin oligomerization and channel formation [145].

4.1.2. Pore Formation

Pore-forming toxins can be classified as α- or β-PFTs depending on the composition of their membrane-spanning regions, i.e., out of α-helices or β-barrels [9,147]. Two alternative pore-forming mechanisms for α- and β-PFTs exist. Most α-PFTs bind to the membrane, and once a critical concentration is reached, PFT subunits insert concomitantly into the membrane and oligomerize to form the final pore. The formation of incomplete but functional pores is possible. β-PFTs concentrate at the membrane interface into the pre-pore, and once oligomerization is completed, conformational change of the toxin molecule inserts it into the membrane [131]. However, the details of the pore formation process by PFTs, including VacA, are still not completely understood [147].

Activation of the Toxin

When added to mammalian cells culture, water-soluble VacA exhibits little or no activity [109,112,138,148], and it needs to be activated either by acid treatment [109,112,138,148] or alkaline activation [109,112]. This phenomenon likely results from the quaternary structure of the VacA molecules. The acid or alkaline treatment results in the disassembly of VacA oligomers into monomers [91,112,138,149], influencing physical properties of VacA [148]. VacA monomers can subsequently reassemble into oligomers when in contact with membranes [102] or in solution at neutral pH [91,112,117] (Figure 4).

Changes accompanying monomer formation and reassembly into oligomers enhance interaction with and insertion into artificial membranes [117,150,151] or of target cells [112,125,142]. Therefore, it was expected that the pH treatment leads to conformational changes that facilitate membrane insertion of the toxin. However, it was demonstrated that the VacA subunit structure does not undergo gross rearrangements, and it was proposed that the dissociation of the high-order oligomeric forms is the main result of VacA acid activation [102]. The monomers can then bind cell surface via specific receptors with increased affinity compared to that of oligomerized toxin [112]. The hydrophobic regions exposed in monomers (but inaccessible in dodecamers and tetradecamers) insert into the membrane, oligomerize, and form a pore.

The ability of VacA to undergo reversible oligomerization and disassembly seems to be crucial for cytotoxic activity. The non-oligomerizing mutant variants of VacA (Δ28–108, Δ49–57, Δ56–83, Δ85–127, Δ112–196, Δ114–294, and Δ346–347) lack cytotoxic activity [70,94, 106,122,123,152]. In addition, the VacA variants (K44C/E338C, K47C/E338C, K55C/D346C) locked into an oligomeric state (cysteine substitutions at sites of p33-p55 contact) are not toxic [152]. Dominant negative mutant forms of VacA (Δ6–27, Δ49–57, Δ346–347) can inhibit the activity of wild-type VacA through a process that involves the formation of mixed oligomers [70,122] or interference with the reassembly of wild-type VacA oligomers after acid treatment [123].

Models of VacA Interaction with Membrane and Pore Formation

Three different models for VacA functional pore formation can be distinguished based on the temporal order of toxin insertion and oligomerization (Figure 5):

1. Regions of p33 insert into the lipid bilayer simultaneously with p88 oligomerization into a hexamer
2. Regions of p33 insert into the lipid bilayer before p88 oligomerization into a hexamer
3. Regions of p33 insert into the lipid bilayer after p88 oligomerization into a hexamer

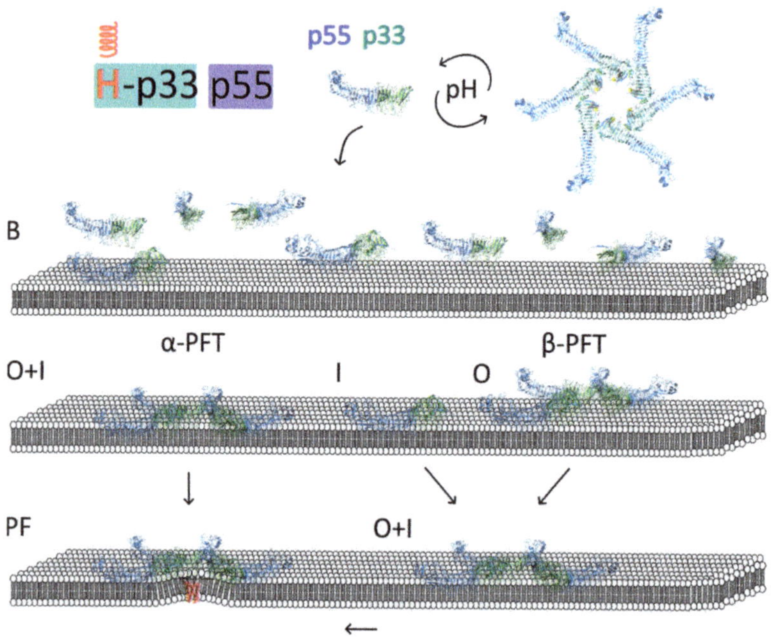

Figure 5. Three different models for acid-induced VacA pore formation can be distinguished based on the temporal order of toxin insertion and oligomerization: (1) Mechanisms characteristic for most of α−PFTs; in this scenario, regions of p33 insert into the lipid bilayer simultaneously with p88 oligomerization into a hexamer; (2) Regions of p33 can also insert into the lipid bilayer before p88 oligomerization into a hexamer; and (3) There is also possibility, as for β−PFTs, that regions of p33 insert into the lipid bilayer after p88 oligomerization into a hexamer and then form a pore. It should

be noted that the position of the hydrophobic stretch (H) in the VacA structure is not known. Therefore, its presence was shown in the membrane-inserted oligomer only. VacA model structure used: PDBid: 6NYG and 1SEW. B, binding, I, insertion, O, oligomerization, PF, pore formation, pore forming α-helices are shown in red, p33 domain—deep teal, p55-marine. Based on [102].

Once bound to the membrane, the p33 regions are expected to undergo structural rearrangement(s) that allow VacA to insert into the bilayer and organize into stable hexamers to form membrane pores [102]. The amino terminus of the p33 subunit contains a predicted hydrophobic region of 32 uncharged residues responsible for insertion into membrane [3,60]. This region contains several GXXXG motifs, amino acid sequences that are predicted to mediate transmembrane dimerization [126,153]. Moreover, it was demonstrated that this part of p33 is relatively unstructured in the monomeric form of VacA and becomes organized during oligomer assembly to form α-helix. This feature may facilitate insertion into the lipid bilayer and pore formation [70,126]. Substitutions of amino acids within this region, including the GXXXG motifs, abolish membrane channel-forming activity and vacuolating toxin activity [93,126,154].

However, VacA oligomers were also shown to exhibit a weak vacuolating activity [109, 112,138,148] and ability to bind to cell membranes. The membrane binding of oligomers was significantly weaker when compared to acid-activated VacA, but the protein was still targeted to lipid rafts [13]. The model of VacA interaction with membrane at neutral pH was proposed with the oligomerization occurring before insertion and pore formation [104] (Figure 6). The model assumptions are as follows:

1. Regions important for insertion into the membrane and channel formation are located at the N-terminal region of the p33 subdomain. These include helices V6–G26 and W30–E37, with the P40 residue at the loop connected to the W30–E37 helix. The W30–E37 helix is hidden in a protomer–protomer interface in the water-soluble hexamer.
2. The interaction of the VacA hexamer with the membrane initially occurs via a cluster of tryptophan residues (W49, W80, W82, W90, and W96) located near the inner rim of the hexamer and the helix W30–E37. Association with the membrane can also be mediated by electrostatic interactions between positively charged amino acid residues of the bottom side of the hexamer (facing another hexamer in dodecamer) and anionic phospholipids.
3. Interaction with membrane lipids induces a change in the position of the W30–E37 helix, while P40 acts as a hinge for this movement. This leads to the exposure of the hydrophobic N-terminus (p33 domain), which forms a helix bundle within the membrane. Importantly, the structure of most of the hexamer's elements remains unchanged.

Roles of Specific VacA Regions in Membrane Binding, Oligomerization, and Membrane Insertion

The participation of individual regions of the mature toxin in processes related to membrane binding, insertion into lipid bilayer, oligomerization, pore formation, and, consequently, in virulence and induction of disease symptoms has been intensively studied by many research teams. Both subunits of the mature toxin, p33 and p55, are required for cell vacuolation and host cell binding [92,97,99,155,156]. However, each domain is responsible for different VacA effects. It is generally accepted that p33 is responsible for VacA pore formation, while p55 is believed to mediate VacA binding to receptors on host cells [155,157]. However, studies performed on various VacA toxin isoforms and mutants suggest that this functional separation is not so clear, and that both subunits, p33 and p55, participate in cell binding.

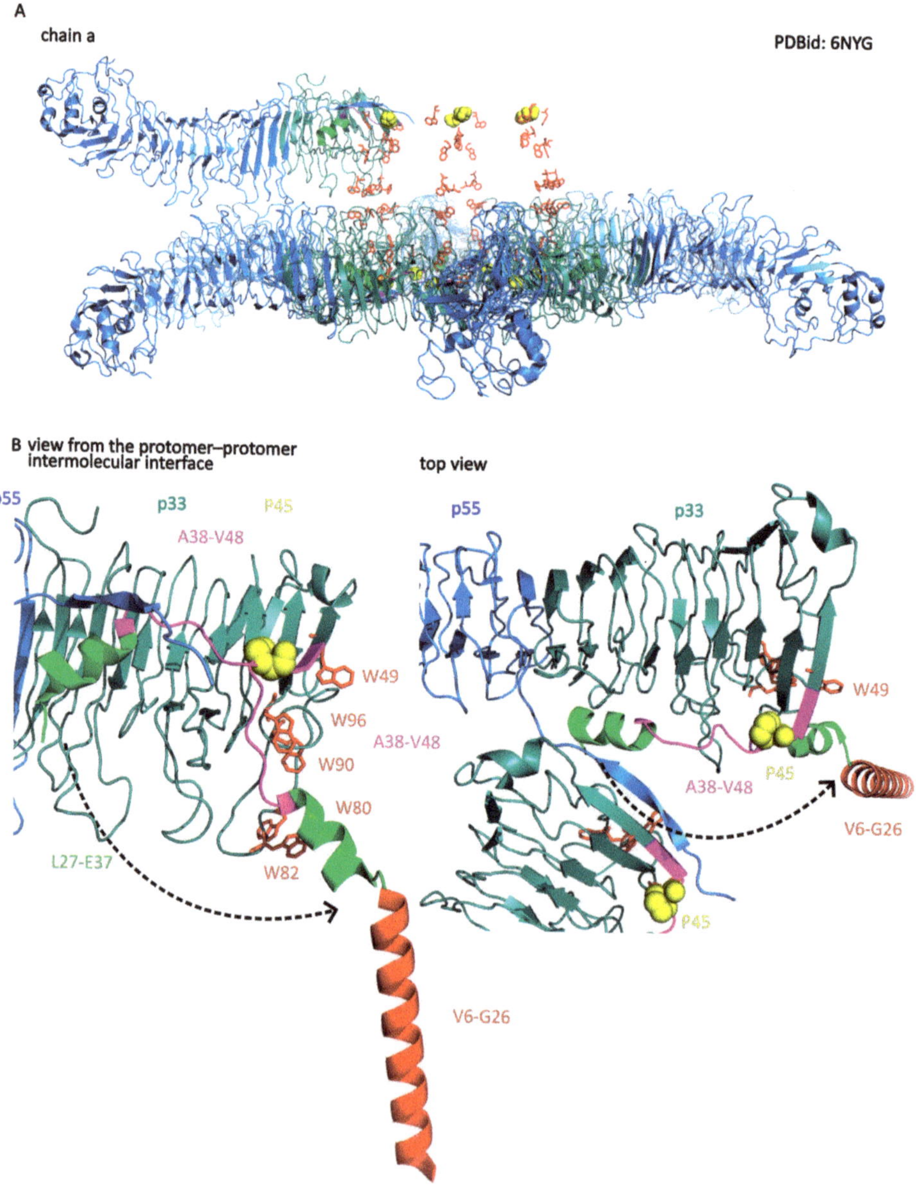

Figure 6. Model of the VacA dodecamer (**A**) and membrane insertion of the VacA hexamer (**B**). (**A**) Structure of water-soluble VacA dodecamer (PDBid: 6NYG). For clarity, five out of six protein chains (b–f) were removed. Structural elements were colored as follows: p33 domain—deepteal, p55-marine, L27-E37 α-helix—green, linker from helix L27-E37 to W49—magenta, tryptophan residues of the tryptophan-rich region—red, Proline 45—yellow. (**B**) After membrane binding aided by tryptophan-rich region (W49, W80, W82, W90, and W96), the N-terminal segment (helix L27–E37 and linker A38–V48 with channel-forming helix V6–G26 not visible in the structure) swings out from the protomer–protomer interface using P45 as a hinge and then exposes the hydrophobic N terminus to form the helix within the membrane (helix V6–G26 modelled in the structure in "after insertion" state). VacA model structure used: PDBid: 6NYG and 1SEW. Based on [104].

As mentioned in Section 3.1, various forms of mature toxin may arise as a result of the *vacA* gene polymorphism. There are three regions related to the degree of virulence of bacteria and the effects they cause on host cells: "s," "i," and "m".

1. The "s" region (signal region), localized at the N-terminus of the VacA preprotein, is identical to the signal sequence in the s1 subtype (and is completely removed during export from the cytoplasm), while in the s2 subtype, the cleavage site of the signal sequence is different. As a result, the VacA protein is longer by 12 amino acid residues at the N-terminus. Its processing during VacA export affects the ability of membrane insertion and pore formation by the secreted toxin. Consequently, the N-terminal sequence of the mature toxin in the s1 type is different than that of the s2 type. This is associated with the different properties of both forms. The s1 type fully exhibits vacuolating activity, while the s2 type lacks detectable cytotoxic activity [158].
2. The "i" region (intermediate region) placed at the C-terminus of the p33 domain has been reported to be involved in both cell vacuolating and binding to various cell types [27,159].
3. The "m" region (middle region) located within the p55 domain with the most common types (m1 or m2) is associated with differences in the ability of VacA to bind to distinct cell types [111,160,161] and exhibit cytotoxin activity [16,111,160–162]. The m1 and m2 forms seem to have different cell-binding specificity. The m2 form induces vacuolization in the primary gastric cells of the RK-13 cell line, but, contrary to the m1 form, it is not able to cause vacuolization in HeLa cells [160]. Several different chimeric variants in the m1/m2 mid-region (R460-G793) of VacA were tested for vacuolating activity and confirmed differences between the m1 and m2 forms of VacA in inducing vacuole formation in RK-13 and HeLa cells [163–165]. The experiments allowed for the specification of amino acids 460–569 within the p55 domain to be responsible for cell binding [164] and showed no significant role of the m2 variant 21-amino acid insert on vacuolization activity of VacA [163]. Different posttranslational modifications of RPTPα from HeLa may be responsible for the reduced susceptibility to m2 VacA [111]. These data favor a protein receptor-mediated interaction of VacA with plasma membrane.

There are multiple studies showing the correlation of VacA polymorphic variants with the severity of diseases attributed to *H. pylori*. This correlation was shown for gastritis [166,167], peptic ulcer [16,166,168–173], gastric cancer [27,33,169,172–180], and intestinal metaplasia [167,177,179,181].

VacA Pore Structure and Function

The formation of a membrane pore by VacA is dependent on p33. However, both domains, p33 and p55, are necessary for efficient cell binding and insertion into the membrane [155]. The structural model of the VacA hexameric anion-selective pore was generated. According to the model, the N-terminus of each VacA hexamer subunit, VacA-transmembrane (TM) (Residues 1–32) hydrophobic stretch, is expected to traverse the membrane as an α-helix. Six of the Gly residues in three GXXXG motifs pack against small Ala or Val side chains to generate the pore [182]. The importance of GXXXG motifs ($V^{12}xGxxxGxxxGxxxGxL^{28}$) was demonstrated using deletion mutants Δ6–27 [70,94], Δ1–17, Δ14–17 [93], Δ1–23/Δ673–913 [183] within the hydrophobic stretch, resulting in the loss of vacuolating activity of VacA. The G14 [93,126,183] and G18 [126] residues were demonstrated as essential for membrane channel formation and VacA cytotoxicity. The VacA Δ6–27 mutant also fails to form membrane channels. Compared to WT VacA, it shows a lack of an organized p33 central core [101]. The importance of the proper N-terminus of the mature VacA toxin is additionally underlined by the fact that the "s2" variant with the 12 amino acid N-terminal extension shows alterations in pore formation and vacuolation activity [158]. The structural model of VacA pore proposed by (Kim et al., 2004) [182] was partially confirmed by cryo–EM structural study where it was possible to trace the N terminus of p33 starting from residue L27. The visible N-terminal region of p33 consists of an α-helix W30–E37, followed by a long loop connected to the β-helix p33 domain central core [104] (Figure 6). The formation of the α-helix by the residues V6–G26 of each VacA oligomer subunit is also expected, but there is still a lack of experimental evidence of the functional VacA pore structure.

Modelling the structure of the VacA pore and studying its properties revealed similarity to other known channels. (Kim et al., 2004) [182] used the structure of the MscS (mechanosensitive channel of small conductance) protein from *Escherichia coli* containing the $A^{98}xxGAxGxAxGxA^{110}$ motif, similar to the VacA's hydrophobic stretch $V^{12}xGxAxGxAxGxV^{24}$, to conduct a quality test of the modeling algorithm used in VacA–TM, which allowed for the validation of the model [182].

The VacA channel exhibits electrophysiological properties of the host chloride channels (ClC). Similar magnitudes of conductance, ion selectivity, and localization within eukaryotic cells largely mimic the electrophysiological behavior of channels in the host cells with a difference only in the membrane potential at which it closes. This feature allows VacA to perturb the homeostatic ionic balance across a membrane without necessarily jeopardizing vitality [6]. Interestingly, VacA is also similar to the cystic fibrosis transmembrane conductance regulator (CFTR). Both proteins form anion-selective (Cl^-), low-conductance pores. Moreover, the conductance of VacA channel can be in the same range as those produced by CFTR stimulation. The similar biophysical properties of VacA and CFTR, as well as the ability of both proteins to penetrate the plasma membrane of respiratory cells and their presence in the endosomal compartment, suggested the use of the VacA protein as a tool to elucidate the different roles of CFTR in the pathogenesis of the CF lung disease [184].

The VacA-induced permeabilization of cells is attributed to the formation of VacA channels in the plasma membrane [118–120]. Channel formation is accompanied by several phenomena such as membrane depolarization [118,125–127], increased current detected in the cell plasma membranes [118] or lipid bilayers [117,119,120,185], and ion conductivity. The conductivity of the VacA channel is very low, and, depending on the experimental conditions, values between 10 and 30 pS were obtained [107,118–120]. VacA channels can conduct chloride [118–120,124], bicarbonate [120,124], and small organic molecules [120], including the passive transport of urea [186]. The VacA-dependent increase of current conduction was effectively inhibited by the chloride channel blockers [118,119,121,185,187] or the chemical modification of VacA [120]. The inhibitors of anion-selective channels [118,121,141] or chemical modification [120] block VacA cytotoxicity without affecting cell binding and endocytosis. Chloride channel blockers also effectively inhibited toxin-induced urea flux [186].

The results mentioned above collectively support the dependence of VacA cytotoxicity on membrane channel formation and indicate that the amino-terminal hydrophobic region of VacA plays an essential role in both membrane channel formation and cytotoxicity.

4.1.3. Internalization

Once bound and inserted into the plasma membrane, VacA becomes efficiently internalized [90,109,141,188,189]. VacA can be released to the extracellular environment not only as a soluble protein but also as a component of the *H. pylori* outer membrane vesicles (OMVs), including particles 20–500 nm in size that are derived from the OM of the Gram-negative bacteria [190] (Figure 7). VacA was detected numerous times in OMVs from different *H. pylori* strains (Table 1; [73,88]), vesicles from bacterial culture and infected cell lines, vesicles from patients infected with *H. pylori* [88], and in gastric juice samples of infected individuals [81]. It was established that approximately 25% of the produced VacA are packaged into OMVs [191]. It has been shown that *H. pylori* vesicles attach to, and are rapidly consumed by, epithelial cells [43,74,78,192,193]. VacA can also be delivered to cells using OMVs as a carrier [74,81,192–195], and the presence of VacA in the OMVs increases rate of their internalization [193]. The OMV's VacA was shown to be biologically active as vacuolation was observed after incubation of cells with OMVs containing VacA [75]. However, OMV's VacA is less effective in this respect compared to the free soluble form [191].

VacA delivery to cells from OMVs was confirmed by several experiments. Orally administered OMVs can enter gastric epithelial cells in the stomach of mice [81], as well as attach to and internalize into primary human antrum cells [192] or human gastric tubular adenocarcinoma [74]. The process of *H. pylori* OMVs internalization is relatively fast as all associated OMVs localize to cells within 20 min. Interestingly, the presence of VacA in the OMVs stimulates their uptake [193]. The exact mechanism of OMV internalization is not known. OMVs can be internalized via clathrin-dependent [193] and clathrin-independent mechanisms [193,194]. Cholesterol is hypothesized to be involved in the internalization of *H. pylori* OMVs by AGS cells, as OMV uptake was significantly reduced due to the depletion of cholesterol in the cell membrane or disruption of cholesterol-rich lipid rafts [194,195].

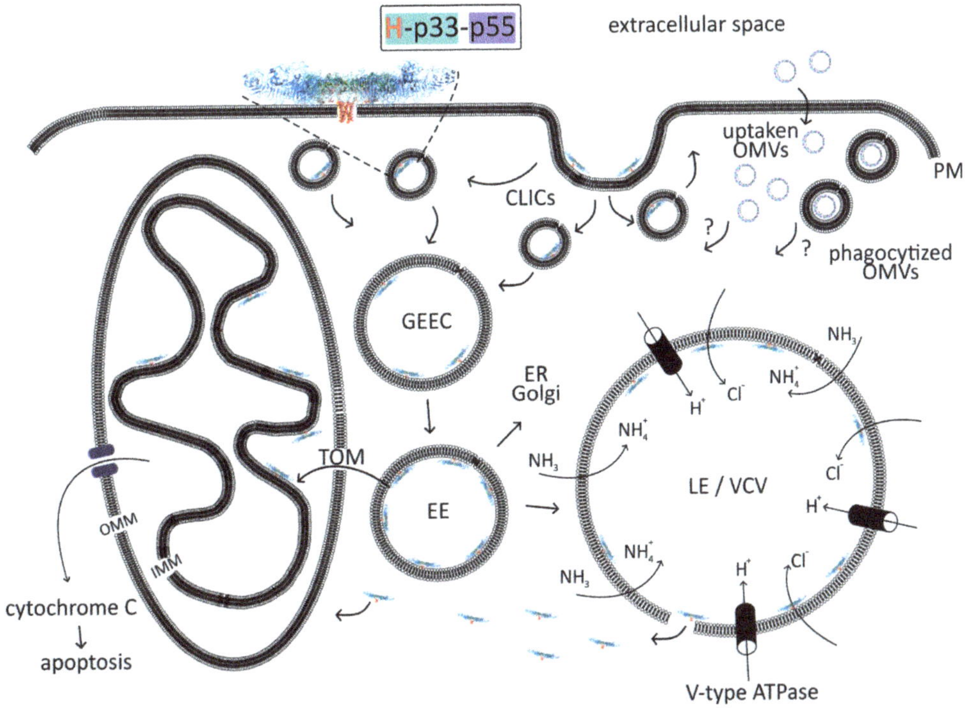

Figure 7. VacA intracellular trafficking. VacA binds to the cell surface and forms a channel that is subsequently endocytosed. The source of intracellular VacA can be also uptaken or phagocytized OMVs, but mechanisms of OMV VacA delivery to target sites are unknown. The first vesicles derived directly from the cell surface and exploited by VacA are termed primary uncoated clathrin-independent tubulovesicular carriers (CLICs). Endocytosis is dependent on Cdc42 and independent of clathrin. VacA accumulates in an early endosomal compartment enriched in GPI-anchored proteins (GEEC). Those compartments are formed by fusion of CLICs. GEECs mature into early endosomes (EE), and then to late endosomes (LE). VacA in cooperation with V-type ATPases causes swelling of the endosomes and, thereby, a formation of vacuoles. VacA can be released from the internal vesicular compartment by rupture of vacuoles and reach a further target, e.g., mitochondria. Another way to reach mitochondria may be the transfer of VacA from EEs during their attachment to the mitochondrial surface. VacA eventually accumulates in the mitochondrial inner membrane. Dissipation of the mitochondrial membrane potential by VacA forming chloride channels causes recruitment of Bax and Bak, release of cytochrome c, and apoptosis. Pore-forming α-helices are shown in red, p33 domain—deepteal, p55 domain—marine. VacA model structure used: PDBid: 6NYG & 1SEW. The fate of internalized or phagocytized OMVs is the subject of ongoing research and, as such, is marked by question marks. Based on [12,196].

Soluble VacA internalization depends on the presence of lipid rafts in the plasma membrane [114,125,135,142–144]. Raft-dependent endocytosis encompasses various pathways but can be generally defined as the cholesterol-sensitive, clathrin-independent (CI) internalization of ligands and receptors from the plasma membrane [197]. It was demonstrated that VacA is composed of host epithelial cells via the clathrin-independent mechanism [129,145,198], while both cholesterol [140] and sphingomyelin [129,135,145,198] are important for binding to cell membranes. Other factors were also recognized as necessary for the uptake of VacA, i.e., actin, calpain, or various PKC [11]. The involvement of actin in the internalization appears to be a common feature of all clathrin-independent endocytosis pathways. VacA was not uniformly associated with the cell surface but was mostly found at the cell's leading edges on filamentous actin (F-actin)–rich membrane extensions whose formation was controlled by the small GTPase Rac1 [129]. Rac1 regulates the reorganization of the actin cytoskeleton and intracellular signal transduction and was shown to regulate VacA activity [129,198,199]. Actin-dependent internalization of the toxin is required for vacuolization because the disruption of the actin cytoskeleton retains VacA on the cell surface [129] or inhibits vacuole formation [129,141,145,200]. VacA internalization does not differ significantly between epithelial cells and lymphocytes and involves clathrin-independent endocytosis, Rac1, and Cdc42 [198].

4.2. Spread of VacA in the Host Cell

A detailed analysis of VacA internalization revealed that the CLIC/GEEC endocytic pathway is involved in this process [129,201] (Figure 7). The CLIC/GEEC pathway was discovered relatively recently. It is clathrin-independent, dynamin-independent, actin-dependent, and it involves Cdc42 to form CLICs, i.e., primary uncoated clathrin-independent tubulovesicular carriers [202]. CLICs are the first vesicles derived directly from the cell surface and accommodate lipid-anchored proteins such as GPI–APs (glycosylphosphatidylinositol–anchored proteins) and a major fraction of the internalized fluid phase. Once formed, CLICs fuse to form a specialized early endosomal compartment called the GPI–AP enriched endosomal compartments (GEECs) [202].

4.2.1. Endosomes

CLIC/GEEC (GPI-AP) Endocytic Pathway of VacA

The VacA exploits the CLIC/GEEC endocytic pathway to reach endosomes and its site of action (Figure 7). VacA was shown to accumulate in GEECs [129,136,201]. A lipid-based sorting mechanism has been proposed for GEECs as cholesterol and sphingolipid levels affect endocytosis via this pathway [202]. Cholesterol and sphingomyelin are important not only for VacA binding to cell membranes [135,140], but they also are suggested to participate in the regulation of the trafficking of VacA within cells [136,144]. For example, normal cells containing a high proportion of long acyl chain sphingomyelin (C18) direct VacA to the GEEC pathway, while cells artificially enriched in short-acyl-chain sphingomyelin (C2) recycle VacA back to the plasma membrane in a Cdc42-independent fashion [136]. Additionally, VacA-induced cell vacuolation is inhibited by the treatment of the cells with either the cholesterol-depleting or cholesterol-binding agents [125,142,144,145].

Endosome maturation is accompanied not only with changes in phosphatidylinositol phospholipids but with the differential recruitment and activation of Rab family GTPases, which control membrane identity, function, and trafficking [203].

Early Endosomes

Most GPI-anchored proteins are recycled back to the plasma membrane [204]. VacA avoids this step, and within 30 min, the toxin reaches early endosomes (EEs) [129] characterized by coating with the small GTPase Rab5. Once VacA reaches EE, VacA-containing vesicles become bridged to filamentous actin (F-actin) structures with the participation of CD2-associated protein (CD2AP). This allows them to exhibit high mobility and be distributed throughout the cell until they finally reach late endosomes (LE) [201]. There are also results suggesting that VacA can be directed from EE to mitochondria by the mechanism that relies on F-actin-driven vesicular motility mechanism [205]. The inhibition of F-actin by cytochalasin D blocks VacA-induced alteration in mitochondrial morphology and significantly decreases apoptosis [200], further supporting the reliance of VacA delivery to mitochondria on F-actin.

Late Endosomes/Lysosomes

Within 120 min, VacA is found in late endosomes (LEs) [129] where it forms anion-selective channels [117–121] and exhibits its hallmark feature i.e., the ability to induce the development of large membrane-bound vacuoles [4,5], termed VacA-containing vacuoles (VCVs) [11]. The membranes of the VacA-induced vacuoles contain small GTPase Rab7, a marker typically found in membranes of late endosomes (LEs) [11,206–211], which suggests that the vacuoles arise from late endosomal compartments [212]. The vacuoles also contain lysosomal (LY) markers, i.e., LAMP1 (Lgp120) [209,211] and LAMP2 (Lgp110) [208], but they do not contain markers for early endocytic compartments [206,207,209]. Therefore, the VacA vacuole is hypothesized to be a hybrid endolysosomal compartment [11]. Another protein colocalizing with Rab7 on vacuolar membranes is vacuolar H^+ ATPase (vacuolar-type ATPase, V-ATPase), which is essential for the VacA-driven vacuole formation [207,213–216]. During endosome maturation, the lumens become increasingly acidic due to the activity of the membrane-embedded V-ATPase. V-ATPases pump hydrogen ions into the vacuole lumen, leading to a pH decrease [203]. The activity of V-ATPase is stimulated by an increase in the intraluminal chloride concentration [216,217], which is dependent on the capacity of VacA to form channels in LE [70,93,118,121,126,216]. The transmembrane pH gradient is needed for the formation and growth of vacuoles [207,213]. A strongly acidic environment leads to the protonation of membrane-permeant weak bases that diffuse into the LEs where they become trapped. This results in water influx, an increase of the osmotic pressure in these compartments, and cell swelling [39,120,188,217,218]. The presence of the weak base NH_4Cl in the culture medium significantly increases toxin vacuolation activity [188,217,219,220]. Vacuoles virtually identical to those observed in VacA-treated cultured cells also occur in superficial gastric epithelial cells biopsied from *H. pylori*-infected patients [221–223] and can be observed in mice infused with VacA from surgically implanted intragastric catheters in parietal cells within gastric tissue [224].

VacA induces considerable vacuole growth until most of the cell cytoplasm is occupied by a few large vacuoles. This growth can be only sustained by the continuous addition of the membrane. One possibility is the use of endosome and lysosome interaction pathways, leading to direct fusion [225]. This scenario is supported by the recognition of several factors known to regulate fusion events between vesicles in the late endocytic pathway, e.g., Rab7, which is required for vacuole formation [209,211,226]. In addition, functional interactions between Rab7 and its downstream effector, Rab-interacting lysosomal protein (RILP), are also necessary for the formation of these bacterial compartments [211]. Another group of proteins involved in fusion processes are SNARE proteins (soluble N-ethylmaleimide-sensitive factor (NSF) attachment protein receptor). One of them, syntaxin-7, which is localized on both late endosome and lysosome, plays a crucial role in their heterotypic fusion and was suggested to be involved in the intracellular vacuolation induced by VacA [227]. Another SNARE protein, VAMP7, is localized to the VacA-induced vacuoles. The expression level of VAMP7 was enhanced in VacA-intoxicated cells, and the downregulation of VAMP7 resulted in the inhibition of VacA-induced vacuolation. VAMP7 was identified as a partner of syntaxin-7 and VacA [228]. The role of SNARE proteins in VacA-induced vacuolization should be further investigated, as another study excluded the involvement of α-SNAP and syntaxin-7 in this process. [212]. Another source of membrane required for the gradual enlargement of vacuoles may be the inner membrane of the organelle, which would fuse with the LE/nascent vacuole membrane [212,216]. This view is supported by the observation that vacuoles are largely devoid of intravesicular membrane structures, characterizing the late endosomal and lysosomal compartments from which vacuoles originate [212,229,230]. They also lack lyso–bisphosphatidic acid, a key component of such internal membrane structures [212].

The physiological role of VacA-induced vacuolation in the pathogenesis of *H. pylori* infection is still unclear. Nevertheless, vacuolation affects a number of cellular functions and may, therefore, be beneficial to bacteria. First of all, VacA leads to the partial neutralization of the acidic pH of the lumen of endosomes and lysosomes. The internal pH of vacuoles was estimated to be at least 0.4 pH units higher than that of LEs and lysosomes in the toxin free cells [231]. This change may result in the inhibition of the degradative power of late endosomal/prelysosomal cargo (e.g., EGF) and mistargeting of acidic hydrolases, e.g., procathepsin D. [231]. VacA itself is also not degraded in the lysosomes [74]. *H. pylori* is a facultative intracellular pathogen and can invade gastric epithelial cells. Intracellular *H. pylori* has been reported within VacA-dependent large vacuoles in a manner associated with an increased survival of bacteria in the cells. Hence, the vacuoles may provide an intracellular niche [232–234]. The suitability of these vacuoles as an intracellular niche for *H. pylori* may be due, at least in part, to the occurrence of VacA-dependent changes in the sorting of lysosomal hydrolases during vacuole biogenesis. This creates nondegradable compartments, preferentially improving the long-term intracellular survival of *H. pylori* [208,211,231]. Additionally, VacA impairs an endolysosomal calcium channel TRPML1 (transient receptor potential membrane channel mucolipin 1) activity, leading to the inhibition of lysosome and autophagy killing, which promotes bacterial colonization [233]. TRPML1 activation in the *H. pylori*-infected gastric cells or human gastric organoids reverses the toxic effects of VacA and leads to the efficient killing of intracellular *H. pylori* [233,235]. Vacuolation has also been suggested to be a simple mechanism for the release of VacA from the internal vesicular compartment. In this way, the toxin can enter the cytosol and reach a further target, e.g., mitochondria [236]. Finally, vacuolation can be considered just as a side effect of the toxin activity routed through affected compartments, as in the case of classical A–B toxins [188]. VacA exhibits a pore-forming activity, and this results in the enlargement of resulting hybrid endosomes and lysosomes before it reaches its destination [130].

4.2.2. Mitochondria/Apoptosis

The presence of *H. pylori* in the gastric mucosa causes cell apoptosis [237], and VacA is one of the factors that induces this process [113,200,238–245]. A significant proportion of VacA can be detected in mitochondria [107,189,200,210,238,246], suggesting VacA might act directly on these organelles to release proapoptotic signals. The details of the pathway for targeting VacA to mitochondria are not fully understood. It was proposed that VacA can reach mitochondria after vacuole rupture and release into the cytosol, or it can be transferred to the mitochondrial surface during direct contact of endosomes with this organelle. Both pathways should eventually lead VacA to reach the mitochondrial inner membrane, where it can form anion-selective channels. VacA can also act on mitochondria indirectly by influencing the activity of the Bcl-2 family proteins, e.g., Bcl-2 (B-cell lymphoma 2) [210,238,243,247,248], Bcl-XL (B-cell lymphoma-extra large) [247,248], Bax (Bcl-2-associated X protein) [210,241,243,246,249], or Bak (Bcl-2 homologous antagonist/killer) [210,246]. Bax and Bak may also play important, direct roles in facilitating the transfer of VacA from endosomes to mitochondria. It is known that VacA leads to the activation of Bax and Bak pro-apoptotic factors [210,246]. The mechanism by which VacA induces Bax or Bak activation is not fully understood. However, it was found that VacA channel formation is necessary for this activity. The mutant variants in the channel-forming region (P9A or G14A) do not induce the accumulation of Bax on endosomes. VacA and Bax colocalization leads to an exposure of the Bax C-terminal and hydrophobic membrane-penetrating α-helice, which then becomes inserted into the outer mitochondrial membrane (OMM). The presence of Bax on endosomes, together with the physical proximity of VacA-containing vesicles and mitochondria, allows for the endosome–mitochondria juxtaposition and transfer of VacA from endosomes to mitochondria. Bax- and Bak-deficient cells were unable to cause the endosome–mitochondria juxtaposition, and the VacA was not retrieved in mitochondria [246].

For proteins to cross the outer mitochondrial membrane, they must pass through the Translocase of the Outer Mitochondrial membrane (TOM complex) [250]. The TOM complex forms a channel that has a diameter insufficient for the translocation of folded proteins, and VacA must be unfolded before or during translocation [128]. Transferring to the mitochondria is possible only if VacA contains the intact N-terminus of 32 residues [107,246], which probably acts as a mitochondrial-targeting signal for interactions with the TOM complex. VacA (at least p33 domain) is eventually integrated into the mitochondrial inner membrane (IMM) [107]. Those findings are supported by the fact that the expression of the p33 domain in cells triggers apoptosis [107,238,251]. However, (Foo et al., 2010) [252] showed that the deletion of residues 6–27 from the p33 domain did not hinder the import of the p33 subunit into mitochondria but was important for the stable integration of the subunits into the inner mitochondrial membrane [252]. There is some indication that the subunit p33 is usually accompanied by p55 [238,246,252], although the p55 subunit can be imported into mitochondria alone via internal targeting signals. However, p55 remains stably associated with mitochondrial membranes only when p33 is present [252].

VacA induces multiple negative effects on mitochondria, including mitochondrial fragmentation [251], reduction [189,210,253,254], or the irreversible loss of mitochondrial transmembrane potential [12], as well as the depletion of ATP [253]. Collectively, it leads to serious disturbances in cell metabolism. The VacA effects on mitochondria are most probably due to the presence of VacA channels in mitochondrial membranes as the VacA variants unable to form channels (Δ6–27, chimeric s2/m1 toxin [240], and VacA P9A [246]), did not induce apoptosis. Additionally, VacA mutants P9A, G14A, Δ6–27 [189,255], and S2M2 variant [189] were also defective in mediating cytochrome c release [189,255] and did not reduce the mitochondrial transmembrane potential [189]. Similarly, it was also shown that chloride channel blockers inhibit the VacA-induced cytochrome c release [189,255] and reduction of mitochondrial transmembrane potential [189]. The disruption of the mitochondrial transmembrane potential occurs prior to mitochondrial outer membrane permeabilization (MOMP) [189]. The dissipation of the mitochondrial transmembrane electrical potential results in mitochondrial recruitment and activation of dynamin-related protein 1 (Drp1), which is a crucial regulator of mitochondrial fission through its GTPase activity and causes mitochondrial fragmentation (with a disappearance of the typical spaghetti-like morphology). In cells infected with an *H. pylori vacA* knockout strain, no mitochondrial fragmentation was visible. The inhibition of Drp1-dependent mitochondrial fission within VacA-intoxicated cells prevents the activation of Bax, MOMP, and, consequently, cell death [251]. As mentioned earlier, VacA can also act on mitochondria indirectly by influencing the activity of the Bcl-2 family proteins. VacA was shown to upregulate the expression of Bax (a multi-domain pro-apoptotic protein) and VDAC1 (voltage-dependent anion-selective channel 1, an endogenous outer-mitochondrial membrane channel), resulting in the VacA-induced MOMP and subsequent release of cytochrome c [243]. Factors like the increased cytoplasmic pool of the apoptosis-inducing factor (AIF), seen after *H. pylori* exposure [256] or expression of cellular inhibitor of apoptosis protein (c-IAP)-2 upregulated during the early period of VacA stimulation [249], may also significantly influence the fate of the cell. Ultimately, *H. pylori* VacA causes the release of cytochrome c [189,210,238,243,255] and induces cell death.

5. Conclusions

The VacA protein is a very important virulence factor of the human pathogen, the *H. pylori* bacterium. It is a toxin secreted outside the bacterium that enters host cells and forms transmembrane pores in both the plasma membrane and organelle membranes. This review presents the route that VacA must follow from its synthesis in the bacterial cytoplasm to its incorporation into the membranes of host cell organelles. Reaching the target location requires overcoming several membrane barriers, which occurs using various mechanisms: classical export via the SEC translocon, T5SS secretion, penetration into the lipid bilayer thanks to regions capable of forming a membrane channel structure, endocytosis, and TOM complex translocation. The individual stages of this journey are accompanied by changes in the primary sequence of the VacA protein, which undergoes the sequential post-translational modifications, consisting in cutting off fragments of the polypeptide chain, which ultimately leads to the formation of the mature toxin. The release of the toxin into the extracellular environment enables it to adopt structures capable of interacting with the host cell membranes, and further conformational changes lead to the formation of transmembrane pores. Not all processes involved in the VacA transport have been understood. For example, the mechanism of pore formation in the membrane has not been characterized in detail to date. It is also unknown how exactly VacA enters the inner membrane of mitochondria. The proposed functions of VacA as a virulence factor are related to its ability to form transmembrane pores. These pores have the characteristics of chloride channels, which causes disturbances in the homeostasis of ion exchanges in the cell. Such channel activity leads to the formation of large cytoplasmic vacuoles and also causes disruptions in the functioning of endosomes, lysosomes, and mitochondria. In this way, VacA is thought to facilitate the successful infection of the gastric epithelium by *H. pylori*.

Author Contributions: Conceptualization, M.J. and J.S.-G.; funding acquisition: M.J.; writing—original draft preparation, M.J. and J.S.-G.; writing—review and editing, J.S.-G. and M.J.; visualization, M.J.; figures M.J. All authors have read and agreed to the published version of the manuscript.

Funding: This research was funded by University of Gdansk, grant number 1220.6010.157.2023.

Institutional Review Board Statement: Not applicable.

Informed Consent Statement: Not applicable.

Data Availability Statement: Data sharing is not applicable.

Conflicts of Interest: The authors declare no conflict of interest.

References

1. Murray, P.R.; Rosenthal, K.S.; Pfaller, M.A. *Medical Microbiology*; Elsevier: Amsterdam, The Netherlands, 2020.
2. Foegeding, N.J.; Caston, R.R.; McClain, M.S.; Ohi, M.D.; Cover, T.L. An Overview of *Helicobacter pylori* VacA Toxin Biology. *Toxins* **2016**, *8*, 173. [CrossRef] [PubMed]
3. Cover, T.L.; Blanke, S.R. *Helicobacter pylori* VacA, a paradigm for toxin multifunctionality. *Nat. Rev. Microbiol.* **2005**, *3*, 320–332. [CrossRef] [PubMed]
4. Leunk, R.D.; Johnson, P.T.; David, B.C.; Kraft, W.G.; Morgan, D.R. Cytotoxic activity in broth-culture filtrates of Campylobacter pylori. *J. Med. Microbiol.* **1988**, *26*, 93–99. [CrossRef]
5. Cover, T.L.; Halter, S.A.; Blaser, M.J. Characterization of HeLa cell vacuoles induced by *Helicobacter pylori* broth culture supernatant. *Hum. Pathol.* **1992**, *23*, 1004–1010. [CrossRef] [PubMed]
6. Czajkowsky, D.M.; Iwamoto, H.; Szabo, G.; Cover, T.L.; Shao, Z. Mimicry of a host anion channel by a *Helicobacter pylori* pore-forming toxin. *Biophys. J.* **2005**, *89*, 3093–3101. [CrossRef] [PubMed]
7. Wayne, R. Chapter 2—Plasma Membrane. In *Plant Cell Biology*, 2nd ed.; Wayne, R., Ed.; Academic Press: Cambridge, MA, USA, 2019; pp. 17–53.
8. Ma, Y.; Poole, K.; Goyette, J.; Gaus, K. Introducing Membrane Charge and Membrane Potential to T Cell Signaling. *Front. Immunol.* **2017**, *8*, 1513. [CrossRef] [PubMed]
9. Dal Peraro, M.; van der Goot, F.G. Pore-forming toxins: Ancient, but never really out of fashion. *Nat. Rev. Microbiol.* **2016**, *14*, 77–92. [CrossRef]
10. Schmitt, W.; Haas, R. Genetic analysis of the Helicobacter pylori vacuolating cytotoxin: Structural similarities with the IgA protease type of exported protein. *Mol. Microbiol.* **1994**, *12*, 307–319. [CrossRef]
11. Kern, B.; Jain, U.; Utsch, C.; Otto, A.; Busch, B.; Jiménez-Soto, L.; Becher, D.; Haas, R. Characterization of *Helicobacter pylori* VacA-containing vacuoles (VCVs), VacA intracellular trafficking and interference with calcium signalling in T lymphocytes. *Cell Microbiol.* **2015**, *17*, 1811–1832. [CrossRef]
12. Rassow, J. Helicobacter pylori vacuolating toxin A and apoptosis. *Cell Commun. Signal* **2011**, *9*, 26. [CrossRef]
13. Raghunathan, K.; Foegeding, N.J.; Campbell, A.M.; Cover, T.L.; Ohi, M.D.; Kenworthy, A.K. Determinants of Raft Partitioning of the *Helicobacter pylori* Pore-Forming Toxin VacA. *Infect. Immun.* **2018**, *86*, e00872-17. [CrossRef] [PubMed]

14. Telford, J.L.; Ghiara, P.; Dell'Orco, M.; Comanducci, M.; Burroni, D.; Bugnoli, M.; Tecce, M.F.; Censini, S.; Covacci, A.; Xiang, Z.; et al. Gene structure of the *Helicobacter pylori* cytotoxin and evidence of its key role in gastric disease. *J. Exp. Med.* **1994**, *179*, 1653–1658. [CrossRef] [PubMed]
15. Cover, T.L.; Tummuru, M.K.; Cao, P.; Thompson, S.A.; Blaser, M.J. Divergence of genetic sequences for the vacuolating cytotoxin among *Helicobacter pylori* strains. *J. Biol. Chem.* **1994**, *269*, 10566–10573. [CrossRef]
16. Atherton, J.C.; Cao, P.; Peek, R.M., Jr.; Tummuru, M.K.; Blaser, M.J.; Cover, T.L. Mosaicism in vacuolating cytotoxin alleles of Helicobacter pylori. Association of specific vacA types with cytotoxin production and peptic ulceration. *J. Biol. Chem.* **1995**, *270*, 17771–17777. [CrossRef] [PubMed]
17. Cover, T.L.; Peek, R.M., Jr. Diet, microbial virulence, and *Helicobacter pylori*-induced gastric cancer. *Gut Microbes* **2013**, *4*, 482–493. [CrossRef] [PubMed]
18. Forsyth, M.H.; Cover, T.L. Mutational analysis of the vacA promoter provides insight into gene transcription in *Helicobacter pylori*. *J. Bacteriol.* **1999**, *181*, 2261–2266. [CrossRef] [PubMed]
19. Amilon, K.R.; Letley, D.P.; Winter, J.A.; Robinson, K.; Atherton, J.C. Expression of the *Helicobacter pylori* virulence factor vacuolating cytotoxin A (vacA) is influenced by a potential stem-loop structure in the 5′ untranslated region of the transcript. *Mol. Microbiol.* **2015**, *98*, 831–846. [CrossRef]
20. Barnard, F.M.; Loughlin, M.F.; Fainberg, H.P.; Messenger, M.P.; Ussery, D.W.; Williams, P.; Jenks, P.J. Global regulation of virulence and the stress response by CsrA in the highly adapted human gastric pathogen *Helicobacter pylori*. *Mol. Microbiol.* **2004**, *51*, 15–32. [CrossRef]
21. Merrell, D.S.; Thompson, L.J.; Kim, C.C.; Mitchell, H.; Tompkins, L.S.; Lee, A.; Falkow, S. Growth phase-dependent response of *Helicobacter pylori* to iron starvation. *Infect. Immun.* **2003**, *71*, 6510–6525. [CrossRef]
22. van Amsterdam, K.; van Vliet, A.H.; Kusters, J.G.; Feller, M.; Dankert, J.; van der Ende, A. Induced *Helicobacter pylori* vacuolating cytotoxin VacA expression after initial colonisation of human gastric epithelial cells. *FEMS Immunol. Med. Microbiol.* **2003**, *39*, 251–256. [CrossRef]
23. Gancz, H.; Jones, K.R.; Merrell, D.S. Sodium chloride affects *Helicobacter pylori* growth and gene expression. *J. Bacteriol.* **2008**, *190*, 4100–4105. [CrossRef] [PubMed]
24. Caston, R.R.; Loh, J.T.; Voss, B.J.; McDonald, W.H.; Scholz, M.B.; McClain, M.S.; Cover, T.L. Effect of environmental salt concentration on the *Helicobacter pylori* exoproteome. *J. Proteom.* **2019**, *202*, 103374. [CrossRef] [PubMed]
25. van Doorn, L.J.; Figueiredo, C.; Sanna, R.; Pena, S.; Midolo, P.; Ng, E.K.; Atherton, J.C.; Blaser, M.J.; Quint, W.G. Expanding allelic diversity of *Helicobacter pylori* vacA. *J. Clin. Microbiol.* **1998**, *36*, 2597–2603. [CrossRef] [PubMed]
26. Delgado-Sapién, G.; Cerritos, R.; González-Valencia, G.; Méndez, J.L.; Cravioto, A.; Morales-Espinosa, R. Novel VacA S1d Signal Sequence Found in Helicobacter Pylori from Mexican Children with Recurrent Abdominal Pain. *SOJ Microbiol. Infect. Dis.* **2017**, *5*, 1–7. [CrossRef]
27. Rhead, J.L.; Letley, D.P.; Mohammadi, M.; Hussein, N.; Mohagheghi, M.A.; Eshagh Hosseini, M.; Atherton, J.C. A new *Helicobacter pylori* vacuolating cytotoxin determinant, the intermediate region, is associated with gastric cancer. *Gastroenterology* **2007**, *133*, 926–936. [CrossRef]
28. Chung, C.; Olivares, A.; Torres, E.; Yilmaz, O.; Cohen, H.; Perez-Perez, G. Diversity of VacA intermediate region among *Helicobacter pylori* strains from several regions of the world. *J. Clin. Microbiol.* **2010**, *48*, 690–696. [CrossRef]
29. Han, S.R.; Schreiber, H.J.; Bhakdi, S.; Loos, M.; Maeurer, M.J. vacA genotypes and genetic diversity in clinical isolates of *Helicobacter pylori*. *Clin. Diagn. Lab. Immunol.* **1998**, *5*, 139–145. [CrossRef]
30. Strobel, S.; Bereswill, S.; Balig, P.; Allgaier, P.; Sonntag, H.G.; Kist, M. Identification and analysis of a new vacA genotype variant of *Helicobacter pylori* in different patient groups in Germany. *J. Clin. Microbiol.* **1998**, *36*, 1285–1289. [CrossRef]
31. Pan, Z.J.; Berg, D.E.; van der Hulst, R.W.; Su, W.W.; Raudonikiene, A.; Xiao, S.D.; Dankert, J.; Tytgat, G.N.; van der Ende, A. Prevalence of vacuolating cytotoxin production and distribution of distinct vacA alleles in *Helicobacter pylori* from China. *J. Infect. Dis.* **1998**, *178*, 220–226. [CrossRef]
32. Mukhopadhyay, A.K.; Kersulyte, D.; Jeong, J.Y.; Datta, S.; Ito, Y.; Chowdhury, A.; Chowdhury, S.; Santra, A.; Bhattacharya, S.K.; Azuma, T.; et al. Distinctiveness of genotypes of *Helicobacter pylori* in Calcutta, India. *J. Bacteriol.* **2000**, *182*, 3219–3227. [CrossRef]
33. Ogiwara, H.; Sugimoto, M.; Ohno, T.; Vilaichone, R.K.; Mahachai, V.; Graham, D.Y.; Yamaoka, Y. Role of deletion located between the intermediate and middle regions of the *Helicobacter pylori* vacA gene in cases of gastroduodenal diseases. *J. Clin. Microbiol.* **2009**, *47*, 3493–3500. [CrossRef] [PubMed]
34. Soyfoo, D.M.; Doomah, Y.H.; Xu, D.; Zhang, C.; Sang, H.M.; Liu, Y.Y.; Zhang, G.X.; Jiang, J.X.; Xu, S.F. New genotypes of *Helicobacter pylori* VacA d-region identified from global strains. *BMC Mol. Cell Biol.* **2021**, *22*, 4. [CrossRef] [PubMed]
35. Bakhti, S.Z.; Latifi-Navid, S.; Mohammadi, S.; Zahri, S.; Bakhti, F.S.; Feizi, F.; Yazdanbod, A.; Siavoshi, F. Relevance of *Helicobacter pylori* vacA 3′-end Region Polymorphism to Gastric Cancer. *Helicobacter* **2016**, *21*, 305–316. [CrossRef] [PubMed]
36. Thi Huyen Trang, T.; Thanh Binh, T.; Yamaoka, Y. Relationship between vacA Types and Development of Gastroduodenal Diseases. *Toxins* **2016**, *8*, 182. [CrossRef]
37. Junaid, M.; Linn, A.K.; Javadi, M.B.; Al-Gubare, S.; Ali, N.; Katzenmeier, G. Vacuolating cytotoxin A (VacA)—A multi-talented pore-forming toxin from *Helicobacter pylori*. *Toxicon* **2016**, *118*, 27–35. [CrossRef] [PubMed]
38. Kim, I.J.; Blanke, S.R. Remodeling the host environment: Modulation of the gastric epithelium by the *Helicobacter pylori* vacuolating toxin (VacA). *Front. Cell Infect. Microbiol.* **2012**, *2*, 37. [CrossRef] [PubMed]

39. Cover, T.L.; Dooley, C.P.; Blaser, M.J. Characterization of and human serologic response to proteins in *Helicobacter pylori* broth culture supernatants with vacuolizing cytotoxin activity. *Infect. Immun.* **1990**, *58*, 603–610. [CrossRef]
40. Covacci, A.; Censini, S.; Bugnoli, M.; Petracca, R.; Burroni, D.; Macchia, G.; Massone, A.; Papini, E.; Xiang, Z.; Figura, N.; et al. Molecular characterization of the 128-kDa immunodominant antigen of *Helicobacter pylori* associated with cytotoxicity and duodenal ulcer. *Proc. Natl. Acad. Sci. USA* **1993**, *90*, 5791–5795. [CrossRef]
41. Phadnis, S.H.; Ilver, D.; Janzon, L.; Normark, S.; Westblom, T.U. Pathological significance and molecular characterization of the vacuolating toxin gene of *Helicobacter pylori*. *Infect. Immun.* **1994**, *62*, 1557–1565. [CrossRef]
42. Rapisarda, C.; Fronzes, R. Secretion Systems Used by Bacteria to Subvert Host Functions. *Curr. Issues Mol. Biol.* **2018**, *25*, 1–42. [CrossRef]
43. Fiocca, R.; Necchi, V.; Sommi, P.; Ricci, V.; Telford, J.; Cover, T.L.; Solcia, E. Release of Helicobacter pylori vacuolating cytotoxin by both a specific secretion pathway and budding of outer membrane vesicles. Uptake of released toxin and vesicles by gastric epithelium. *J. Pathol.* **1999**, *188*, 220–226. [CrossRef]
44. Cover, T.L.; Blaser, M.J. Purification and characterization of the vacuolating toxin from *Helicobacter pylori*. *J. Biol. Chem.* **1992**, *267*, 10570–10575. [CrossRef] [PubMed]
45. Troman, L.; Collinson, I. Pushing the Envelope: The Mysterious Journey Through the Bacterial Secretory Machinery, and Beyond. *Front. Microbiol.* **2021**, *12*, 782900. [CrossRef] [PubMed]
46. Smets, D.; Loos, M.S.; Karamanou, S.; Economou, A. Protein Transport Across the Bacterial Plasma Membrane by the Sec Pathway. *Protein J.* **2019**, *38*, 262–273. [CrossRef] [PubMed]
47. Liechti, G.; Goldberg, J.B. Outer membrane biogenesis in *Escherichia coli*, *Neisseria meningitidis*, and *Helicobacter pylori*: Paradigm deviations in H. pylori. *Front. Cell Infect. Microbiol.* **2012**, *2*, 29. [CrossRef] [PubMed]
48. Leyton, D.L.; Rossiter, A.E.; Henderson, I.R. From self sufficiency to dependence: Mechanisms and factors important for autotransporter biogenesis. *Nat. Rev. Microbiol.* **2012**, *10*, 213–225. [CrossRef]
49. Ambroziak, P.; Rzepka, I.; Skorko-Glonek, J. SecA—A multidomain and multitask bacterial export protein. *Acta Biochim. Pol.* **2021**, *68*, 427–436. [CrossRef]
50. Macošek, J.; Mas, G.; Hiller, S. Redefining Molecular Chaperones as Chaotropes. *Front. Mol. Biosci.* **2021**, *8*, 683132. [CrossRef]
51. Troman, L.; Alvira, S.; Daum, B.; Gold, V.A.M.; Collinson, I. Interaction of the periplasmic chaperone SurA with the inner membrane protein secretion (SEC) machinery. *Biochem. J.* **2023**, *480*, 283–296. [CrossRef]
52. Schiffrin, B.; Machin, J.M.; Karamanos, T.K.; Zhuravleva, A.; Brockwell, D.J.; Radford, S.E.; Calabrese, A.N. Dynamic interplay between the periplasmic chaperone SurA and the BAM complex in outer membrane protein folding. *Commun. Biol.* **2022**, *5*, 560. [CrossRef]
53. Figaj, D.; Ambroziak, P.; Rzepka, I.; Skórko-Glonek, J. SurA-like and Skp-like Proteins as Important Virulence Determinants of the Gram Negative Bacterial Pathogens. *Int. J. Mol. Sci.* **2023**, *24*, 295. [CrossRef] [PubMed]
54. Dautin, N. Folding Control in the Path of Type 5 Secretion. *Toxins* **2021**, *13*, 341. [CrossRef] [PubMed]
55. Meuskens, I.; Saragliadis, A.; Leo, J.C.; Linke, D. Type V Secretion Systems: An Overview of Passenger Domain Functions. *Front. Microbiol.* **2019**, *10*, 1163. [CrossRef] [PubMed]
56. Klauser, T.; Pohlner, J.; Meyer, T.F. The secretion pathway of IgA protease-type proteins in gram-negative bacteria. *Bioessays* **1993**, *15*, 799–805. [CrossRef] [PubMed]
57. Loveless, B.J.; Saier, M.H., Jr. A novel family of channel-forming, autotransporting, bacterial virulence factors. *Mol. Membr. Biol.* **1997**, *14*, 113–123. [CrossRef] [PubMed]
58. Wang, H.J.; Chang, P.C.; Kuo, C.H.; Tzeng, C.S.; Wang, W.C. Characterization of the C-terminal domain of *Helicobacter pylori* vacuolating toxin and its relationship with extracellular toxin production. *Biochem. Biophys. Res. Commun.* **1998**, *250*, 397–402. [CrossRef]
59. Fischer, W.; Buhrdorf, R.; Gerland, E.; Haas, R. Outer membrane targeting of passenger proteins by the vacuolating cytotoxin autotransporter of *Helicobacter pylori*. *Infect. Immun.* **2001**, *69*, 6769–6775. [CrossRef]
60. Nguyen, V.Q.; Caprioli, R.M.; Cover, T.L. Carboxy-terminal proteolytic processing of *Helicobacter pylori* vacuolating toxin. *Infect. Immun.* **2001**, *69*, 543–546. [CrossRef]
61. Bumann, D.; Aksu, S.; Wendland, M.; Janek, K.; Zimny-Arndt, U.; Sabarth, N.; Meyer, T.F.; Jungblut, P.R. Proteome analysis of secreted proteins of the gastric pathogen *Helicobacter pylori*. *Infect. Immun.* **2002**, *70*, 3396–3403. [CrossRef]
62. Smith, T.G.; Lim, J.M.; Weinberg, M.V.; Wells, L.; Hoover, T.R. Direct analysis of the extracellular proteome from two strains of *Helicobacter pylori*. *Proteomics* **2007**, *7*, 2240–2245. [CrossRef]
63. Voss, B.J.; Gaddy, J.A.; McDonald, W.H.; Cover, T.L. Analysis of surface-exposed outer membrane proteins in *Helicobacter pylori*. *J. Bacteriol.* **2014**, *196*, 2455–2471. [CrossRef] [PubMed]
64. Snider, C.A.; Voss, B.J.; McDonald, W.H.; Cover, T.L. Growth phase-dependent composition of the *Helicobacter pylori* exoproteome. *J. Proteom.* **2016**, *130*, 94–107. [CrossRef] [PubMed]
65. Kim, N.; Weeks, D.L.; Shin, J.M.; Scott, D.R.; Young, M.K.; Sachs, G. Proteins released by *Helicobacter pylori* in vitro. *J. Bacteriol.* **2002**, *184*, 6155–6162. [CrossRef] [PubMed]
66. Dautin, N.; Barnard, T.J.; Anderson, D.E.; Bernstein, H.D. Cleavage of a bacterial autotransporter by an evolutionarily convergent autocatalytic mechanism. *EMBO J.* **2007**, *26*, 1942–1952. [CrossRef]
67. Tajima, N.; Kawai, F.; Park, S.Y.; Tame, J.R. A novel intein-like autoproteolytic mechanism in autotransporter proteins. *J. Mol. Biol.* **2010**, *402*, 645–656. [CrossRef]

68. Rossetto, O.; de Bernard, M.; Pellizzari, R.; Vitale, G.; Caccin, P.; Schiavo, G.; Montecucco, C. Bacterial toxins with intracellular protease activity. *Clin. Chim. Acta* **2000**, *291*, 189–199. [CrossRef]
69. Reyrat, J.M.; Lanzavecchia, S.; Lupetti, P.; de Bernard, M.; Pagliaccia, C.; Pelicic, V.; Charrel, M.; Ulivieri, C.; Norais, N.; Ji, X.; et al. 3D imaging of the 58 kDa cell binding subunit of the *Helicobacter pylori* cytotoxin. *J. Mol. Biol.* **1999**, *290*, 459–470. [CrossRef]
70. Vinion-Dubiel, A.D.; McClain, M.S.; Czajkowsky, D.M.; Iwamoto, H.; Ye, D.; Cao, P.; Schraw, W.; Szabo, G.; Blanke, S.R.; Shao, Z.; et al. A dominant negative mutant of *Helicobacter pylori* vacuolating toxin (VacA) inhibits VacA-induced cell vacuolation. *J. Biol. Chem.* **1999**, *274*, 37736–37742. [CrossRef]
71. Sabarth, N.; Hurvitz, R.; Schmidt, M.; Zimny-Arndt, U.; Jungblut, P.R.; Meyer, T.F.; Bumann, D. Identification of *Helicobacter pylori* surface proteins by selective proteinase K digestion and antibody phage display. *J. Microbiol. Methods* **2005**, *62*, 345–349. [CrossRef]
72. Ilver, D.; Barone, S.; Mercati, D.; Lupetti, P.; Telford, J.L. *Helicobacter pylori* toxin VacA is transferred to host cells via a novel contact-dependent mechanism. *Cell Microbiol.* **2004**, *6*, 167–174. [CrossRef]
73. Jarzab, M.; Posselt, G.; Meisner-Kober, N.; Wessler, S. *Helicobacter pylori*-Derived Outer Membrane Vesicles (OMVs): Role in Bacterial Pathogenesis? *Microorganisms* **2020**, *8*, 1328. [CrossRef] [PubMed]
74. Sommi, P.; Ricci, V.; Fiocca, R.; Necchi, V.; Romano, M.; Telford, J.L.; Solcia, E.; Ventura, U. Persistence of *Helicobacter pylori* VacA toxin and vacuolating potential in cultured gastric epithelial cells. *Am. J. Physiol.* **1998**, *275*, G681–G688. [CrossRef] [PubMed]
75. Keenan, J.; Day, T.; Neal, S.; Cook, B.; Perez-Perez, G.; Allardyce, R.; Bagshaw, P. A role for the bacterial outer membrane in the pathogenesis of *Helicobacter pylori* infection. *FEMS Microbiol. Lett.* **2000**, *182*, 259–264. [CrossRef] [PubMed]
76. Ayala, G.; Torres, L.; Espinosa, M.; Fierros-Zarate, G.; Maldonado, V.; Melendez-Zajgla, J. External membrane vesicles from *Helicobacter pylori* induce apoptosis in gastric epithelial cells. *FEMS Microbiol. Lett.* **2006**, *260*, 178–185. [CrossRef] [PubMed]
77. Mullaney, E.; Brown, P.A.; Smith, S.M.; Botting, C.H.; Yamaoka, Y.Y.; Terres, A.M.; Kelleher, D.P.; Windle, H.J. Proteomic and functional characterization of the outer membrane vesicles from the gastric pathogen *Helicobacter pylori*. *Proteom. Clin. Appl.* **2009**, *3*, 785–796. [CrossRef] [PubMed]
78. Olofsson, A.; Vallstrom, A.; Petzold, K.; Tegtmeyer, N.; Schleucher, J.; Carlsson, S.; Haas, R.; Backert, S.; Wai, S.N.; Grobner, G.; et al. Biochemical and functional characterization of *Helicobacter pylori* vesicles. *Mol. Microbiol.* **2010**, *77*, 1539–1555. [CrossRef] [PubMed]
79. Winter, J.; Letley, D.; Rhead, J.; Atherton, J.; Robinson, K. *Helicobacter pylori* membrane vesicles stimulate innate pro- and anti-inflammatory responses and induce apoptosis in Jurkat T cells. *Infect. Immun.* **2014**, *82*, 1372–1381. [CrossRef]
80. Turner, L.; Praszkier, J.; Hutton, M.L.; Steer, D.; Ramm, G.; Kaparakis-Liaskos, M.; Ferrero, R.L. Increased Outer Membrane Vesicle Formation in a *Helicobacter pylori* tolB Mutant. *Helicobacter* **2015**, *20*, 269–283. [CrossRef]
81. Choi, H.I.; Choi, J.P.; Seo, J.; Kim, B.J.; Rho, M.; Han, J.K.; Kim, J.G. *Helicobacter pylori*-derived extracellular vesicles increased in the gastric juices of gastric adenocarcinoma patients and induced inflammation mainly via specific targeting of gastric epithelial cells. *Exp. Mol. Med.* **2017**, *49*, e330. [CrossRef]
82. Turner, L.; Bitto, N.J.; Steer, D.L.; Lo, C.; D'Costa, K.; Ramm, G.; Shambrook, M.; Hill, A.F.; Ferrero, R.L.; Kaparakis-Liaskos, M. *Helicobacter pylori* Outer Membrane Vesicle Size Determines Their Mechanisms of Host Cell Entry and Protein Content. *Front. Immunol.* **2018**, *9*, 1466. [CrossRef]
83. Liu, Q.; Li, X.; Zhang, Y.; Song, Z.; Li, R.; Ruan, H.; Huang, X. Orally-administered outer-membrane vesicles from *Helicobacter pylori* reduce *H. pylori* infection via Th2-biased immune responses in mice. *Pathog. Dis.* **2019**, *77*, ftz050. [CrossRef] [PubMed]
84. Zavan, L.; Bitto, N.J.; Johnston, E.L.; Greening, D.W.; Kaparakis-Liaskos, M. *Helicobacter pylori* Growth Stage Determines the Size, Protein Composition, and Preferential Cargo Packaging of Outer Membrane Vesicles. *Proteomics* **2019**, *19*, e1800209. [CrossRef] [PubMed]
85. Chew, Y.; Chung, H.Y.; Lin, P.Y.; Wu, D.C.; Huang, S.K.; Kao, M.C. Outer Membrane Vesicle Production by *Helicobacter pylori* Represents an Approach for the Delivery of Virulence Factors CagA, VacA and UreA into Human Gastric Adenocarcinoma (AGS) Cells. *Int. J. Mol. Sci.* **2021**, *22*, 3942. [CrossRef] [PubMed]
86. Melo, J.; Pinto, V.; Fernandes, T.; Malheiro, A.R.; Osório, H.; Figueiredo, C.; Leite, M. Isolation Method and Characterization of Outer Membranes Vesicles of *Helicobacter pylori* Grown in a Chemically Defined Medium. *Front. Microbiol.* **2021**, *12*, 654193. [CrossRef] [PubMed]
87. Wei, S.; Li, X.; Wang, J.; Wang, Y.; Zhang, C.; Dai, S.; Wang, X.; Deng, X.; Zhao, L.; Shan, B. Outer Membrane Vesicles Secreted by *Helicobacter pylori* Transmitting Gastric Pathogenic Virulence Factors. *ACS Omega* **2022**, *7*, 240–258. [CrossRef] [PubMed]
88. Saberi, S.; Esmaeili, M.; Saghiri, R.; Shekari, F.; Mohammadi, M. Assessment of the mixed origin of the gastric epithelial extracellular vesicles in acellular transfer of *Helicobacter pylori* toxins and a systematic review. *Microb. Pathog.* **2023**, *177*, 106024. [CrossRef] [PubMed]
89. Palacios, E.; Lobos-González, L.; Guerrero, S.; Kogan, M.J.; Shao, B.; Heinecke, J.W.; Quest, A.F.G.; Leyton, L.; Valenzuela-Valderrama, M. *Helicobacter pylori* outer membrane vesicles induce astrocyte reactivity through nuclear factor-κappa B activation and cause neuronal damage in vivo in a murine model. *J. Neuroinflammation* **2023**, *20*, 66. [CrossRef] [PubMed]
90. Garner, J.A.; Cover, T.L. Binding and internalization of the *Helicobacter pylori* vacuolating cytotoxin by epithelial cells. *Infect. Immun.* **1996**, *64*, 4197–4203. [CrossRef]
91. Cover, T.L.; Hanson, P.I.; Heuser, J.E. Acid-induced dissociation of VacA, the *Helicobacter pylori* vacuolating cytotoxin, reveals its pattern of assembly. *J. Cell Biol.* **1997**, *138*, 759–769. [CrossRef]

92. Ye, D.; Willhite, D.C.; Blanke, S.R. Identification of the minimal intracellular vacuolating domain of the *Helicobacter pylori* vacuolating toxin. *J. Biol. Chem.* **1999**, *274*, 9277–9282. [CrossRef]
93. Ye, D.; Blanke, S.R. Mutational analysis of the *Helicobacter pylori* vacuolating toxin amino terminus: Identification of amino acids essential for cellular vacuolation. *Infect. Immun.* **2000**, *68*, 4354–4357. [CrossRef] [PubMed]
94. Torres, V.J.; McClain, M.S.; Cover, T.L. Interactions between p-33 and p-55 domains of the *Helicobacter pylori* vacuolating cytotoxin (VacA). *J. Biol. Chem.* **2004**, *279*, 2324–2331. [CrossRef] [PubMed]
95. Vaidya, M.; Panchal, H. In silico investigation and structural characterization of virulent factor and a metallo peptidase present in *Helicobacter pylori* strain J99. *Interdiscip. Sci.* **2012**, *4*, 302–309. [CrossRef] [PubMed]
96. Lupetti, P.; Heuser, J.E.; Manetti, R.; Massari, P.; Lanzavecchia, S.; Bellon, P.L.; Dallai, R.; Rappuoli, R.; Telford, J.L. Oligomeric and subunit structure of the *Helicobacter pylori* vacuolating cytotoxin. *J. Cell Biol.* **1996**, *133*, 801–807. [CrossRef] [PubMed]
97. Ye, D.; Blanke, S.R. Functional complementation reveals the importance of intermolecular monomer interactions for *Helicobacter pylori* VacA vacuolating activity. *Mol. Microbiol.* **2002**, *43*, 1243–1253. [CrossRef] [PubMed]
98. Adrian, M.; Cover, T.L.; Dubochet, J.; Heuser, J.E. Multiple oligomeric states of the *Helicobacter pylori* vacuolating toxin demonstrated by cryo-electron microscopy. *J. Mol. Biol.* **2002**, *318*, 121–133. [CrossRef] [PubMed]
99. El-Bez, C.; Adrian, M.; Dubochet, J.; Cover, T.L. High resolution structural analysis of *Helicobacter pylori* VacA toxin oligomers by cryo-negative staining electron microscopy. *J. Struct. Biol.* **2005**, *151*, 215–228. [CrossRef] [PubMed]
100. Lanzavecchia, S.; Bellon, P.L.; Lupetti, P.; Dallai, R.; Rappuoli, R.; Telford, J.L. Three-dimensional reconstruction of metal replicas of the *Helicobacter pylori* vacuolating cytotoxin. *J. Struct. Biol.* **1998**, *121*, 9–18. [CrossRef]
101. Chambers, M.G.; Pyburn, T.M.; González-Rivera, C.; Collier, S.E.; Eli, I.; Yip, C.K.; Takizawa, Y.; Lacy, D.B.; Cover, T.L.; Ohi, M.D. Structural analysis of the oligomeric states of *Helicobacter pylori* VacA toxin. *J. Mol. Biol.* **2013**, *425*, 524–535. [CrossRef]
102. Pyburn, T.M.; Foegeding, N.J.; González-Rivera, C.; McDonald, N.A.; Gould, K.L.; Cover, T.L.; Ohi, M.D. Structural organization of membrane-inserted hexamers formed by *Helicobacter pylori* VacA toxin. *Mol. Microbiol.* **2016**, *102*, 22–36. [CrossRef]
103. Su, M.; Erwin, A.L.; Campbell, A.M.; Pyburn, T.M.; Salay, L.E.; Hanks, J.L.; Lacy, D.B.; Akey, D.L.; Cover, T.L.; Ohi, M.D. Cryo-EM Analysis Reveals Structural Basis of *Helicobacter pylori* VacA Toxin Oligomerization. *J. Mol. Biol.* **2019**, *431*, 1956–1965. [CrossRef] [PubMed]
104. Zhang, K.; Zhang, H.; Li, S.; Pintilie, G.D.; Mou, T.C.; Gao, Y.; Zhang, Q.; van den Bedem, H.; Schmid, M.F.; Au, S.W.N.; et al. Cryo-EM structures of *Helicobacter pylori* vacuolating cytotoxin A oligomeric assemblies at near-atomic resolution. *Proc. Natl. Acad. Sci. USA* **2019**, *116*, 6800–6805. [CrossRef] [PubMed]
105. Gangwer, K.A.; Mushrush, D.J.; Stauff, D.L.; Spiller, B.; McClain, M.S.; Cover, T.L.; Lacy, D.B. Crystal structure of the *Helicobacter pylori* vacuolating toxin p55 domain. *Proc. Natl. Acad. Sci. USA* **2007**, *104*, 16293–16298. [CrossRef] [PubMed]
106. González-Rivera, C.; Campbell, A.M.; Rutherford, S.A.; Pyburn, T.M.; Foegeding, N.J.; Barke, T.L.; Spiller, B.W.; McClain, M.S.; Ohi, M.D.; Lacy, D.B.; et al. A Nonoligomerizing Mutant Form of *Helicobacter pylori* VacA Allows Structural Analysis of the p33 Domain. *Infect. Immun.* **2016**, *84*, 2662–2670. [CrossRef] [PubMed]
107. Domańska, G.; Motz, C.; Meinecke, M.; Harsman, A.; Papatheodorou, P.; Reljic, B.; Dian-Lothrop, E.A.; Galmiche, A.; Kepp, O.; Becker, L.; et al. *Helicobacter pylori* VacA toxin/subunit p34: Targeting of an anion channel to the inner mitochondrial membrane. *PLoS Pathog.* **2010**, *6*, e1000878. [CrossRef] [PubMed]
108. Seto, K.; Hayashi-Kuwabara, Y.; Yoneta, T.; Suda, H.; Tamaki, H. Vacuolation induced by cytotoxin from *Helicobacter pylori* is mediated by the EGF receptor in HeLa cells. *FEBS Lett.* **1998**, *431*, 347–350. [CrossRef]
109. McClain, M.S.; Schraw, W.; Ricci, V.; Boquet, P.; Cover, T.L. Acid activation of *Helicobacter pylori* vacuolating cytotoxin (VacA) results in toxin internalization by eukaryotic cells. *Mol. Microbiol.* **2000**, *37*, 433–442. [CrossRef]
110. Yahiro, K.; Wada, A.; Nakayama, M.; Kimura, T.; Ogushi, K.; Niidome, T.; Aoyagi, H.; Yoshino, K.; Yonezawa, K.; Moss, J.; et al. Protein-tyrosine phosphatase alpha, RPTP alpha, is a *Helicobacter pylori* VacA receptor. *J. Biol. Chem.* **2003**, *278*, 19183–19189. [CrossRef]
111. De Guzman, B.B.; Hisatsune, J.; Nakayama, M.; Yahiro, K.; Wada, A.; Yamasaki, E.; Nishi, Y.; Yamazaki, S.; Azuma, T.; Ito, Y.; et al. Cytotoxicity and recognition of receptor-like protein tyrosine phosphatases, RPTPalpha and RPTPbeta, by *Helicobacter pylori* m2VacA. *Cell Microbiol.* **2005**, *7*, 1285–1293. [CrossRef]
112. Yahiro, K.; Niidome, T.; Kimura, M.; Hatakeyama, T.; Aoyagi, H.; Kurazono, H.; Imagawa, K.; Wada, A.; Moss, J.; Hirayama, T. Activation of *Helicobacter pylori* VacA toxin by alkaline or acid conditions increases its binding to a 250-kDa receptor protein-tyrosine phosphatase beta. *J. Biol. Chem.* **1999**, *274*, 36693–36699. [CrossRef]
113. Yahiro, K.; Satoh, M.; Nakano, M.; Hisatsune, J.; Isomoto, H.; Sap, J.; Suzuki, H.; Nomura, F.; Noda, M.; Moss, J.; et al. Low-density lipoprotein receptor-related protein-1 (LRP1) mediates autophagy and apoptosis caused by *Helicobacter pylori* VacA. *J. Biol. Chem.* **2012**, *287*, 31104–31115. [CrossRef] [PubMed]
114. Nakayama, M.; Hisatsune, J.; Yamasaki, E.; Nishi, Y.; Wada, A.; Kurazono, H.; Sap, J.; Yahiro, K.; Moss, J.; Hirayama, T. Clustering of *Helicobacter pylori* VacA in lipid rafts, mediated by its receptor, receptor-like protein tyrosine phosphatase beta, is required for intoxication in AZ-521 Cells. *Infect. Immun.* **2006**, *74*, 6571–6580. [CrossRef] [PubMed]
115. Fujikawa, A.; Shirasaka, D.; Yamamoto, S.; Ota, H.; Yahiro, K.; Fukada, M.; Shintani, T.; Wada, A.; Aoyama, N.; Hirayama, T.; et al. Mice deficient in protein tyrosine phosphatase receptor type Z are resistant to gastric ulcer induction by VacA of *Helicobacter pylori*. *Nat. Genet.* **2003**, *33*, 375–381. [CrossRef] [PubMed]

116. Sewald, X.; Gebert-Vogl, B.; Prassl, S.; Barwig, I.; Weiss, E.; Fabbri, M.; Osicka, R.; Schiemann, M.; Busch, D.H.; Semmrich, M.; et al. Integrin subunit CD18 Is the T-lymphocyte receptor for the *Helicobacter pylori* vacuolating cytotoxin. *Cell Host Microbe* **2008**, *3*, 20–29. [CrossRef] [PubMed]
117. Czajkowsky, D.M.; Iwamoto, H.; Cover, T.L.; Shao, Z. The vacuolating toxin from *Helicobacter pylori* forms hexameric pores in lipid bilayers at low pH. *Proc. Natl. Acad. Sci. USA* **1999**, *96*, 2001–2006. [CrossRef] [PubMed]
118. Szabò, I.; Brutsche, S.; Tombola, F.; Moschioni, M.; Satin, B.; Telford, J.L.; Rappuoli, R.; Montecucco, C.; Papini, E.; Zoratti, M. Formation of anion-selective channels in the cell plasma membrane by the toxin VacA of *Helicobacter pylori* is required for its biological activity. *EMBO J.* **1999**, *18*, 5517–5527. [CrossRef] [PubMed]
119. Iwamoto, H.; Czajkowsky, D.M.; Cover, T.L.; Szabo, G.; Shao, Z. VacA from *Helicobacter pylori*: A hexameric chloride channel. *FEBS Lett.* **1999**, *450*, 101–104. [CrossRef] [PubMed]
120. Tombola, F.; Carlesso, C.; Szabò, I.; de Bernard, M.; Reyrat, J.M.; Telford, J.L.; Rappuoli, R.; Montecucco, C.; Papini, E.; Zoratti, M. *Helicobacter pylori* vacuolating toxin forms anion-selective channels in planar lipid bilayers: Possible implications for the mechanism of cellular vacuolation. *Biophys. J.* **1999**, *76*, 1401–1409. [CrossRef]
121. Tombola, F.; Oregna, F.; Brutsche, S.; Szabò, I.; Del Giudice, G.; Rappuoli, R.; Montecucco, C.; Papini, E.; Zoratti, M. Inhibition of the vacuolating and anion channel activities of the VacA toxin of *Helicobacter pylori*. *FEBS Lett.* **1999**, *460*, 221–225. [CrossRef]
122. Ivie, S.E.; McClain, M.S.; Torres, V.J.; Algood, H.M.; Lacy, D.B.; Yang, R.; Blanke, S.R.; Cover, T.L. *Helicobacter pylori* VacA subdomain required for intracellular toxin activity and assembly of functional oligomeric complexes. *Infect. Immun.* **2008**, *76*, 2843–2851. [CrossRef]
123. Genisset, C.; Galeotti, C.L.; Lupetti, P.; Mercati, D.; Skibinski, D.A.; Barone, S.; Battistutta, R.; de Bernard, M.; Telford, J.L. A *Helicobacter pylori* vacuolating toxin mutant that fails to oligomerize has a dominant negative phenotype. *Infect. Immun.* **2006**, *74*, 1786–1794. [CrossRef] [PubMed]
124. Debellis, L.; Papini, E.; Caroppo, R.; Montecucco, C.; Curci, S. *Helicobacter pylori* cytotoxin VacA increases alkaline secretion in gastric epithelial cells. *Am. J. Physiol. Gastrointest. Liver Physiol.* **2001**, *281*, G1440–G1448. [CrossRef] [PubMed]
125. Schraw, W.; Li, Y.; McClain, M.S.; van der Goot, F.G.; Cover, T.L. Association of Helicobacter pylori vacuolating toxin (VacA) with lipid rafts. *J. Biol. Chem.* **2002**, *277*, 34642–34650. [CrossRef] [PubMed]
126. McClain, M.S.; Iwamoto, H.; Cao, P.; Vinion-Dubiel, A.D.; Li, Y.; Szabo, G.; Shao, Z.; Cover, T.L. Essential role of a GXXXG motif for membrane channel formation by *Helicobacter pylori* vacuolating toxin. *J. Biol. Chem.* **2003**, *278*, 12101–12108. [CrossRef] [PubMed]
127. Caston, R.R.; Sierra, J.C.; Foegeding, N.J.; Truelock, M.D.; Campbell, A.M.; Frick-Cheng, A.E.; Bimczok, D.; Wilson, K.T.; McClain, M.S.; Cover, T.L. Functional Properties of *Helicobacter pylori* VacA Toxin m1 and m2 Variants. *Infect. Immun.* **2020**, *88*, e00032-20. [CrossRef] [PubMed]
128. Palframan, S.L.; Kwok, T.; Gabriel, K. Vacuolating cytotoxin A (VacA), a key toxin for *Helicobacter pylori* pathogenesis. *Front. Cell Infect. Microbiol.* **2012**, *2*, 92. [CrossRef] [PubMed]
129. Gauthier, N.C.; Monzo, P.; Kaddai, V.; Doye, A.; Ricci, V.; Boquet, P. *Helicobacter pylori* VacA cytotoxin: A probe for a clathrin-independent and Cdc42-dependent pinocytic pathway routed to late endosomes. *Mol. Biol. Cell* **2005**, *16*, 4852–4866. [CrossRef] [PubMed]
130. Boquet, P.; Ricci, V. Intoxication strategy of *Helicobacter pylori* VacA toxin. *Trends Microbiol.* **2012**, *20*, 165–174. [CrossRef]
131. Ulhuq, F.R.; Mariano, G. Bacterial pore-forming toxins. *Microbiology* **2022**, *168*, 001154. [CrossRef]
132. Yahiro, K.; Wada, A.; Yamasaki, E.; Nakayama, M.; Nishi, Y.; Hisatsune, J.; Morinaga, N.; Sap, J.; Noda, M.; Moss, J.; et al. Essential domain of receptor tyrosine phosphatase beta (RPTPbeta) for interaction with *Helicobacter pylori* vacuolating cytotoxin. *J. Biol. Chem.* **2004**, *279*, 51013–51021. [CrossRef]
133. Satoh, K.; Hirayama, T.; Takano, K.; Suzuki-Inoue, K.; Sato, T.; Ohta, M.; Nakagomi, J.; Ozaki, Y. VacA, the vacuolating cytotoxin of *Helicobacter pylori*, binds to multimerin 1 on human platelets. *Thromb. J.* **2013**, *11*, 23. [CrossRef]
134. Utt, M.; Danielsson, B.; Wadström, T. *Helicobacter pylori* vacuolating cytotoxin binding to a putative cell surface receptor, heparan sulfate, studied by surface plasmon resonance. *FEMS Immunol. Med. Microbiol.* **2001**, *30*, 109–113. [CrossRef]
135. Gupta, V.R.; Patel, H.K.; Kostolansky, S.S.; Ballivian, R.A.; Eichberg, J.; Blanke, S.R. Sphingomyelin functions as a novel receptor for *Helicobacter pylori* VacA. *PLoS Pathog.* **2008**, *4*, e1000073. [CrossRef]
136. Gupta, V.R.; Wilson, B.A.; Blanke, S.R. Sphingomyelin is important for the cellular entry and intracellular localization of *Helicobacter pylori* VacA. *Cell Microbiol.* **2010**, *12*, 1517–1533. [CrossRef]
137. Roche, N.; Ilver, D.; Angström, J.; Barone, S.; Telford, J.L.; Teneberg, S. Human gastric glycosphingolipids recognized by *Helicobacter pylori* vacuolating cytotoxin VacA. *Microbes Infect.* **2007**, *9*, 605–614. [CrossRef]
138. Molinari, M.; Galli, C.; de Bernard, M.; Norais, N.; Ruysschaert, J.M.; Rappuoli, R.; Montecucco, C. The acid activation of *Helicobacter pylori* toxin VacA: Structural and membrane binding studies. *Biochem. Biophys. Res. Commun.* **1998**, *248*, 334–340. [CrossRef]
139. Sezgin, E.; Levental, I.; Mayor, S.; Eggeling, C. The mystery of membrane organization: Composition, regulation and roles of lipid rafts. *Nat. Rev. Mol. Cell Biol.* **2017**, *18*, 361–374. [CrossRef]
140. Geisse, N.A.; Cover, T.L.; Henderson, R.M.; Edwardson, J.M. Targeting of *Helicobacter pylori* vacuolating toxin to lipid raft membrane domains analysed by atomic force microscopy. *Biochem. J.* **2004**, *381*, 911–917. [CrossRef]

141. Gauthier, N.C.; Ricci, V.; Gounon, P.; Doye, A.; Tauc, M.; Poujeol, P.; Boquet, P. Glycosylphosphatidylinositol-anchored proteins and actin cytoskeleton modulate chloride transport by channels formed by the *Helicobacter pylori* vacuolating cytotoxin VacA in HeLa cells. *J. Biol. Chem.* **2004**, *279*, 9481–9489. [CrossRef]
142. Kuo, C.H.; Wang, W.C. Binding and internalization of *Helicobacter pylori* VacA via cellular lipid rafts in epithelial cells. *Biochem. Biophys. Res. Commun.* **2003**, *303*, 640–644. [CrossRef]
143. Lai, C.H.; Chang, Y.C.; Du, S.Y.; Wang, H.J.; Kuo, C.H.; Fang, S.H.; Fu, H.W.; Lin, H.H.; Chiang, A.S.; Wang, W.C. Cholesterol depletion reduces *Helicobacter pylori* CagA translocation and CagA-induced responses in AGS cells. *Infect. Immun.* **2008**, *76*, 3293–3303. [CrossRef] [PubMed]
144. Patel, H.K.; Willhite, D.C.; Patel, R.M.; Ye, D.; Williams, C.L.; Torres, E.M.; Marty, K.B.; MacDonald, R.A.; Blanke, S.R. Plasma membrane cholesterol modulates cellular vacuolation induced by the *Helicobacter pylori* vacuolating cytotoxin. *Infect. Immun.* **2002**, *70*, 4112–4123. [CrossRef] [PubMed]
145. Ricci, V.; Galmiche, A.; Doye, A.; Necchi, V.; Solcia, E.; Boquet, P. High cell sensitivity to *Helicobacter pylori* VacA toxin depends on a GPI-anchored protein and is not blocked by inhibition of the clathrin-mediated pathway of endocytosis. *Mol. Biol. Cell* **2000**, *11*, 3897–3909. [CrossRef] [PubMed]
146. Escribá, P.V.; González-Ros, J.M.; Goñi, F.M.; Kinnunen, P.K.; Vigh, L.; Sánchez-Magraner, L.; Fernández, A.M.; Busquets, X.; Horváth, I.; Barceló-Coblijn, G. Membranes: A meeting point for lipids, proteins and therapies. *J. Cell Mol. Med.* **2008**, *12*, 829–875. [CrossRef] [PubMed]
147. Li, Y.; Li, Y.; Mengist, H.M.; Shi, C.; Zhang, C.; Wang, B.; Li, T.; Huang, Y.; Xu, Y.; Jin, T. Structural Basis of the Pore-Forming Toxin/Membrane Interaction. *Toxins* **2021**, *13*, 128. [CrossRef] [PubMed]
148. de Bernard, M.; Papini, E.; de Filippis, V.; Gottardi, E.; Telford, J.; Manetti, R.; Fontana, A.; Rappuoli, R.; Montecucco, C. Low pH activates the vacuolating toxin of *Helicobacter pylori*, which becomes acid and pepsin resistant. *J. Biol. Chem.* **1995**, *270*, 23937–23940. [CrossRef] [PubMed]
149. Reyrat, J.M.; Charrel, M.; Pagliaccia, C.; Burroni, D.; Lupetti, P.; de Bernard, M.; Ji, X.; Norais, N.; Papini, E.; Dallai, R.; et al. Characterisation of a monoclonal antibody and its use to purify the cytotoxin of *Helicobacter pylori*. *FEMS Microbiol. Lett.* **1998**, *165*, 79–84. [CrossRef]
150. Moll, G.; Papini, E.; Colonna, R.; Burroni, D.; Telford, J.; Rappuoli, R.; Montecucco, C. Lipid interaction of the 37-kDa and 58-kDa fragments of the *Helicobacter pylori* cytotoxin. *Eur. J. Biochem.* **1995**, *234*, 947–952. [CrossRef]
151. Pagliaccia, C.; Wang, X.M.; Tardy, F.; Telford, J.L.; Ruysschaert, J.M.; Cabiaux, V. Structure and interaction of VacA of *Helicobacter pylori* with a lipid membrane. *Eur. J. Biochem.* **2000**, *267*, 104–109. [CrossRef]
152. Caso, G.C.; McClain, M.S.; Erwin, A.L.; Truelock, M.D.; Campbell, A.M.; Leasure, C.S.; Nagel, M.; Schey, K.L.; Lacy, D.B.; Ohi, M.D.; et al. Functional Properties of Oligomeric and Monomeric Forms of *Helicobacter pylori* VacA Toxin. *Infect. Immun.* **2021**, *89*, e0034821. [CrossRef]
153. McClain, M.S.; Cao, P.; Cover, T.L. Amino-terminal hydrophobic region of Helicobacter pylori vacuolating cytotoxin (VacA) mediates transmembrane protein dimerization. *Infect. Immun.* **2001**, *69*, 1181–1184. [CrossRef] [PubMed]
154. McClain, M.S.; Czajkowsky, D.M.; Torres, V.J.; Szabo, G.; Shao, Z.; Cover, T.L. Random mutagenesis of *Helicobacter pylori* vacA to identify amino acids essential for vacuolating cytotoxic activity. *Infect. Immun.* **2006**, *74*, 6188–6195. [CrossRef] [PubMed]
155. Torres, V.J.; Ivie, S.E.; McClain, M.S.; Cover, T.L. Functional properties of the p33 and p55 domains of the *Helicobacter pylori* vacuolating cytotoxin. *J. Biol. Chem.* **2005**, *280*, 21107–21114. [CrossRef] [PubMed]
156. Torres, V.J.; McClain, M.S.; Cover, T.L. Mapping of a domain required for protein-protein interactions and inhibitory activity of a *Helicobacter pylori* dominant-negative VacA mutant protein. *Infect. Immun.* **2006**, *74*, 2093–2101. [CrossRef] [PubMed]
157. Yahiro, K.; Hirayama, T.; Moss, J.; Noda, M. New Insights into VacA Intoxication Mediated through Its Cell Surface Receptors. *Toxins* **2016**, *8*, 152. [CrossRef] [PubMed]
158. McClain, M.S.; Cao, P.; Iwamoto, H.; Vinion-Dubiel, A.D.; Szabo, G.; Shao, Z.; Cover, T.L. A 12-amino-acid segment, present in type s2 but not type s1 *Helicobacter pylori* VacA proteins, abolishes cytotoxin activity and alters membrane channel formation. *J. Bacteriol.* **2001**, *183*, 6499–6508. [CrossRef]
159. González-Rivera, C.; Algood, H.M.; Radin, J.N.; McClain, M.S.; Cover, T.L. The intermediate region of *Helicobacter pylori* VacA is a determinant of toxin potency in a Jurkat T cell assay. *Infect. Immun.* **2012**, *80*, 2578–2588. [CrossRef]
160. Pagliaccia, C.; de Bernard, M.; Lupetti, P.; Ji, X.; Burroni, D.; Cover, T.L.; Papini, E.; Rappuoli, R.; Telford, J.L.; Reyrat, J.M. The m2 form of the *Helicobacter pylori* cytotoxin has cell type-specific vacuolating activity. *Proc. Natl. Acad. Sci. USA* **1998**, *95*, 10212–10217. [CrossRef]
161. Wang, W.C.; Wang, H.J.; Kuo, C.H. Two distinctive cell binding patterns by vacuolating toxin fused with glutathione S-transferase: One high-affinity m1-specific binding and the other lower-affinity binding for variant m forms. *Biochemistry* **2001**, *40*, 11887–11896. [CrossRef]
162. Atherton, J.C.; Peek, R.M., Jr.; Tham, K.T.; Cover, T.L.; Blaser, M.J. Clinical and pathological importance of heterogeneity in vacA, the vacuolating cytotoxin gene of *Helicobacter pylori*. *Gastroenterology* **1997**, *112*, 92–99. [CrossRef]
163. Ji, X.; Fernandez, T.; Burroni, D.; Pagliaccia, C.; Atherton, J.C.; Reyrat, J.M.; Rappuoli, R.; Telford, J.L. Cell specificity of *Helicobacter pylori* cytotoxin is determined by a short region in the polymorphic midregion. *Infect. Immun.* **2000**, *68*, 3754–3757. [CrossRef] [PubMed]

164. Skibinski, D.A.; Genisset, C.; Barone, S.; Telford, J.L. The cell-specific phenotype of the polymorphic vacA midregion is independent of the appearance of the cell surface receptor protein tyrosine phosphatase beta. *Infect. Immun.* **2006**, *74*, 49–55. [CrossRef] [PubMed]
165. Letley, D.P.; Rhead, J.L.; Twells, R.J.; Dove, B.; Atherton, J.C. Determinants of non-toxicity in the gastric pathogen *Helicobacter pylori*. *J. Biol. Chem.* **2003**, *278*, 26734–26741. [CrossRef] [PubMed]
166. Korona-Glowniak, I.; Cichoz-Lach, H.; Siwiec, R.; Andrzejczuk, S.; Glowniak, A.; Matras, P.; Malm, A. Antibiotic Resistance and Genotypes of *Helicobacter pylori* Strains in Patients with Gastroduodenal Disease in Southeast Poland. *J. Clin. Med.* **2019**, *8*, 71. [CrossRef]
167. Jouimyi, M.R.; Bounder, G.; Essaidi, I.; Boura, H.; Badre, W.; Benomar, H.; Zerouali, K.; Lebrazi, H.; Kettani, A.; Maachi, F. Association of *Helicobacter pylori* vacA polymorphisms with the risk of gastric precancerous lesions in a Moroccan population. *J. Infect. Dev. Ctries.* **2021**, *15*, 1124–1132. [CrossRef] [PubMed]
168. Keikha, M.; Ali-Hassanzadeh, M.; Karbalaei, M. Association of *Helicobacter pylori* vacA genotypes and peptic ulcer in Iranian population: A systematic review and meta-analysis. *BMC Gastroenterol.* **2020**, *20*, 266. [CrossRef] [PubMed]
169. Basso, D.; Zambon, C.F.; Letley, D.P.; Stranges, A.; Marchet, A.; Rhead, J.L.; Schiavon, S.; Guariso, G.; Ceroti, M.; Nitti, D.; et al. Clinical relevance of *Helicobacter pylori* cagA and vacA gene polymorphisms. *Gastroenterology* **2008**, *135*, 91–99. [CrossRef] [PubMed]
170. Louw, J.A.; Kidd, M.S.; Kummer, A.F.; Taylor, K.; Kotze, U.; Hanslo, D. The relationship between *Helicobacter pylori* infection, the virulence genotypes of the infecting strain and gastric cancer in the African setting. *Helicobacter* **2001**, *6*, 268–273. [CrossRef]
171. Phan, T.N.; Santona, A.; Tran, V.H.; Tran, T.N.H.; Le, V.A.; Cappuccinelli, P.; Rubino, S.; Paglietti, B. Genotyping of *Helicobacter pylori* shows high diversity of strains circulating in central Vietnam. *Infect. Genet. Evol.* **2017**, *52*, 19–25. [CrossRef]
172. Memon, A.A.; Hussein, N.R.; Miendje Deyi, V.Y.; Burette, A.; Atherton, J.C. Vacuolating cytotoxin genotypes are strong markers of gastric cancer and duodenal ulcer-associated *Helicobacter pylori* strains: A matched case-control study. *J. Clin. Microbiol.* **2014**, *52*, 2984–2989. [CrossRef]
173. Chang, W.L.; Yeh, Y.C.; Sheu, B.S. The impacts of *H. pylori* virulence factors on the development of gastroduodenal diseases. *J. Biomed. Sci.* **2018**, *25*, 68. [CrossRef] [PubMed]
174. Figueiredo, C.; Machado, J.C.; Pharoah, P.; Seruca, R.; Sousa, S.; Carvalho, R.; Capelinha, A.F.; Quint, W.; Caldas, C.; van Doorn, L.J.; et al. *Helicobacter pylori* and interleukin 1 genotyping: An opportunity to identify high-risk individuals for gastric carcinoma. *J. Natl. Cancer Inst.* **2002**, *94*, 1680–1687. [CrossRef] [PubMed]
175. Njenga, P.; Njau, A.; Moloo, Z.; Revathi, G.; Tshibangu, E.; Yamaoka, Y. Pattern and trends of *Helicobacter pylori* genotypes in gastric cancer: A Kenyan 8-year study. *Front. Med.* **2023**, *10*, 1119513. [CrossRef]
176. Lima, V.P.; Silva-Fernandes, I.J.; Alves, M.K.; Rabenhorst, S.H. Prevalence of *Helicobacter pylori* genotypes (vacA, cagA, cagE and virB11) in gastric cancer in Brazilian's patients: An association with histopathological parameters. *Cancer Epidemiol.* **2011**, *35*, e32–e37. [CrossRef] [PubMed]
177. El Khadir, M.; Boukhris Alaoui, S.; Benajah, D.A.; Ibrahimi, S.A.; Chbani, L.; El Abkari, M.; Bennani, B. VacA genotypes and cagA-EPIYA-C motifs of *Helicobacter pylori* and gastric histopathological lesions. *Int. J. Cancer* **2020**, *147*, 3206–3214. [CrossRef] [PubMed]
178. Ferreira, R.M.; Machado, J.C.; Letley, D.; Atherton, J.C.; Pardo, M.L.; Gonzalez, C.A.; Carneiro, F.; Figueiredo, C. A novel method for genotyping the *Helicobacter pylori* vacA intermediate region directly in gastric biopsy specimens. *J. Clin. Microbiol.* **2012**, *50*, 3983–3989. [CrossRef] [PubMed]
179. Douraghi, M.; Talebkhan, Y.; Zeraati, H.; Ebrahimzadeh, F.; Nahvijoo, A.; Morakabati, A.; Ghafarpour, M.; Esmaili, M.; Bababeik, M.; Oghalaie, A.; et al. Multiple gene status in *Helicobacter pylori* strains and risk of gastric cancer development. *Digestion* **2009**, *80*, 200–207. [CrossRef]
180. Miehlke, S.; Kirsch, C.; Agha-Amiri, K.; Günther, T.; Lehn, N.; Malfertheiner, P.; Stolte, M.; Ehninger, G.; Bayerdörffer, E. The *Helicobacter pylori* vacA s1, m1 genotype and cagA is associated with gastric carcinoma in Germany. *Int. J. Cancer* **2000**, *87*, 322–327. [CrossRef]
181. Winter, J.A.; Letley, D.P.; Cook, K.W.; Rhead, J.L.; Zaitoun, A.A.; Ingram, R.J.; Amilon, K.R.; Croxall, N.J.; Kaye, P.V.; Robinson, K.; et al. A role for the vacuolating cytotoxin, VacA, in colonization and Helicobacter pylori-induced metaplasia in the stomach. *J. Infect. Dis.* **2014**, *210*, 954–963. [CrossRef]
182. Kim, S.; Chamberlain, A.K.; Bowie, J.U. Membrane channel structure of *Helicobacter pylori* vacuolating toxin: Role of multiple GXXXG motifs in cylindrical channels. *Proc. Natl. Acad. Sci. USA* **2004**, *101*, 5988–5991. [CrossRef]
183. de Bernard, M.; Burroni, D.; Papini, E.; Rappuoli, R.; Telford, J.; Montecucco, C. Identification of the *Helicobacter pylori* VacA toxin domain active in the cell cytosol. *Infect. Immun.* **1998**, *66*, 6014–6016. [CrossRef]
184. Campello, S.; Tombola, F.; Cabrini, G.; Zoratti, M. The vacuolating toxin of *Helicobacter pylori* mimicks the CFTR-mediated chloride conductance. *FEBS Lett.* **2002**, *532*, 237–240. [CrossRef]
185. Tombola, F.; Pagliaccia, C.; Campello, S.; Telford, J.L.; Montecucco, C.; Papini, E.; Zoratti, M. How the loop and middle regions influence the properties of *Helicobacter pylori* VacA channels. *Biophys. J.* **2001**, *81*, 3204–3215. [CrossRef]
186. Tombola, F.; Morbiato, L.; Del Giudice, G.; Rappuoli, R.; Zoratti, M.; Papini, E. The *Helicobacter pylori* VacA toxin is a urea permease that promotes urea diffusion across epithelia. *J. Clin. Investig.* **2001**, *108*, 929–937. [CrossRef]

187. Tombola, F.; Del Giudice, G.; Papini, E.; Zoratti, M. Blockers of VacA provide insights into the structure of the pore. *Biophys. J.* **2000**, *79*, 863–873. [CrossRef]
188. Ricci, V.; Sommi, P.; Fiocca, R.; Romano, M.; Solcia, E.; Ventura, U. Helicobacter pylori vacuolating toxin accumulates within the endosomal-vacuolar compartment of cultured gastric cells and potentiates the vacuolating activity of ammonia. *J. Pathol.* **1997**, *183*, 453–459. [CrossRef]
189. Willhite, D.C.; Blanke, S.R. *Helicobacter pylori* vacuolating cytotoxin enters cells, localizes to the mitochondria, and induces mitochondrial membrane permeability changes correlated to toxin channel activity. *Cell Microbiol.* **2004**, *6*, 143–154. [CrossRef]
190. Ahmed, A.A.Q.; Besio, R.; Xiao, L.; Forlino, A. Outer Membrane Vesicles (OMVs) as Biomedical Tools and Their Relevance as Immune-Modulating Agents against *H. pylori* Infections: Current Status and Future Prospects. *Int. J. Mol. Sci.* **2023**, *24*, 8542. [CrossRef]
191. Ricci, V.; Chiozzi, V.; Necchi, V.; Oldani, A.; Romano, M.; Solcia, E.; Ventura, U. Free-soluble and outer membrane vesicle-associated VacA from *Helicobacter pylori*: Two forms of release, a different activity. *Biochem. Biophys. Res. Commun.* **2005**, *337*, 173–178. [CrossRef]
192. Heczko, U.; Smith, V.C.; Mark Meloche, R.; Buchan, A.M.; Finlay, B.B. Characteristics of *Helicobacter pylori* attachment to human primary antral epithelial cells. *Microbes Infect.* **2000**, *2*, 1669–1676. [CrossRef]
193. Parker, H.; Chitcholtan, K.; Hampton, M.B.; Keenan, J.I. Uptake of *Helicobacter pylori* outer membrane vesicles by gastric epithelial cells. *Infect. Immun.* **2010**, *78*, 5054–5061. [CrossRef] [PubMed]
194. Olofsson, A.; Nygard Skalman, L.; Obi, I.; Lundmark, R.; Arnqvist, A. Uptake of *Helicobacter pylori* vesicles is facilitated by clathrin-dependent and clathrin-independent endocytic pathways. *mBio* **2014**, *5*, e00979-14. [CrossRef] [PubMed]
195. Kaparakis, M.; Turnbull, L.; Carneiro, L.; Firth, S.; Coleman, H.A.; Parkington, H.C.; Le Bourhis, L.; Karrar, A.; Viala, J.; Mak, J.; et al. Bacterial membrane vesicles deliver peptidoglycan to NOD1 in epithelial cells. *Cell Microbiol.* **2010**, *12*, 372–385. [CrossRef] [PubMed]
196. Utsch, C.; Haas, R. VacA's Induction of VacA-Containing Vacuoles (VCVs) and Their Immunomodulatory Activities on Human T Cells. *Toxins* **2016**, *8*, 190. [CrossRef] [PubMed]
197. Lajoie, P.; Nabi, I.R. Regulation of raft-dependent endocytosis. *J. Cell Mol. Med.* **2007**, *11*, 644–653. [CrossRef]
198. Sewald, X.; Jiménez-Soto, L.; Haas, R. PKC-dependent endocytosis of the *Helicobacter pylori* vacuolating cytotoxin in primary T lymphocytes. *Cell Microbiol.* **2011**, *13*, 482–496. [CrossRef] [PubMed]
199. Hotchin, N.A.; Cover, T.L.; Akhtar, N. Cell vacuolation induced by the VacA cytotoxin of *Helicobacter pylori* is regulated by the Rac1 GTPase. *J. Biol. Chem.* **2000**, *275*, 14009–14012. [CrossRef]
200. Oldani, A.; Cormont, M.; Hofman, V.; Chiozzi, V.; Oregioni, O.; Canonici, A.; Sciullo, A.; Sommi, P.; Fabbri, A.; Ricci, V.; et al. *Helicobacter pylori* counteracts the apoptotic action of its VacA toxin by injecting the CagA protein into gastric epithelial cells. *PLoS Pathog.* **2009**, *5*, e1000603. [CrossRef]
201. Gauthier, N.C.; Monzo, P.; Gonzalez, T.; Doye, A.; Oldani, A.; Gounon, P.; Ricci, V.; Cormont, M.; Boquet, P. Early endosomes associated with dynamic F-actin structures are required for late trafficking of *H. pylori* VacA toxin. *J. Cell Biol.* **2007**, *177*, 343–354. [CrossRef]
202. Mayor, S.; Parton, R.G.; Donaldson, J.G. Clathrin-independent pathways of endocytosis. *Cold Spring Harb. Perspect. Biol.* **2014**, *6*, a016758. [CrossRef]
203. Elkin, S.R.; Lakoduk, A.M.; Schmid, S.L. Endocytic pathways and endosomal trafficking: A primer. *Wien. Med. Wochenschr.* **2016**, *166*, 196–204. [CrossRef]
204. Sabharanjak, S.; Sharma, P.; Parton, R.G.; Mayor, S. GPI-anchored proteins are delivered to recycling endosomes via a distinct cdc42-regulated, clathrin-independent pinocytic pathway. *Dev. Cell* **2002**, *2*, 411–423. [CrossRef]
205. Ricci, V.; Romano, M.; Boquet, P. Molecular cross-talk between *Helicobacter pylori* and human gastric mucosa. *World J. Gastroenterol.* **2011**, *17*, 1383–1399. [CrossRef]
206. Papini, E.; de Bernard, M.; Milia, E.; Bugnoli, M.; Zerial, M.; Rappuoli, R.; Montecucco, C. Cellular vacuoles induced by *Helicobacter pylori* originate from late endosomal compartments. *Proc. Natl. Acad. Sci. USA* **1994**, *91*, 9720–9724. [CrossRef]
207. Papini, E.; Gottardi, E.; Satin, B.; de Bernard, M.; Massari, P.; Telford, J.; Rappuoli, R.; Sato, S.B.; Montecucco, C. The vacuolar ATPase proton pump is present on intracellular vacuoles induced by *Helicobacter pylori*. *J. Med. Microbiol.* **1996**, *45*, 84–89. [CrossRef]
208. Molinari, M.; Galli, C.; Norais, N.; Telford, J.L.; Rappuoli, R.; Luzio, J.P.; Montecucco, C. Vacuoles induced by *Helicobacter pylori* toxin contain both late endosomal and lysosomal markers. *J. Biol. Chem.* **1997**, *272*, 25339–25344. [CrossRef]
209. Li, Y.; Wandinger-Ness, A.; Goldenring, J.R.; Cover, T.L. Clustering and redistribution of late endocytic compartments in response to *Helicobacter pylori* vacuolating toxin. *Mol. Biol. Cell* **2004**, *15*, 1946–1959. [CrossRef]
210. Yamasaki, E.; Wada, A.; Kumatori, A.; Nakagawa, I.; Funao, J.; Nakayama, M.; Hisatsune, J.; Kimura, M.; Moss, J.; Hirayama, T. *Helicobacter pylori* vacuolating cytotoxin induces activation of the proapoptotic proteins Bax and Bak, leading to cytochrome c release and cell death, independent of vacuolation. *J. Biol. Chem.* **2006**, *281*, 11250–11259. [CrossRef]
211. Terebiznik, M.R.; Vazquez, C.L.; Torbicki, K.; Banks, D.; Wang, T.; Hong, W.; Blanke, S.R.; Colombo, M.I.; Jones, N.L. *Helicobacter pylori* VacA toxin promotes bacterial intracellular survival in gastric epithelial cells. *Infect. Immun.* **2006**, *74*, 6599–6614. [CrossRef]
212. de Bernard, M.; Moschioni, M.; Habermann, A.; Griffiths, G.; Montecucco, C. Cell vacuolization induced by *Helicobacter pylori* VacA cytotoxin does not depend on late endosomal SNAREs. *Cell Microbiol.* **2002**, *4*, 11–18. [CrossRef]

213. Papini, E.; Bugnoli, M.; De Bernard, M.; Figura, N.; Rappuoli, R.; Montecucco, C. Bafilomycin A1 inhibits *Helicobacter pylori*-induced vacuolization of HeLa cells. *Mol. Microbiol.* **1993**, *7*, 323–327. [CrossRef] [PubMed]
214. Papini, E.; de Bernard, M.; Bugnoli, M.; Milia, E.; Rappuoli, R.; Montecucco, C. Cell vacuolization induced by *Helicobacter pylori*: Inhibition by bafilomycins A1, B1, C1 and D. *FEMS Microbiol. Lett.* **1993**, *113*, 155–159. [CrossRef] [PubMed]
215. Cover, T.L.; Reddy, L.Y.; Blaser, M.J. Effects of ATPase inhibitors on the response of HeLa cells to *Helicobacter pylori* vacuolating toxin. *Infect. Immun.* **1993**, *61*, 1427–1431. [CrossRef] [PubMed]
216. Genisset, C.; Puhar, A.; Calore, F.; de Bernard, M.; Dell'Antone, P.; Montecucco, C. The concerted action of the *Helicobacter pylori* cytotoxin VacA and of the v-ATPase proton pump induces swelling of isolated endosomes. *Cell Microbiol.* **2007**, *9*, 1481–1490. [CrossRef] [PubMed]
217. Morbiato, L.; Tombola, F.; Campello, S.; Del Giudice, G.; Rappuoli, R.; Zoratti, M.; Papini, E. Vacuolation induced by VacA toxin of *Helicobacter pylori* requires the intracellular accumulation of membrane permeant bases, Cl(−) and water. *FEBS Lett.* **2001**, *508*, 479–483. [CrossRef] [PubMed]
218. Montecucco, C. Protein toxins and membrane transport. *Curr. Opin. Cell Biol.* **1998**, *10*, 530–536. [CrossRef]
219. Cover, T.L.; Puryear, W.; Perez-Perez, G.I.; Blaser, M.J. Effect of urease on HeLa cell vacuolation induced by *Helicobacter pylori* cytotoxin. *Infect. Immun.* **1991**, *59*, 1264–1270. [CrossRef]
220. Cover, T.L.; Vaughn, S.G.; Cao, P.; Blaser, M.J. Potentiation of *Helicobacter pylori* vacuolating toxin activity by nicotine and other weak bases. *J. Infect. Dis.* **1992**, *166*, 1073–1078. [CrossRef]
221. Necchi, V.; Sommi, P.; Ricci, V.; Solcia, E. In vivo accumulation of *Helicobacter pylori* products, NOD1, ubiquitinated proteins and proteasome in a novel cytoplasmic structure. *PLoS ONE* **2010**, *5*, e9716. [CrossRef]
222. Tricottet, V.; Bruneval, P.; Vire, O.; Camilleri, J.P.; Bloch, F.; Bonte, N.; Roge, J. Campylobacter-like organisms and surface epithelium abnormalities in active, chronic gastritis in humans: An ultrastructural study. *Ultrastruct. Pathol.* **1986**, *10*, 113–122. [CrossRef]
223. Fiocca, R.; Luinetti, O.; Villani, L.; Chiaravalli, A.M.; Capella, C.; Solcia, E. Epithelial cytotoxicity, immune responses, and inflammatory components of *Helicobacter pylori* gastritis. *Scand. J. Gastroenterol. Suppl.* **1994**, *205*, 11–21. [CrossRef] [PubMed]
224. Holland, R.L.; Bosi, K.D.; Harpring, G.H.; Luo, J.; Wallig, M.; Phillips, H.; Blanke, S.R. Chronic in vivo exposure to *Helicobacter pylori* VacA: Assessing the efficacy of automated and long-term intragastric toxin infusion. *Sci. Rep.* **2020**, *10*, 9307. [CrossRef] [PubMed]
225. Jeger, J.L. Endosomes, lysosomes, and the role of endosomal and lysosomal biogenesis in cancer development. *Mol. Biol. Rep.* **2020**, *47*, 9801–9810. [CrossRef] [PubMed]
226. Papini, E.; Satin, B.; Bucci, C.; de Bernard, M.; Telford, J.L.; Manetti, R.; Rappuoli, R.; Zerial, M.; Montecucco, C. The small GTP binding protein rab7 is essential for cellular vacuolation induced by *Helicobacter pylori* cytotoxin. *EMBO J.* **1997**, *16*, 15–24. [CrossRef] [PubMed]
227. Suzuki, J.; Ohnishi, H.; Wada, A.; Hirayama, T.; Ohno, H.; Ueda, N.; Yasuda, H.; Iiri, T.; Wada, Y.; Futai, M.; et al. Involvement of syntaxin 7 in human gastric epithelial cell vacuolation induced by the *Helicobacter pylori*-produced cytotoxin VacA. *J. Biol. Chem.* **2003**, *278*, 25585–25590. [CrossRef] [PubMed]
228. Mashima, H.; Suzuki, J.; Hirayama, T.; Yoshikumi, Y.; Ohno, H.; Ohnishi, H.; Yasuda, H.; Fujita, T.; Omata, M. Involvement of vesicle-associated membrane protein 7 in human gastric epithelial cell vacuolation induced by *Helicobacter pylori*-produced VacA. *Infect. Immun.* **2008**, *76*, 2296–2303. [CrossRef] [PubMed]
229. Montecucco, C.; Papini, E.; de Bernard, M.; Zoratti, M. Molecular and cellular activities of *Helicobacter pylori* pathogenic factors. *FEBS Lett.* **1999**, *452*, 16–21. [CrossRef] [PubMed]
230. Ricci, V.; Sommi, P.; Boquet, P. 19—Helicobacter pylori vacuolating toxin. In *The Comprehensive Sourcebook of Bacterial Protein Toxins*, 4th ed.; Alouf, J., Ladant, D., Popoff, M.R., Eds.; Academic Press: Boston, MA, USA, 2015; pp. 515–557.
231. Satin, B.; Norais, N.; Telford, J.; Rappuoli, R.; Murgia, M.; Montecucco, C.; Papini, E. Effect of *Helicobacter pylori* vacuolating toxin on maturation and extracellular release of procathepsin D and on epidermal growth factor degradation. *J. Biol. Chem.* **1997**, *272*, 25022–25028. [CrossRef]
232. Huang, Y.; Wang, Q.L.; Cheng, D.D.; Xu, W.T.; Lu, N.H. Adhesion and Invasion of Gastric Mucosa Epithelial Cells by *Helicobacter pylori*. *Front. Cell Infect. Microbiol.* **2016**, *6*, 159. [CrossRef]
233. Capurro, M.I.; Greenfield, L.K.; Prashar, A.; Xia, S.; Abdullah, M.; Wong, H.; Zhong, X.Z.; Bertaux-Skeirik, N.; Chakrabarti, J.; Siddiqui, I.; et al. VacA generates a protective intracellular reservoir for *Helicobacter pylori* that is eliminated by activation of the lysosomal calcium channel TRPML1. *Nat. Microbiol.* **2019**, *4*, 1411–1423. [CrossRef]
234. Sit, W.Y.; Chen, Y.A.; Chen, Y.L.; Lai, C.H.; Wang, W.C. Cellular evasion strategies of *Helicobacter pylori* in regulating its intracellular fate. *Semin. Cell Dev. Biol.* **2020**, *101*, 59–67. [CrossRef] [PubMed]
235. Capurro, M.I.; Prashar, A.; Jones, N.L. MCOLN1/TRPML1 inhibition—A novel strategy used by *Helicobacter pylori* to escape autophagic killing and antibiotic eradication therapy in vivo. *Autophagy* **2020**, *16*, 169–170. [CrossRef] [PubMed]
236. Boquet, P.; Ricci, V.; Galmiche, A.; Gauthier, N.C. Gastric cell apoptosis and *H. pylori*: Has the main function of VacA finally been identified? *Trends Microbiol.* **2003**, *11*, 410–413. [CrossRef] [PubMed]
237. Blanke, S.R. Micro-managing the executioner: Pathogen targeting of mitochondria. *Trends Microbiol.* **2005**, *13*, 64–71. [CrossRef] [PubMed]

238. Galmiche, A.; Rassow, J.; Doye, A.; Cagnol, S.; Chambard, J.C.; Contamin, S.; de Thillot, V.; Just, I.; Ricci, V.; Solcia, E.; et al. The N-terminal 34 kDa fragment of *Helicobacter pylori* vacuolating cytotoxin targets mitochondria and induces cytochrome c release. *EMBO J.* **2000**, *19*, 6361–6370. [CrossRef] [PubMed]
239. Kuck, D.; Kolmerer, B.; Iking-Konert, C.; Krammer, P.H.; Stremmel, W.; Rudi, J. Vacuolating cytotoxin of *Helicobacter pylori* induces apoptosis in the human gastric epithelial cell line AGS. *Infect. Immun.* **2001**, *69*, 5080–5087. [CrossRef]
240. Cover, T.L.; Krishna, U.S.; Israel, D.A.; Peek, R.M., Jr. Induction of gastric epithelial cell apoptosis by *Helicobacter pylori* vacuolating cytotoxin. *Cancer Res.* **2003**, *63*, 951–957.
241. Cho, S.J.; Kang, N.S.; Park, S.Y.; Kim, B.O.; Rhee, D.K.; Pyo, S. Induction of apoptosis and expression of apoptosis related genes in human epithelial carcinoma cells by *Helicobacter pylori* VacA toxin. *Toxicon* **2003**, *42*, 601–611. [CrossRef]
242. Manente, L.; Perna, A.; Buommino, E.; Altucci, L.; Lucariello, A.; Citro, G.; Baldi, A.; Iaquinto, G.; Tufano, M.A.; De Luca, A. The *Helicobacter pylori*'s protein VacA has direct effects on the regulation of cell cycle and apoptosis in gastric epithelial cells. *J. Cell Physiol.* **2008**, *214*, 582–587. [CrossRef]
243. Lan, C.H.; Sheng, J.Q.; Fang, D.C.; Meng, Q.Z.; Fan, L.L.; Huang, Z.R. Involvement of VDAC1 and Bcl-2 family of proteins in VacA-induced cytochrome c release and apoptosis of gastric epithelial carcinoma cells. *J. Dig. Dis.* **2010**, *11*, 43–49. [CrossRef]
244. Yahiro, K.; Akazawa, Y.; Nakano, M.; Suzuki, H.; Hisatune, J.; Isomoto, H.; Sap, J.; Noda, M.; Moss, J.; Hirayama, T. *Helicobacter pylori* VacA induces apoptosis by accumulation of connexin 43 in autophagic vesicles via a Rac1/ERK-dependent pathway. *Cell Death Discov.* **2015**, *1*, 15035. [CrossRef] [PubMed]
245. Kim, J.M.; Kim, J.S.; Kim, N.; Ko, S.H.; Jeon, J.I.; Kim, Y.J. Helicobacter pylori vacuolating cytotoxin induces apoptosis via activation of endoplasmic reticulum stress in dendritic cells. *J. Gastroenterol. Hepatol.* **2015**, *30*, 99–108. [CrossRef] [PubMed]
246. Calore, F.; Genisset, C.; Casellato, A.; Rossato, M.; Codolo, G.; Esposti, M.D.; Scorrano, L.; de Bernard, M. Endosome-mitochondria juxtaposition during apoptosis induced by *H. pylori* VacA. *Cell Death Differ.* **2010**, *17*, 1707–1716. [CrossRef] [PubMed]
247. Ganten, T.M.; Aravena, E.; Sykora, J.; Koschny, R.; Mohr, J.; Rudi, J.; Stremmel, W.; Walczak, H. *Helicobacter pylori*-induced apoptosis in T cells is mediated by the mitochondrial pathway independent of death receptors. *Eur. J. Clin. Investig.* **2007**, *37*, 117–125. [CrossRef] [PubMed]
248. Matsumoto, A.; Isomoto, H.; Nakayama, M.; Hisatsune, J.; Nishi, Y.; Nakashima, Y.; Matsushima, K.; Kurazono, H.; Nakao, K.; Hirayama, T.; et al. *Helicobacter pylori* VacA reduces the cellular expression of STAT3 and pro-survival Bcl-2 family proteins, Bcl-2 and Bcl-XL, leading to apoptosis in gastric epithelial cells. *Dig. Dis. Sci.* **2011**, *56*, 999–1006. [CrossRef]
249. Kim, J.M.; Kim, J.S.; Lee, J.Y.; Sim, Y.S.; Kim, Y.J.; Oh, Y.K.; Yoon, H.J.; Kang, J.S.; Youn, J.; Kim, N.; et al. Dual effects of *Helicobacter pylori* vacuolating cytotoxin on human eosinophil apoptosis in early and late periods of stimulation. *Eur. J. Immunol.* **2010**, *40*, 1651–1662. [CrossRef]
250. Chacinska, A.; Koehler, C.M.; Milenkovic, D.; Lithgow, T.; Pfanner, N. Importing mitochondrial proteins: Machineries and mechanisms. *Cell* **2009**, *138*, 628–644. [CrossRef]
251. Jain, P.; Luo, Z.Q.; Blanke, S.R. *Helicobacter pylori* vacuolating cytotoxin A (VacA) engages the mitochondrial fission machinery to induce host cell death. *Proc. Natl. Acad. Sci. USA* **2011**, *108*, 16032–16037. [CrossRef]
252. Foo, J.H.; Culvenor, J.G.; Ferrero, R.L.; Kwok, T.; Lithgow, T.; Gabriel, K. Both the p33 and p55 subunits of the *Helicobacter pylori* VacA toxin are targeted to mammalian mitochondria. *J. Mol. Biol.* **2010**, *401*, 792–798. [CrossRef]
253. Kimura, M.; Goto, S.; Wada, A.; Yahiro, K.; Niidome, T.; Hatakeyama, T.; Aoyagi, H.; Hirayama, T.; Kondo, T. Vacuolating cytotoxin purified from *Helicobacter pylori* causes mitochondrial damage in human gastric cells. *Microb. Pathog.* **1999**, *26*, 45–52. [CrossRef]
254. Wang, L.; Yi, J.; Yin, X.Y.; Hou, J.X.; Chen, J.; Xie, B.; Chen, G.; Wang, Q.F.; Wang, L.N.; Wang, X.Y.; et al. Vacuolating Cytotoxin A Triggers Mitophagy in *Helicobacter pylori*-Infected Human Gastric Epithelium Cells. *Front. Oncol.* **2022**, *12*, 881829. [CrossRef] [PubMed]
255. Willhite, D.C.; Cover, T.L.; Blanke, S.R. Cellular vacuolation and mitochondrial cytochrome c release are independent outcomes of *Helicobacter pylori* vacuolating cytotoxin activity that are each dependent on membrane channel formation. *J. Biol. Chem.* **2003**, *278*, 48204–48209. [CrossRef] [PubMed]
256. Ashktorab, H.; Dashwood, R.H.; Dashwood, M.M.; Zaidi, S.I.; Hewitt, S.M.; Green, W.R.; Lee, E.L.; Daremipouran, M.; Nouraie, M.; Malekzadeh, R.; et al. *H. pylori*-induced apoptosis in human gastric cancer cells mediated via the release of apoptosis-inducing factor from mitochondria. *Helicobacter* **2008**, *13*, 506–517. [CrossRef] [PubMed]

membranes

MDPI

Article

Real-Time Observation of Capsaicin-Induced Intracellular Domain Dynamics of TRPV1 Using the Diffracted X-ray Tracking Method

Kazuhiro Mio [1,2,*], Tatsunari Ohkubo [1,2], Daisuke Sasaki [3], Tatsuya Arai [3], Mayui Sugiura [1], Shoko Fujimura [1], Shunsuke Nozawa [4], Hiroshi Sekiguchi [5], Masahiro Kuramochi [6] and Yuji C. Sasaki [1,3,5,*]

1 AIST-UTokyo Advanced Operando-Measurement Technology Open Innovation Laboratory (OPERANDO-OIL), National Institute of Advanced Industrial Science and Technology (AIST), 6-2-3 Kashiwanoha, Chiba 277-0882, Japan
2 Graduate School of Medical Life Science, Yokohama City University, 1-7-29 Suehiro-cho, Tsurumi-ku, Yokohama 230-0045, Japan
3 Graduate School of Frontier Sciences, The University of Tokyo, 5-1-5 Kashiwanoha, Chiba 277-8561, Japan; t.arai@edu.k.u-tokyo.ac.jp (T.A.)
4 Photon Factory, Institute of Materials Structure Science, High Energy Accelerator Research Organization, 1-1 Oho, Tsukuba 305-0801, Japan; noz@post.kek.jp
5 Center for Synchrotron Radiation Research, Japan Synchrotron Radiation Research Institute, 1-1-1 Kouto, Sayo-cho 679-5198, Japan
6 Graduate School of Science and Engineering, Ibaraki University, Hitachi 316-8511, Japan
* Correspondence: kazu.mio@aist.go.jp (K.M.); ycsasaki@edu.k.u-tokyo.ac.jp (Y.C.S.)

Abstract: The transient receptor potential vanilloid type 1 (TRPV1) is a multimodal receptor which responds to various stimuli, including capsaicin, protons, and heat. Recent advances in cryo-electron microscopy have revealed the structures of TRPV1. However, due to the large size of TRPV1 and its structural complexity, the detailed process of channel gating has not been well documented. In this study, we applied the diffracted X-ray tracking (DXT) technique to analyze the intracellular domain dynamics of the TRPV1 protein. DXT enables the capture of intramolecular motion through the analysis of trajectories of Laue spots generated from attached gold nanocrystals. Diffraction data were recorded at two different frame rates: 100 μs/frame and 12.5 ms/frame. The data from the 100 μs/frame recording were further divided into two groups based on the moving speed, using the lifetime filtering technique, and they were analyzed separately. Capsaicin increased the slope angle of the MSD curve of the C-terminus in 100 μs/frame recording, which accompanied a shifting of the rotational bias toward the counterclockwise direction, as viewed from the cytoplasmic side. This capsaicin-induced fluctuation was not observed in the 12.5 ms/frame recording, indicating that it is a high-frequency fluctuation. An intrinsiccounterclockwise twisting motion was observed in various speed components at the N-terminus, regardless of the capsaicin administration. Additionally, the competitive inhibitor AMG9810 induced a clockwise twisting motion, which is the opposite direction to capsaicin. These findings contribute to our understanding of the activation mechanisms of the TRPV1 channel.

Keywords: diffracted X-ray tracking technique; TRP channel; conformation dynamics; single molecule analysis; TRPV1

1. Introduction

The transient receptor potential vanilloid 1 (TRPV1) channel is activated by various stimuli, including heat, protons, and vanilloids, and plays a key role in nociception [1–3]. TRPV1 is a homotetramer channel, with each subunit containing six transmembrane segments and a pore-forming region. The pore is formed by the assembly of transmembrane segment 5 (S5), a pore loop, and transmembrane segment 6 (S6), surrounded by a voltage sensor-like domain composed of a bundle of four transmembrane helices (S1–S4). Both the N- and C-termini are intracellular, and an ankyrin repeat domain is prominently present at the N-terminus. Cryo-electron microscopy (cryo-EM) has revealed near-atomic resolution structures of TRPV1 [4–6], providing a fundamental framework for TRP activation.

Cryo-EM data of various gating stages, along with simulation studies [7–9], suggest the occurrence of rotational and twisting motion during gating, but direct detection of these movements has proven technically challenging.

The most controversial issue regarding the cryo-EM structures is that the capsaicin-bound structure was presented as a "closed form", with the lower gate open but the upper gate closed [5]. Subsequent studies demonstrated the "open form" by applying capsaicin and heat simultaneously [10], but raised a puzzling question of why capsaicin alone fails to induce the open conformation. To address this enigma, we have focused on the investigation of intramolecular motions.

In our previous study, we employed the diffracted X-ray tracking (DXT) technique to measure the dynamics of the extracellular domains of TRPV1. Capsaicin induced rotational bias towards the clockwise (CW) direction, which was speculated to be a channel opening movement, and this bias was sustained in motions with longer lifetime groups [11]. Based on this finding, it was hypothesized that the primary effect of capsaicin involves the opening of the lower gate and an increase in the oscillation frequency of the upper gate, thereby facilitating calcium influx [11].

In the present study, to elucidate the dynamics associate with ligand binding and channel gating, we investigated the motion of the N- and C-terminal domains using a high-speed time-resolved DXT technique with different recording speeds.

2. Methods

2.1. Plasmid Construct and Protein Purification

Full-length human TRPV1 was subcloned into pcDNA3.1 and tagged with $(His)_6$ sequence at the N-terminus and the FLAG sequence (DYKDDDDK) at the C-terminus. The TRPV1 protein was expressed in HEK293F cells and cell surface biotinylation was performed using sulfo-NHS-LC-biotin (Thermo Fisher, MA, USA) according to manufacturer's instructions. The TRPV1 protein was purified with two-step purification using an anti-FLAG affinity chromatography (Sigma-Aldrich, MO, USA) and a Superdex 200 Increase size exclusion chromatography (Cytiva, MA, USA). The peak fractions were concentrated to 0.1 mg/mL for DXT.

2.2. Sample Preparation for DXT Assays

Gold nanocrystals, which were grown on a KCl (100) surface via epitaxial growth under a 10^{-4} Pa vacuum, were dissolved in double-distilled water, and conjugated with anti $(His)_6$ monoclonal antibody (clone 9C11, FUJIFILM-Wako, Osaka, Japan) or anti FLAG monoclonal antibody (clone M2, Sigma-Aldrich) under borate buffer, pH 9.0. Buffer component was substituted with recording buffer (PBS (pH 7.4) with 5 mM n-decyl-β-D-maltoside (Sigma-Aldrich)) before applying to TRPV1. For the C-terminal analysis of the 12.5 ms/frame recording, ZnO crystal was used for labeling antibodies.

To immobilize biotinylated TRPV1 proteins to the basement surface, 12.5 μm thick polyimide film (Du Pont-Toray, Tokyo, Japan) was coated with cadmium and chromium (Cd/Cr) by evaporation, then biotin-functionalized self-assembled monolayers (Biotin-SAM) were formed using Biotin-SAM Formation Reagent (Dojindo, Kumamoto, Japan). After binding streptavidin on the Biotin-SAM membrane, twenty microliters of biotinylated TRPV1 (N-terminally $(His)_6$-tagged and C-terminally FLAG tagged, ~0.1 mg/mL) were applied and incubated at 4 °C for 6 h. Excess protein was washed out with buffer, and the gold- or ZnO-conjugated antibodies (dispersed in a recording buffer) were applied. After incubation for 20 min at 4 °C, excess antibodies were washed out. Ten microliter of recording buffer containing 10 μM capsaicin with or without 10 μM AMG9810 was applied. The sample chamber was covered with another layer of polyimide film, sandwiched by stainless steel frames and screw clamped. Capsaicin and AMG9810 were purchased from Sigma-Aldrich.

2.3. DXT Measurements

DXT measurements were conducted using the SPring-8 BL40XU and the Photon Factory Advanced Ring AR-NW14A beamlines. The X-ray beam size at SPring-8 BL40XU beamline was adjusted to 50 μm in diameter by inserting a pinhole aperture upstream of the sample, and time-resolved diffraction images from the gold nanocrystals were recorded by an X-ray image intensifier (V7739P, Hamamatsu photonics, Hamamatsu, Japan) and a CMOS camera (Phantom V2511, Vision Research, NJ, USA). For each sample, diffractions at 36 positions (6×6) within 1 mm^2 (1 mm $\times$ 1 mm) were recorded. The distance between the sample and the detector was set to 45.6 mm. X-rays with an energy bandwidth of 0.1 (15.8 keV in peak energy and photon flux of 10^{13} photons/s) were used for the DXT. Data were recorded with 100 μs/frame intervals and measured for 450 frames per spot. Time-resolved diffraction images at the Photon Factory Advanced Ring AR-NW14A beamline were recorded using a 2D photon-counting detector Pilatus 100 K (Dectris, Baden-Dättwil, Switzerland). The gap of an insertion device U20 was set to 15 mm and the white X-ray with 17.7 keV in peak energy with an energy bandwidth of 0.1 (photon flux of 10^{13} photons/s) was used [12]. The distance between the sample and the detector was set at 95 mm, with beam size of 100 μm $\times$ 250 μm. Data were recorded with 12.5 ms/frame intervals and measured for 5000 frames per spot. Recording was performed in triplicate for each sample, with 0.2 mm distance. Temperature at the sample chamber was kept at 25 °C using the Peltier cooling-and-heating stage (Type 10084L, Japan Hightech, Fukuoka, Japan). The concentration of the antibody-gold (or ZnO) conjugate was optimized to obtain approximately 1000 trajectories per recording. X-ray radiation damage was assessed by comparing the MSD curves calculated from the two groups: the first half and the second half.

2.4. Image Analysis for DXT

The obtained data were calibrated with the concurrently recorded intensity of incident X-ray (I_0). Each diffraction spot was tracked by TrackPy (v0.3.2 https://doi.org/10.5281/zenodo.60550) after correcting the background. Trajectories were analyzed using a custom software written within IGOR Pro version 9 (Wavemetrics, Lake Oswego, OR, USA). Diffraction spots of Au (111), ZnO (100), (002), (101) were used for the analysis. The diffractions from the fast-moving proteins have a short time duration between the appearance and disappearance in the observation area, while those of slow-moving proteins have longer duration. For the lifetime filtering technique, trajectories with short lifetime (LT < 2.5 ms) and medium lifetime (2.5 ms $\leq$ LT < 4.0 ms) were extracted from the 100 μs/frame recording data, and they were analyzed separately.

2.5. Mean Squared Displacement Curve

Protein dynamics were analyzed using a mean squared displacement (MSD) algorithm to extract the local behavior of the protein as a function of time. The MSD curves were fitted by the following function: $\delta^2(t) = D_\alpha t^\alpha + 2\beta^2$. D_α is the anomalous diffusion constant, a non-linear relationship to time. α represents subdiffusion ($1 >> \alpha > 0$) or superdiffusion ($\alpha > 1$), and β is a measurement error.

3. Results

We analyzed the intracellular domain dynamics of TRPV1 using the DXT technique (Figure 1a). DXT uses X-rays with the wavelength of 0.01 nm to 0.1 nm, enabling positioning accuracy at the picometer level. DXT can also measure dynamics to the order of microseconds to milliseconds, due to the high-brightness X-rays. The motions of the N- and C-termini were investigated separately, using distinct peptide tags and their corresponding antibodies. Gold or ZnO nanocrystals were bound to the antibodies in advance, to obtain diffractions. Diffraction recording was conducted at 100 μs per frame for fast motion and

12.5 ms per frame for slow motion (Figure 1b). The utilization of different frame rates allowed the extraction of distinct motion components. DXT recording with 100 μs/frame was conducted at the SPring-8 BL40XU beamline, and 12.5 ms/frame at the Photon Factory AR-NW14A beamline. The resulting diffraction spots were analyzed in two axes of tilt (θ) (Supplementary Figures S1 and S2 and Supplementary Tables S1 and S2) and twist (χ) angles. For each experimental condition, we analyzed 844–5222 trajectory points, providing a comprehensive assessment of the system.

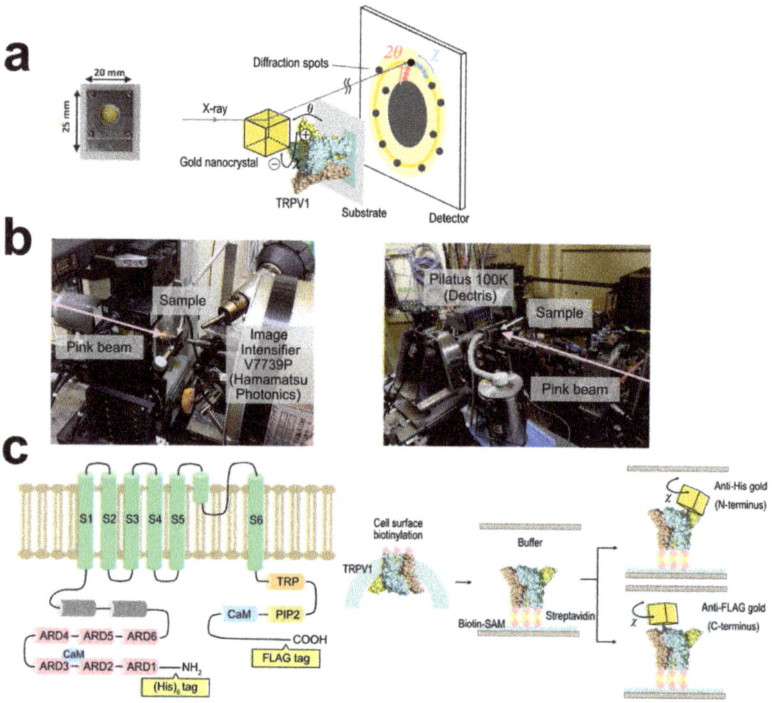

Figure 1. Diffracted X-ray tracking (DXT) measurements. (**a**) Schematic of the DXT measurement. Gold nanocrystals were labeled at specific positions of the target proteins, and their motion was determined by the analysis of the diffraction spots. Pink-beam X-rays elicited trackable diffraction spots from the gold nanocrystals. (**b**) **Left**: Experimental setup for the DXT measurement at the SPring-8 BL40XU beamline. Time-resolved diffraction images from the gold nanocrystals were recorded by an X-ray image intensifier (V7739P, Hamamatsu photonics) and a CMOS camera (Phantom V2511, Vision Research) with 100 μs/frame. **Right**: Photon Factory Advanced Ring AR-NW14A beamline. Data were recorded using a 2D photon-counting detector (Pilatus 100 K, Dectris) with 12.5 ms/frame. (**c**) **Left**: Structure of TRPV1 monomer. The $(His)_6$-tag and FLAG-tag were introduced at the N- and C-terminus, respectively. ARD: ankyrin repeat domain, TRP: TRP-box domain, PIP2: PIP2 binding site, CaM: calmodulin binding site. **Right**: Labeling of TRPV1 with nanocrystal-conjugated antibodies and sample preparation. N-terminally $(His)_6$-tagged, C-terminally FLAG tagged, and biotinylated TRPV1 was immobilized on a biotin-SAM formed polyimide films.

3.1. N-Terminal Dynamics of TRPV1

The N-terminal domain of TRPV1 comprises a typical ankyrin repeat domain (ARD), which is considered to play a major function in thermo-sensation [13,14] as well as modulating channel function by binding with multiple ligands [15–17]. The TRPV1 proteins were transiently expressed in HEK293F cells in suspension culture, and their extracellular domain was biotinylated using the surface biotinylation technique (Figure 1c). The TRPV1 protein was purified, and bound to Cd/Cr-coated polyimide films through a biotin-streptavidin reaction, on which biotin-SAM was generated in advance. The motions of the N-terminus were monitored using gold nanocrystal-conjugated antibodies specific to the hexahistidine $(His)_6$-tag introduced at the N-terminus.

Mean square displacement (MSD) curves were calculated for each experiment. MSD curves of 100 µs/frame recording revealed a slight increase in the slope of the curve by capsaicin (5 mrad^2/ms for control and 10 mrad^2/ms for capsaicin) (Figure 2a). However, further enhancement of the motion to 40 mrad^2/ms was observed upon application of AMG9810 (Figure 2a). In contrast, the MSD curves of 12.5 ms/frame recording showed comparable values between control and AMG9810 of 0.04 mrad^2/ms, but which were suppressed to 0.02 mrad^2/ms by capsaicin (Figure 2b).

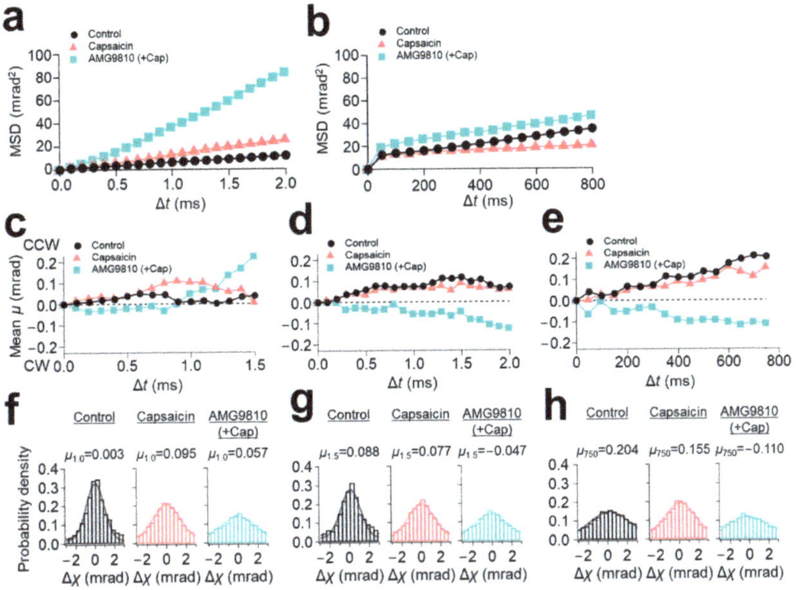

Figure 2. N-terminal domain movement of TRPV1 (χ axis). Mean squared displacement (MSD) curves of TRPV1 N-terminus for the χ axis recorded at (**a**) 100 µs/frame and (**b**) 12.5 ms/frame. (**c**) Rotational angle analysis for the χ axis of short-lifetime group (LT < 2.5 ms, left), (**d**) medium-lifetime group (2.5 ms ≤ LT < 4 ms) at 100 µs/frame recording and (**e**) 12.5 ms/frame recording. (**f**) The distribution of angular displacement of short-lifetime group, (**g**) medium-lifetime group at 100 µs/frame recording, and (**h**) 12.5 ms/frame recording.

To characterize the motion properties, we performed a directional analysis by segregating the rotational (χ) data into CW and counterclockwise (CCW) directions. The rotational bias was determined from the cytoplasmic side by comparing it to the anchored extracellular domains.

The calculation from the entire dataset did not show an apparent rotational bias, due to the closed sample chamber conditions. Therefore, we applied a lifetime filtering technique [11]. Lifetime was defined as the time between the appearance and disappearance of a diffraction spot in the observation area. Diffraction spots can be recorded within the energy range of synchrotron radiation; the fast-moving proteins have a short time duration while slow-moving proteins have longer duration. Therefore, the duration of diffraction spots (lifetime) represents the overall movement of the target protein, providing ligand-induced internal motility of TRPV1. Trajectories with lifetimes below 2.5 ms (LT < 2.5 ms) represent very fast movements, and trajectories with lifetimes between 2.5 ms and 4.0 ms (2.5 ms ≤ LT < 4.0 ms) represent medium-speed movements. They were extracted from the original dataset and analyzed separately.

We successfully extracted capsaicin-induced rotational bias in the domains. The short-lifetime group (LT < 2.5 ms) exhibited a slight CCW bias for all three conditions (Figure 2c). The medium-lifetime group (2.5 ms $\leq$ LT < 4.0 ms) demonstrated a sustained CCW bias for the control and capsaicin, but represented the opposite direction of CW bias for AMG9810 (Figure 2d). The rotational bias obtained from the 12.5 ms/frame recording was similar to those of the medium-speed movement group (2.5 ms $\leq$ LT < 4.0 ms): CCW rotational bias for both control and capsaicin, and CW rotational bias for AMG9810 (Figure 2e).

The rotation bias analysis demonstrated that the N-terminal movement of control and capsaicin comprises various moving speed components having CCW rotational biases (Figure 2c–e). It should be noted that these movements were recorded within the limited recording frame rate windows, and there may exist other motion components outside the windows; probably, much slower untwisting movements may be occurring as well.

The distribution of movement along the χ axis was displayed by histograms and fitted by Gaussian curves. Peak positions of all three curves of the fast-movement group (LT < 2.5 ms) were slightly positive (0.003 for control, 0.095 for capsaicin, and 0.057 for AMG9810) (Figure 2f). The distributions were broadened by capsaicin and further broadened by AMG9810 (full width at half maximum (FWHM) were: 2.191 for control, 3.316 for capsaicin, and 4.270 for AMG9810; Table 1), reflecting an increase in motion for capsaicin and AMG9810, as observed in the Figure 2a. The histograms were similar for the medium-speed movement group (2.5 ms $\leq$ LT < 4.0 ms), but the peak position of AMG9810 was shifted to the negative value (0.088 for control, 0.077 for capsaicin, and −0.047 for AMG9810).

Table 1. Gaussian fitting parameters for the N-terminus (χ).

	Control	Capsaicin	AMG9810 (+Cap)
LT < 2.5 ms (100 µs/frame recording)			
-Location (mrad)	0.003 ± 0.032	0.095 ± 0.034	0.057 ± 0.054
-FWHM (mrad)	2.191 ± 0.081	3.316 ± 0.090	4.270 ± 0.156
-Peak area	0.713 ± 0.026	0.674 ± 0.019	0.546 ± 0.022
2.5 ms ≤ LT < 4.0 ms (100 µs/frame recording)			
-Location (mrad)	0.088 ± 0.044	0.077 ± 0.038	−0.047 ± 0.058
-FWHM (mrad)	2.317 ± 0.110	3.135 ± 0.101	3.837 ± 0.160
-Peak area	0.637 ± 0.030	0.618 ± 0.020	0.508 ± 0.023
12.5 ms/frame recording			
-Location (mrad)	0.204 ± 0.037	0.155 ± 0.037	−0.110 ± 0.055
-FWHM (mrad)	4.653 ± 0.112	3.740 ± 0.101	5.011 ± 0.173
-Peak area	0.671 ± 0.019	0.723 ± 0.021	0.554 ± 0.023

As for the slow-speed motion (12.5 ms/frame recording), the peak position was not significantly influenced by capsaicin (0.204 for control and 0.155 for capsaicin), but the FWHM was decreased (4.653 for control and 3.740 for capsaicin), suggesting that capsaicin reduced the movement at the N-terminus, as observed in the Figure 2b. The peak position of the AMG9810 was shifted to −0.110.

In conclusion, capsaicin did not induce significant motion change to the N-terminal domain, and, regardless of the frame rates, both the control and capsaicin intrinsically had a CCW rotational bias. On the other hand, AMG9810 exhibited a clear increase in the motion with opposite directional movement compared to the control and capsaicin. This trend was typically observed in the motion of the medium- and slow-speed groups. It is speculated that the competitive inhibitor AMG9810 inhibits the endogenous rotational movement in TRPV1, resulting in the extraction of only the CW directional movement.

3.2. C-Terminal Dynamics of TRPV1

The C-terminal domain of TRPV1 locates much closer to the gate in the sequence. The S6 segment undergoes conformational changes upon capsaicin binding [5]. The S6 segment leads to the TRP-box domain, which forms a long helix parallel to the membrane plane and to the cytoplasmic C-terminus. The C-terminal domain contains calmodulin and phosphatidylinositol binding sites, which are considered to modulate channel function [18,19]. The C-terminal domain is mostly intrinsically disordered. The MSD curves from the 100 μs/frame recording demonstrated that capsaicin increased the slope angle from 10 mrad^2/ms (control) to 20 mrad^2/ms at the C-terminus (Figure 3a). However, in contrast to the N-terminus, AMG9810 did not enhance the slope angle. As for the 12.5 ms/frame recording, capsaicin did not change the slope angle of the MSD curve (Figure 3b); instead, AMG9810 increased the slope angle, as observed in the N-terminus.

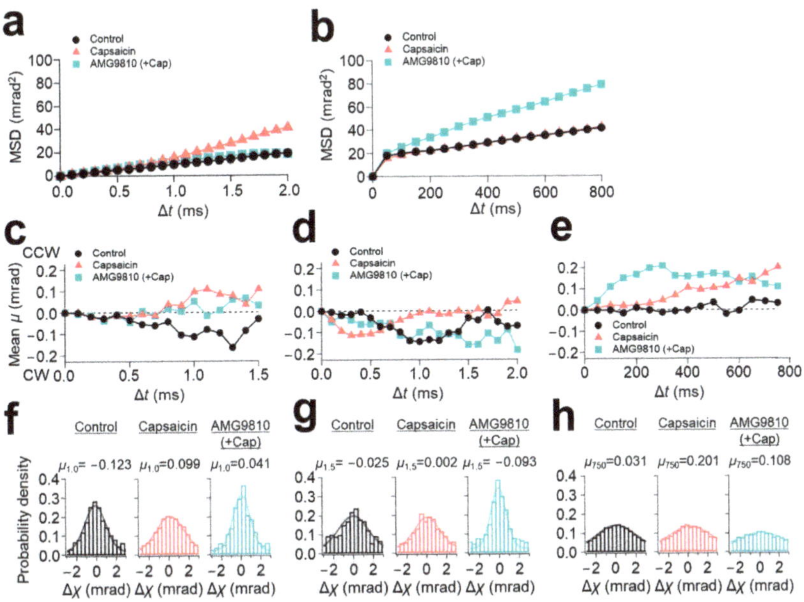

Figure 3. C-terminal domain movement of TRPV1 (χ axis). Mean squared displacement (MSD) curves of TRPV1 C-terminus for the χ axis recorded at (**a**) 100 μs/frame and (**b**) 12.5 ms/frame. (**c**) Rotational angle analysis for the χ axis of short-lifetime group (LT < 2.5 ms, left), (**d**) medium-lifetime group (2.5 ≤ LT < 4 ms) at 100 μs/frame recording and (**e**) 12.5 ms/frame recording. (**f**) The distribution of angular displacement of short-lifetime group, (**g**) medium-lifetime group at 100 μs/frame recording, and (**h**) 12.5 ms/frame recording.

These dynamics were further analyzed using the lifetime filtering technique and the rotational bias analysis. Capsaicin induced CCW rotational bias for the fast-movement group (LT < 2.5 ms) (Figure 3c), consequently shifting the peak position of capsaicin in the histograms from −0.123 (control) to 0.099 (Figure 3f). In contrast to the broadening of the histograms at the N-terminus, the histograms of AMG9810 in both the short-lifetime group (LT < 2.5 ms) and medium-lifetime group (2.5 ms ≤ LT < 4.0 ms) were narrowed, representing a decrease in motion dynamics (Figure 3f,g and Table 2).

As for the slow-speed motion (12.5 ms/frame recording), all three conditions showed CCW rotational biases (Figure 3e). Gaussian fitting for these conditions did not show a significant change in distributions (Figure 3h). The change in the peak position was also limited (0.031 for control, 0.201 for capsaicin, and 0.108 for AMG9810).

Table 2. Gaussian fitting parameters for the C-terminus (χ).

	Control	Capsaicin	AMG9810 (+Cap)
LT < 2.5 ms (100 μs/frame recording)			
-Location (mrad)	−0.123 ± 0.037	0.099 ± 0.028	0.041 ± 0.040
-FWHM (mrad)	2.456 ± 0.095	3.237 ± 0.075	2.093 ± 0.101
-Peak area	0.639 ± 0.024	0.647 ± 0.016	0.656 ± 0.031
2.5 ms ≤ LT < 4.0 ms (100 μs/frame recording)			
-Location (mrad)	−0.025 ± 0.081	0.002 ± 0.053	−0.093 ± 0.042
-FWHM (mrad)	3.537 ± 0.221	3.007 ± 0.139	1.827 ± 0.104
-Peak area	0.690 ± 0.046	0.536 ± 0.025	0.619 ± 0.033
12.5 ms/frame recording			
-Location (mrad)	0.031 ± 0.028	0.201 ± 0.048	0.108 ± 0.058
-FWHM (mrad)	4.492 ± 0.082	5.109 ± 0.152	5.496 ± 0.197
-Peak area	0.598 ± 0.012	0.646 ± 0.023	0.491 ± 0.022

In conclusion, the motion of the C-terminal domain was highly influenced by the ligand. In the 100 μs/frame recording, capsaicin induced an increase in movement, accompanied by a significant shift in the rotational bias towards CCW direction. This movement was considered a very fast oscillatory movement, as it was not observed in other timescale data. Considering the timescale, this movement can be related to the gating dynamics. On the other hand, AMG9810 suppressed the capsaicin-induced motion at the C-terminus in the 100 μs/frame recording. This is in sharp contrast to the AMG9810 at the N-terminus, where it increased movement, accompanied by a clear rotational bias. Histogram analysis also demonstrated that AMG9810 effectively suppressed the C-terminal motion.

3.3. *Subtraction Analysis for Capsaicin-Induced Intracellular Domain Dynamics of TRPV1*

To elucidate the capsaicin-induced rotational (χ) domain movement more clearly, the histogram subtractions between them, with and without capsaicin, were performed and analyzed.

At the N-terminus, capsaicin did not induce significant change in rotational bias, but broadened the population distribution for both the fast- (LT < 2.5 ms) and the medium-moving groups (2.5 ms ≤ LT < 4.0 ms) in the 100 μs/frame recording, and narrowed it in the 12.5 ms/frame recording (Figure 2). Consequently, the subtraction graphs of both the fast- (LT < 2.5 ms) and medium-moving groups (2.5 ms ≤ LT < 4.0 ms) showed a single negative peak around zero mrad, with small positive peaks on both sides, indicating that capsaicin increased the MSD curve angle without changing the rotational bias (Figure 4a, Supplementary Table S3). In contrast to the 100 μs/frame, the subtraction histogram of the 12.5 ms/frame recording showed a single positive peak around the zero mrad, suggesting that capsaicin narrowed the population distribution without changing the rotational bias.

In contrast, capsaicin shifted the rotational bias at the C-terminus in the 100 μs/frame recording (Figure 3c). This shift was typically observed in the fast-moving group (LT < 2.5 ms), where the peak position of capsaicin was shifted CCW from −0.123 mrad (control) to 0.099 mrad (Figure 3f), resulting in the two-peak curve with one negative peak at the −0.28 mrad and one positive peak at the 1.48 mrad (Figure 4b), Supplementary Table S4). The capsaicin-induced rotational bias was not obvious in the subtraction maps of the medium-moving group (2.5 ms ≤ LT < 4.0 ms) and the 12.5 ms/frame recording.

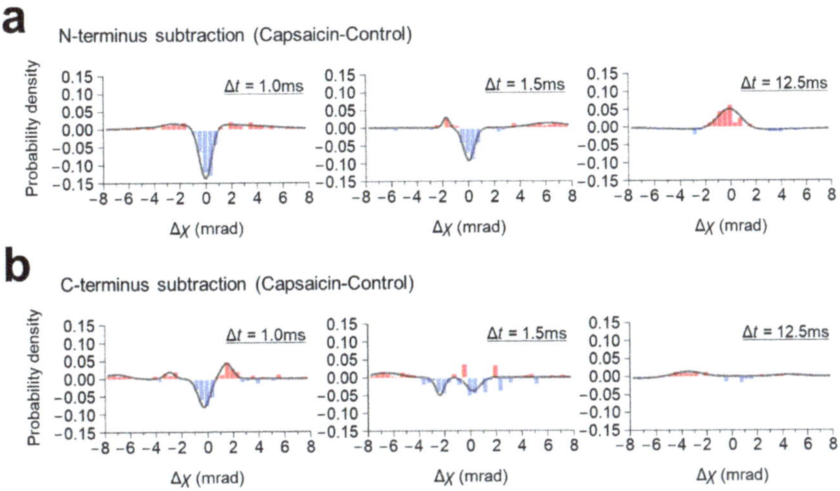

Figure 4. Subtraction maps of the distribution of angular displacement. Maps were generated by subtracting the population distribution histograms of control (without capsaicin) from those of capsaicin. (**a**) Motion difference at the N-terminus, and (**b**) the C-terminus. From left to right: subtraction maps of the short-lifetime group (LT < 2.5 ms), medium-lifetime group (2.5 ms $\leq$ LT < 4 ms) at 100 μs/frame recording, and 12.5 ms/frame recording.

4. Discussion

In this study, we applied the DXT technique to understand the intracellular domain dynamics of the TRPV1 channel. DXT is a very powerful method for clarifying membrane protein dynamics [11,20–22], analyzing the dynamics of functional proteins like chaperonin [23], and analyzing the oligomerization mechanisms of amyloid β (Aβ) [24], the Alzheimer's disease-related protein; it is also able to do this with a high spatiotemporal resolution, even for the intramolecular dynamics in living cells [25,26]. Capsaicin induced high-frequency fluctuational movement at the C-terminus, while CCW twisting motion on a range of various timescales was observed at the N-terminus, regardless of the addition of capsaicin (Figure 5). The rotational bias observed in the control may reduce the energy barrier in gating, which may be beneficial for capsaicin to pull the trigger for gating, with minimum conformational changes.

The N-terminal domain of a TRPV1 monomer comprises six ankyrin repeats; each ankyrin repeat consists of two antiparallel α helices (inner and outer helix), separated by hairpin loops. The ARD of TRPV1 comprises calmodulin and ATP binding sites, and the channel function is modulated by binding with Ca^{2+}, ATP and other multiple ligands [15–17]. Recent studies also suggested that the ARD plays an important role in thermal sensation [13,14]. Point mutations in TRPV1 ARD induce drastic changes in the temperature threshold in tailed amphibians [14].

In our experiment, capsaicin did not show significant influence on the N-terminal domain motion, and, independently, an intrinsic CCW rotational bias was observed for the control and capsaicin. Because recent studies suggested that the N-terminal ARD has a major function in thermal sensation, further studies are expected to apply DXT to analyze the thermo-induced TRPV1 gating mechanisms. We also observed that AMG9810 exhibited a clear increase in the MSD curve, but its rotational bias was in the opposite direction compared to the control and capsaicin. We have confirmed in advance that the AMG9810 inhibits capsaicin-induced calcium entry, using the cell-based calcium influx assay. It is speculated that the competitive inhibitor AMG9810 inhibits the endogenous rotational movement in TRPV1, resulting in the extraction of only the opposing directional movement.

The C-terminal part of a TRPV1 monomer comprises the TRP-box domain, which is conserved in the TRPC, TRPV, and TRPM subfamilies [27]. The TRP-box is a long helix, parallel to the membrane plane [4]. The gating of the TRPV1 channel is regulated by the PIP_2 and calmodulin bindings and by phosphorylation at the TRP-box and other part of the C-terminus [18,19]. However, most of the cytoplasmic C-terminus is intrinsically disordered, and its structure is not well understood.

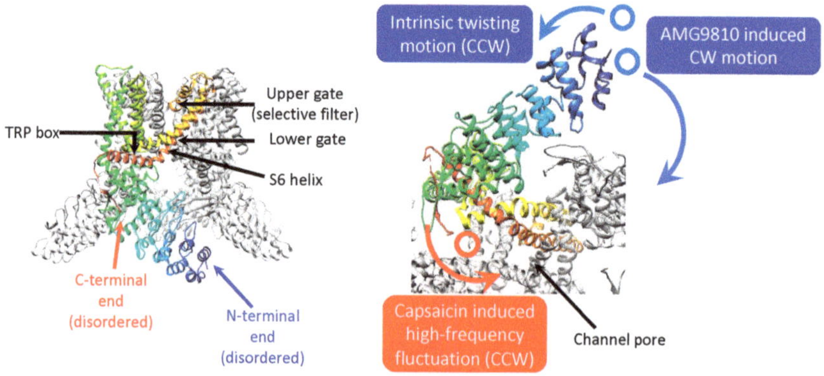

Figure 5. Motion dynamics of TRPV1 cytoplasmic domain activation by capsaicin. **Left**: A monomer subunit from the TRPV1 tetramer is depicted in rainbow color scheme from the N-terminus (blue) to the C-terminus (red). Data are displayed by the UCSF Chimera version 1.15 (PDB: 3J5P). **Right**: DXT demonstrates motion dynamics of the N- and C-terminal domains viewed from the cytoplasmic side.

In our experiment, the C-terminal domain motion was highly influenced by ligands. In the 100 μs/frame recording, capsaicin increased the movement at the C-terminus, accompanied by a significant shift in the rotational bias towards the CCW direction. This movement was considered a very fast oscillatory movement, because it was not obvious in other timescale data. Conformational dynamics of ion channels occur on nanosecond-to-millisecond timescales. Considering the timescale, this movement can be associated with gating-related molecular dynamics. Further studies using functionally modified mutant proteins or introducing protein decoration, including phosphorylation, calmodulin and PIP_2 binding, may be helpful in elucidating the detailed TRPV1 gating mechanisms.

Our previous study using DXT to measure the extracellular domain dynamics of TRPV1 demonstrated that capsaicin induced a CW rotational bias, which was suggested as a gate-opening movement [11]. From these results, capsaicin may primarily open the lower gate by the twisting movement of the transmembrane helices and increase the oscillation frequency of the upper gate as observed in the C-terminal end, facilitating calcium influx. Further study is needed to fully understand TRPV1's gating mechanism and interaction dynamisms by capsaicin and other ligands.

In our analysis, measurements were conducted at different frame rates. We also introduced the lifetime filtering technique, which can successfully classify the motions by their moving speed. Different frame rates captured motion components operating at significantly different speeds, including rapid conformational changes on microsecond-to-millisecond time scales. Patch clamp experiments suggested that a half-activation time of TRPV1 by capsaicin is more than hundred milliseconds, significantly slower than the activation by heat at 51 °C ($t1/2$ = ~6 ms) [28]. Our DXT experiments were performed at frame rates from 100 μs to 12.5 ms, covering the timescale for gate opening expected from the patch clamp experiments. Patch clamp measures ion permeation resulting from gate opening, whereas DXT detects all motions related to gating. In combination with the patch clamp technique, DXT may be a powerful tool for unraveling the mechanisms behind, such as pore dilation (agonist stimulation-induced ionic selectivity changes) [29]. By introduced this technique, we could obtain the capsaicin-induced rotational bias at the intracellular domains, which was not evident in the statistical analysis using the unfiltered data source.

Supplementary Materials: The following supporting information can be downloaded at: https://www.mdpi.com/article/10.3390/membranes13080708/s1, Figure S1: N-terminal domain movement of TRPV1 (θ axis); Figure S2: C-terminal domain movement of TRPV1 (θ axis); Table S1: Gaussian fitting parameters for the N-terminus (θ); Table S2: Gaussian fitting parameters for the C-terminus (θ); Table S3: Fitting parameters for the subtraction analysis (N-terminus, χ); Table S4: Fitting parameters for the subtraction analysis (C-terminus, χ).

Author Contributions: Conceptualization, K.M. and Y.C.S.; formal analysis, T.O., D.S., T.A., M.S., S.F. and M.K.; investigation, K.M., T.O., D.S., T.A., M.S., S.F., S.N., H.S., M.K. and Y.S.; writing—original draft preparation, K.M.; writing—review and editing, K.M., T.O. and Y.C.S. All authors have read and agreed to the published version of the manuscript.

Funding: This work was supported by JSPS KAKENHI Grant Numbers JP 23H01818, JP 21K06515, JP 17H05539 and by JST CREST Grant Number JP18071859, Japan.

Institutional Review Board Statement: Not applicable.

Data Availability Statement: Not applicable.

Acknowledgments: We thank Kayoko Kawaguchi and Takaaki Shiina for technical assistance. DXT and DXB experiments were performed with the approval of the Japan Synchrotron Radiation Research Institute (Proposal No. 2019A1498, and 2019B1294), and of the Photon Factory Program Advisory Committee (Proposal No. 2019G151, 2019G637, and 2021G567).

Conflicts of Interest: The authors declare no conflict of interest.

References

1. Caterina, M.J.; Schumacher, M.A.; Tominaga, M.; Rosen, T.A.; Levine, J.D.; Julius, D. The capsaicin receptor: A heat-activated ion channel in the pain pathway. *Nature* **1997**, *389*, 816–824. [CrossRef] [PubMed]
2. Tominaga, M.; Caterina, M.J.; Malmberg, A.B.; Rosen, T.A.; Gilbert, H.; Skinner, K.; Raumann, B.E.; Basbaum, A.I.; Julius, D. The cloned capsaicin receptor integrates multiple pain-producing stimuli. *Neuron* **1998**, *21*, 531–543. [CrossRef]
3. Julius, D. TRP channels and pain. *Annu. Rev. Cell Dev. Biol.* **2013**, *29*, 355–384. [CrossRef] [PubMed]
4. Liao, M.; Cao, E.; Julius, D.; Cheng, Y. Structure of the TRPV1 ion channel determined by electron cryo-microscopy. *Nature* **2013**, *504*, 107–112. [CrossRef] [PubMed]
5. Cao, E.; Liao, M.; Cheng, Y.; Julius, D. TRPV1 structures in distinct conformations reveal activation mechanisms. *Nature* **2013**, *504*, 113–118. [CrossRef]
6. Diver, M.M.; Lin King, J.V.; Julius, D.; Cheng, Y. Sensory TRP Channels in Three Dimensions. *Annu. Rev. Biochem.* **2022**, *91*, 629–649. [CrossRef]
7. Yang, F.; Xiao, X.; Cheng, W.; Yang, W.; Yu, P.; Song, Z.; Yarov-Yarovoy, V.; Zheng, J. Structural mechanism underlying capsaicin binding and activation of the TRPV1 ion channel. *Nat. Chem. Biol.* **2015**, *11*, 518–524. [CrossRef]
8. Zheng, A.; Qin, F. A combined coarse-grained and all-atom simulation of TRPV1 channel gating and heat activation. *J. Gen. Physiol.* **2015**, *145*, 443–456. [CrossRef]
9. Kasimova, M.A.; Yazici, A.T.; Yudin, Y.; Granata, D.; Klein, M.L.; Rohacs, T.; Carnevale, V. A hypothetical molecular mechanism for TRPV1 activation that invokes rotation of an S6 asparagine. *J. Gen. Physiol.* **2018**, *150*, 1554–1566. [CrossRef]
10. Kwon, D.H.; Zhang, F.; Suo, Y.; Bouvette, J.; Borgnia, M.J.; Lee, S.Y. Heat-dependent opening of TRPV1 in the presence of capsaicin. *Nat. Struct. Mol. Biol.* **2021**, *28*, 554–563. [CrossRef]
11. Fujimura, S.; Mio, K.; Kuramochi, M.; Sekiguchi, H.; Ikezaki, K.; Mio, M.; Hengphasatporn, K.; Shigeta, Y.; Kubo, T.; Sasaki, Y.C. Agonist and Antagonist-Diverted Twisting Motions of a Single TRPV1 Channel. *J. Phys. Chem. B.* **2020**, *124*, 11617–11624. [CrossRef] [PubMed]
12. Nozawa, S.; Adachi, S.I.; Takahashi, J.I.; Tazaki, R.; Guérin, L.; Daimon, M.; Tomita, A.; Sato, T.; Chollet, M.; Collet, E. Developing 100 ps-resolved X-ray structural analysis capabilities on beamline NW14A at the Photon Factory Advanced Ring. *J. Synchrotron. Rad.* **2007**, *14*, 313–319. [CrossRef] [PubMed]
13. Ladrón-De-Guevara, E.; Dominguez, L.; Rangel-Yescas, G.E.; Fernández-Velasco, D.A.; Torres-Larios, A.; Rosenbaum, T.; Islas, L.D. The contribution of the Ankyrin Repeat Domain of TRPV1 as a thermal module. *Biophys. J.* **2020**, *118*, 836–845. [CrossRef] [PubMed]
14. Hori, S.; Tateyama, M.; Shirai, T.; Kubo, Y.; Saitoh, O. Two single-point mutations in Ankyrin Repeat one drastically change the threshold temperature of TRPV1. *Nat. Commun.* **2023**, *14*, 2415. [CrossRef]

15. Rosenbaum, T.; Gordon-Shaag, A.; Munari, M.; Gordon, S.E. Ca^{2+}/calmodulin modulates TRPV1 activation by capsaicin. *J. Gen. Physiol.* **2004**, *123*, 53–62. [CrossRef]
16. Lishko, P.V.; Procko, E.; Jin, X.; Phelps, C.B.; Gaudet, R. The Ankyrin Repeats of TRPV1 Bind Multiple Ligands and Modulate Channel Sensitivity. *Neuron* **2007**, *54*, 905–918. [CrossRef]
17. Phelps, C.B.; Wang, R.R.; Choo, S.S.; Gaudet, R. Differential Regulation of TRPV1, TRPV3, and TRPV4 Sensitivity through a Conserved Binding Site on the Ankyrin Repeat Domain. *J. Biol. Chem.* **2010**, *285*, 731–740. [CrossRef]
18. Numazaki, M.; Tominaga, T.; Takeuchi, K.; Murayama, N.; Toyooka, H.; Tominaga, M. Structural determinant of TRPV1 desensitization interacts with calmodulin. *Proc. Natl. Acad. Sci. USA* **2003**, *100*, 8002–8006. [CrossRef]
19. Yazici, A.T.; Gianti, E.; Kasimova, M.A.; Lee, B.H.; Carnevale, V.; Rohacs, T. Dual regulation of TRPV1 channels by phosphatidylinositol via functionally distinct binding sites. *J. Biol. Chem.* **2021**, *296*, 100573. [CrossRef]
20. Shimizu, H.; Iwamoto, M.; Konno, T.; Nihei, A.; Sasaki, Y.C.; Oiki, S. Global twisting motion of single molecular KcsA potassium channel upon gating. *Cell* **2008**, *132*, 67–78. [CrossRef]
21. Sekiguchi, H.; Suzuki, Y.; Nishino, Y.; Kobayashi, S.; Shimoyama, Y.; Cai, W.; Nagata, K.; Okada, M.; Ichiyanagi, K.; Ohta, N.; et al. Real time ligand-induced motion mappings of AChBP and nAChR using X-ray single molecule tracking. *Sci. Rep.* **2014**, *4*, 6384. [CrossRef]
22. Ohkubo, T.; Shiina, T.; Kawaguchi, K.; Sasaki, D.; Inamasu, R.; Yang, Y.; Li, Z.; Taninaka, K.; Sakaguchi, M.; Fujimura, S.; et al. Visualizing Intramolecular Dynamics of Membrane Proteins. *Int. J. Mol. Sci.* **2022**, *23*, 14539. [CrossRef]
23. Sekiguchi, H.; Nakagawa, A.; Moriya, K.; Makabe, K.; Ichiyanagi, K.; Nozawa, S.; Sato, T.; Adachi, S.; Kuwajima, K.; Yohda, M.; et al. ATP dependent rotational motion of group II chaperonin observed by X-ray single molecule tracking. *PLoS ONE* **2013**, *8*, e64176. [CrossRef] [PubMed]
24. Chang, J.; Arai, T.; Kuramochi, M.; Inamasu, R.; Lee, Z.; Ohkubo, T.; Mio, K.; Sasaki, Y.C. Dynamic observations of various oligomers in amyloid β isoforms using laboratory diffracted X-ray blinking. *Biochem. Biophys. Rep.* **2022**, *31*, 101298. [CrossRef] [PubMed]
25. Mio, K.; Ishihara, M.; Fujimura, S.; Sasaki, D.; Nozawa, S.; Ichiyanagi, K.; Fukaya, R.; Adachi, S.-I.; Kuramochi, M.; Sekiguchi, H.; et al. X-ray-based living-cell motion analysis of individual serotonin receptors. *Biochem. Biophys. Res. Commun.* **2020**, *529*, 306–313. [CrossRef]
26. Mio, K.; Fujimura, S.; Ishihara, M.; Kuramochi, M.; Sekiguchi, H.; Kubo, T.; Sasaki, Y.C. Living-Cell Diffracted X-ray Tracking Analysis Confirmed Internal Salt Bridge Is Critical for Ligand-Induced Twisting Motion of Serotonin Receptors. *Int. J. Mol. Sci.* **2021**, *22*, 5285. [CrossRef] [PubMed]
27. Montell, C. Physiology, phylogeny, and functions of the TRP superfamily of cation channels. *Sci. STKE* **2001**, *90*, re1. [CrossRef]
28. Yao, J.; Liu, B.; Qin, F. Kinetic and energetic analysis of thermally activated TRPV1 channels. *Biophys. J.* **2010**, *99*, 1743–1753. [CrossRef] [PubMed]
29. Chung, M.K.; Güler, A.D.; Caterina, M.J. TRPV1 shows dynamic ionic selectivity during agonist stimulation. *Nat. Neurosci.* **2008**, *11*, 555–564. [CrossRef]

membranes

MDPI

Article

The Alternating Access Mechanism in Mammalian Multidrug Resistance Transporters and Their Bacterial Homologs

Shadi A Badiee , Ugochi H. Isu , Ehsaneh Khodadadi and Mahmoud Moradi *

Department of Chemistry and Biochemistry, University of Arkansas, Fayetteville, AR 72701, USA; sasadian@uark.edu (S.A.B.); uhisu@uark.edu (U.H.I.); ekhodada@uark.edu (E.K.)
* Correspondence: moradi@uark.edu

Abstract: Multidrug resistance (MDR) proteins belonging to the ATP-Binding Cassette (ABC) transporter group play a crucial role in the export of cytotoxic drugs across cell membranes. These proteins are particularly fascinating due to their ability to confer drug resistance, which subsequently leads to the failure of therapeutic interventions and hinders successful treatments. One key mechanism by which multidrug resistance (MDR) proteins carry out their transport function is through alternating access. This mechanism involves intricate conformational changes that enable the binding and transport of substrates across cellular membranes. In this extensive review, we provide an overview of ABC transporters, including their classifications and structural similarities. We focus specifically on well-known mammalian multidrug resistance proteins such as MRP1 and Pgp (MDR1), as well as bacterial counterparts such as Sav1866 and lipid flippase MsbA. By exploring the structural and functional features of these MDR proteins, we shed light on the roles of their nucleotide-binding domains (NBDs) and transmembrane domains (TMDs) in the transport process. Notably, while the structures of NBDs in prokaryotic ABC proteins, such as Sav1866, MsbA, and mammalian Pgp, are identical, MRP1 exhibits distinct characteristics in its NBDs. Our review also emphasizes the importance of two ATP molecules for the formation of an interface between the two binding sites of NBD domains across all these transporters. ATP hydrolysis occurs following substrate transport and is vital for recycling the transporters in subsequent cycles of substrate transportation. Specifically, among the studied transporters, only NBD2 in MRP1 possesses the ability to hydrolyze ATP, while both NBDs of Pgp, Sav1866, and MsbA are capable of carrying out this reaction. Furthermore, we highlight recent advancements in the study of MDR proteins and the alternating access mechanism. We discuss the experimental and computational approaches utilized to investigate the structure and dynamics of MDR proteins, providing valuable insights into their conformational changes and substrate transport. This review not only contributes to an enhanced understanding of multidrug resistance proteins but also holds immense potential for guiding future research and facilitating the development of effective strategies to overcome multidrug resistance, thus improving therapeutic interventions.

Keywords: multidrug resistance; MDR; ATP-binding cassette transporter; ABC transporters; lipid flippase; MsbA; Sav1866; multidrug resistance protein; MRP1; Pgp; P-glycoprotein; nucleotide-binding domains; NBDs; transmembrane domains; TMDs; alternating access mechanism; alternating access model

1. Introduction

Multidrug resistance (MDR) is one of the obstacles in modern medicine since it reduces the efficacy of several drugs and makes treatment less successful. Resistance to antibiotics, anticancer, and antiviral treatments are mediated by these proteins, which are classified as one of the subfamilies in ATP-binding cassette (ABC) transporters. ABC transporters are composed of two distinct parts called the nucleotide-binding domain (NBD) and the transmembrane domain (TMD). The NBD provides energy for the movement of substrates in or out of the cell through having a binding site for ATP, whereas TMD supplies the substrate translocation route. The selective efflux of substrates is made possible by the alternating access mechanism, in which the transporter switches between inward- and outward-facing conformations.

P-glycoprotein (Pgp, also referred to as MDR1 and ABCB1) and Multidrug resistance protein 1 (MRP1) are critical multidrug resistance in mammalian cancer cells, and these transporters can efflux several different types of medicines. MsbA and Sav1866 are bacterial homologs of mammalian Pgp. Based on their TMD fold, these four transporters are all classified as group IV. Although the TMDs are arranged similarly, their features, such as NBDs, ATP hydrolysis, NBD-TMD interface, transporter mechanism details, substrate specificity, substrate binding pocket, etc., may differ, case by case. In this review paper, we attempt to describe the majority of these features using experimental and computational papers. The transporters MRP1, Pgp, Sav1866, and MsbA were chosen for this review paper to include both bacterial (Sav1866, MsbA) and mammalian (Pgp, MRP1) transporters.

Several recent review papers have thoroughly covered various aspects of the ABC superfamily, such as the role of ABC transporters in specific cancer progression and metastasis, the development of novel ABC transporter modulators, and the collection of data on natural and synthetic compounds targeting cancer-related ABC transporters, among others [1–9]. Our review paper, however, specifically focuses on mammalian proteins and their bacterial homologs, aiming to comprehensively gather data on all aspects of these proteins, particularly their alternating access mechanisms. In this review paper, we briefly explain the variety of efflux pumps, highlighting their importance in multidrug resistance (Section 1.1). This is followed by a broad overview of the ATP-binding cassette (ABC) superfamily (Section 1.2), specifically focusing on the multidrug resistance (MDR) subfamily and their general structure and function (Section 1.3). We also discuss their alternating access mechanisms (Section 1.4) and highlight two vital MDR proteins (MRP1 and Pgp) and the homologs of Pgp (Sav1866, MsbA), providing a detailed analysis of their structures and mechanisms (Section 2).

1.1. Efflux Pumps

Efflux pumps are membrane proteins that play a crucial role in removing various compounds from cells [10–12]. Efflux pumps are found in diverse organisms, encompassing both prokaryotes and eukaryotes. However, there are variations in the types and characteristics of efflux pumps between these two groups. Although certain classes of efflux pumps can confer resistance in both prokaryotic and eukaryotic organisms, there are notable differences in their specific functions and mechanisms [5,13]. Prokaryotic efflux pumps are classified into six distinct groups, whereas eukaryotic efflux pumps are classified into five groups. Efflux pumps play crucial roles in a multitude of cellular processes. These include maintaining the proper potential and pH gradient across the cell membrane, participating in intercellular signaling, contributing to microbial virulence, and facilitating the elimination of unwanted metabolites and toxic substances from the cell. Through these functions, efflux pumps actively contribute to the maintenance of cellular homeostasis [10]. Efflux pumps can be classified into five primary families, namely small multidrug resistance (SMR), multi-antimicrobial extrusion (MATE), major facilitator superfamily (MFS), resistance nodulation and cell division (RND), and ATP-binding cassette (ABC). The ABC family transporters harness energy obtained from ATP hydrolysis. In contrast, pumps belonging to the other four families operate as second-order transporters, utilizing a proton gradient for their transport function [10]. Although efflux pumps are commonly linked to multidrug resistance, the ABC efflux pumps stand out as the primary players involved in multidrug resistance [5], with other efflux pumps possessing unique roles and functions extending beyond MDR [14,15]. The structures and roles of the other types of efflux pumps are briefly described below.

1.1.1. Major Facilitator Superfamily

MFS efflux pumps are membrane proteins that actively transport a wide range of substrates across cellular membranes [16]. Examples include Escherichia coli MdfA and TetA [17,18]. As part of the largest superfamily of secondary active transporters, MFS efflux pumps utilize the energy from electrochemical gradients to transport substrates bidirectionally [19]. They play a role in maintaining cellular homeostasis and regulating molecule levels within the cell. These pumps are diverse and can transport organic compounds, ions, and drugs out of the cell, aiding in cellular detoxification [19]. In some bacteria, MFS efflux pumps are associated with multidrug resistance by expelling antibiotics and contributing to drug resistance [19]. The substrate-binding site in MFS transporters is typically located in a central cavity, involving residues from both the N- and C-terminal domains. Multidrug recognition is achieved through a large substrate-binding pocket that can accommodate compounds of different sizes and shapes [17,19].

1.1.2. Multi-Antimicrobial Extrusion

MATE is a diverse family of secondary active transporters [20] composed of 12 transmembrane helices [21]. These transporters are found in all three domains of life [20] and play a crucial role in facilitating the transport of a wide range of compounds across cellular and organellar membranes [22]. Bacterial MATE transporters have a specific affinity for cationic compounds, including important antibiotics. Their presence contributes significantly to the multi-drug resistance observed in pathogenic bacteria, as these transporters actively extrude antibiotics and confer resistance to these drugs [20]. In plants, MATE transporters are involved in the accumulation of various metabolites within organelles, particularly vacuoles. They play a vital role in maintaining cellular homeostasis by actively transporting specific compounds to their designated locations [23]. In humans, MATE transporters are expressed in the brush-border membrane of the kidney. They are responsible for the clearance of cationic drugs from the body, ensuring their elimination through urine [20]. MATE transporters can be classified into different subfamilies based on their amino acid sequence similarity. These subfamilies include NorM, DinF (DNA-damage inducible protein F), and eukaryotic MATE transporters[24,25].

1.1.3. Small Multidrug Resistance

SMR efflux pumps, which are part of the small multidrug resistance (SMR) family, consist of four transmembrane helices and typically form homodimers or sometimes heterodimers [12,25]. These pumps play a crucial role in exporting a diverse range of toxic compounds from bacterial cells. Their compact size, with a short polypeptide length of 100–150 amino acids [26], and the presence of short hydrophilic loops enable them to solubilize and transport various drugs [12]. Examples of SMR efflux pumps include EmrE, QacG, and QacH, found in different bacterial species [27]. These pumps actively export toxic compounds such as quaternary ammonium compounds and dyes, protecting the cells from their harmful effects [25]. By facilitating the removal of these substances, SMR efflux pumps contribute to bacterial resistance against a wide range of compounds [12,25,26].

1.1.4. Resistance-Nodulation-Division

RND efflux pumps form tripartite complexes composed of three components: an inner membrane protein, a periplasmic membrane fusion protein, and an outer membrane protein [28,29]. Together, these complexes facilitate the active efflux of various substrates from bacterial cells. Examples of RND efflux pumps include AcrAB-TolC and MexAB-OprM, found in Gram-negative bacteria such as *E. coli* and *Pseudomonas aeruginosa*, respectively [28,30]. These pumps are known for their ability to actively pump out a diverse range of substrates, including antibiotics, detergents, and heavy metals. By actively removing these substances from the cell, RND efflux pumps play a crucial role in bacterial resistance, allowing the cells to survive and adapt in different environments [25,28]. They serve as a defense mechanism by protecting the cell from harmful compounds and are also involved in the development of multidrug resistance. The structure of RND transporters is distinct from other secondary transporters, characterized by a conserved transmembrane region and a periplasmic and luminal domain that contribute significantly to the protein's mass [31]. The transmembrane region contains an internal structural repeat resulting from gene duplication, while the periplasmic luminal domain can vary among different RND homologs [25].

1.2. ABC Transporter

ATP-binding cassette (ABC) transporters are a vast class of membrane proteins that utilize the energy from ATP hydrolysis (bond breakage between γ and β phosphate in ATP) to carry a number of different substrates such as peptides [32], ions, lipids, vitamins, amino acids, drugs, etc., across the membrane [33,34]. They can be homo or heterodimers comprising two conserved cytoplasmic NBDs and two distinct TMDs [35–38]. NBDs act as gatekeepers and are necessary to pass molecules through the membrane bilayer. In addition to binding and hydrolyzing ATP, they can induce conformational changes in the related TMDs. This alteration allows substrates to traverse the membrane bilayer and pass into or out of the cytoplasm [39,40]. It is known that both prokaryotic and eukaryotic/mammalian organisms are members of the ABC transporter superfamily [41].

ABC transporters in humans are classified into five subfamilies based on their sequence: ABCA, ABCB, ABCC, ABCD, and ABCG [42–45]. There are two types of ABC transporters based on the direction of substrate movement: importers and exporters. Eukaryotes, unlike bacteria that can yield both ABC importers and exporters, have mainly exporters with some exceptions [46]. ABC transporters were also historically divided into three groups: exporters, importers, and non-transporters [47,48]. However, a recent paper by Thomas et al. (2020) [42] proposed a new classification based on the structural homology of TMDs. This new classification of ABC transporters is implemented in response to the wealth of high-resolution structural information available from X-ray crystallography and cryo-electron microscopy. These methods uncovered diverse TMD folds and evolutionary relationships between bacterial and mammalian transporters [49–53]. By integrating this new structural knowledge with previous phylogenetic analyses, a new classification has been introduced. In accordance with this novel classification, ABC transporters can be divided into seven categories (I–VII) based solely on their TMD fold without considering any additional membrane-integrated domains. TMD fold is determined by utilizing TM-scores obtained from comparing the structural alignment of the TMDs. For instance, Class IV (Figure 1) represents a TMD fold with a 6 + 6 transmembrane helix organization. It can exist as a homo- or heterodimer within a single chain. Notably, transporters such as Pgp, MsbA, Sav1866, and MRP1 fall into this class [42].

Table 1 shows a list of available PDB files for Pgp, MRP1, Sav1866, and MsbA transporter proteins. These files were created using advanced techniques such as X-ray crystallography and cryo-electron microscopy. We sorted the accessible PDB files for each protein by resolution, starting with higher-resolution structures and progressing to lower-resolution structures.

Table 1. Available PDB files for MRP1, Pgp, Sav1866, and MsbA transporter proteins.

Name	PDB ID *	Resolution	Method **
MRP1	6BHU [54]	3.14 Å	Cryo-EM
	6UY0 [55]	3.23 Å	Cryo-EM
	8F4B [56]	3.27 Å	Cryo-EM
	5UJA [57]	3.34 Å	Cryo-EM
	5UJ9 [57]	3.49 Å	Cryo-EM
	7M68 [58]	4.04 Å	Cryo-EM

Table 1. *Cont.*

Name	PDB ID *	Resolution	Method **
Pgp	5KO2 [59]	3.30 Å	X-ray
	5KPD [60]	3.35 Å	X-ray
	4Q9H [61]	3.40 Å	X-ray
	6C0V [62]	3.40 Å	Cryo–EM
	5KPJ [59]	3.50 Å	X-ray
	4XWK [63]	3.50 Å	X-ray
	4Q9L [61]	3.80 Å	X-ray
	4M1M [64]	3.80 Å	X-ray
	5KOI [59]	3.85 Å	X-ray
	6UJN [65]	3.98 Å	X-ray
	5KPI [59]	4.01 Å	X-ray
Sav1866	2HYD [66]	3.00 Å	X-ray
	2ONJ [67]	3.40 Å	X-ray
MsbA	6BPL [68]	2.70 Å	X-ray
	6BPP [68]	2.92 Å	X-ray
	6BL6 [69]	2.80 Å	X-ray
	3B60 [70]	3.70 Å	X-ray
	5TV4 [71]	4.20 Å	Cryo-EM
	6UZL [72]	4.40 Å	Cryo-EM

* Data obtained from the protein data bank (PDB) database (https://www.rcsb.org, accessed on 17 May 2023). Citations to the PDB structures are included. ** Cryo-EM Cryogenic Electron Microscopy; X-ray X-ray Diffraction.

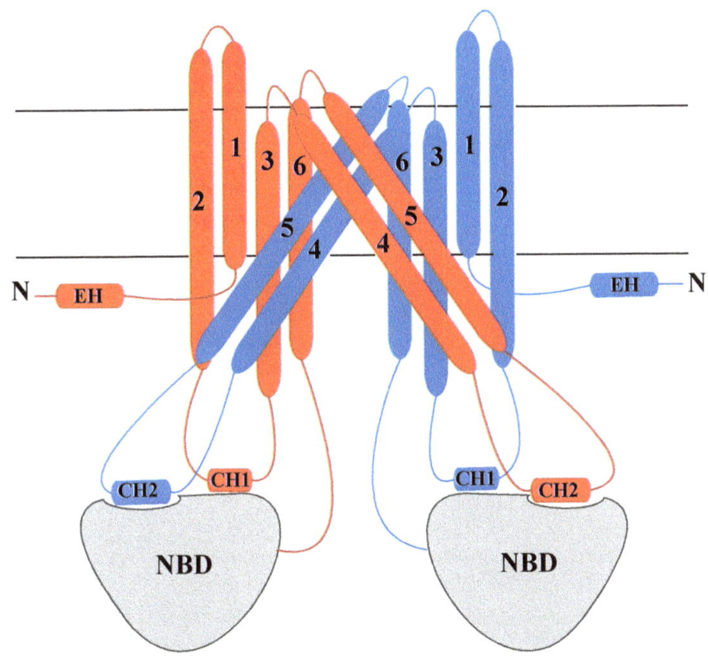

Figure 1. Schematic representation of type IV ABC transporters which are classified based on TMD arrangement. Each TMD (blue and red color) is composed of six transmembrane helices (1–6). CH and EH indicate coupling helix and elbow helix, respectively. In this transporter type, the elbow helix (EH), located at the N-terminus of each transporter half, plays a crucial role in stabilizing the transporter by positioning itself on the membrane-facing surface of the NBD. Each TMD of type IV transporters comprises a conserved core consisting of six transmembrane helices. This structure has two coupling helices (CH1, CH2) that interact with the NBDs.

1.3. MDRs

The ability of cells or organisms to resist the effects of numerous medications or chemotherapeutic treatments is referred to as multidrug resistance (MDR). MDR can be caused by a variety of mechanisms, including the overexpression of efflux pumps, which actively remove drugs from cells, changes in target molecules that reduce drug binding affinity, drug modification or inactivation, decreased drug uptake due to membrane transporter mutations, increased DNA repair mechanisms, and changes in signaling pathways. These mechanisms all contribute to the development of MDR and cause obstacles in disease treatment [73]. Resistance against anticancer medications is one of the significant difficulties related to developing anticancer chemotherapies [74,75]. In addition, resistance to multiple antibiotics is one of the other significant challenges regarding bacterial infections [76,77]. In cancer cells, multidrug resistance (MDR) as a phenomenon is caused by the over-expression of ABC transporters such as P-glycoprotein (Pgp, ABCB1 or MDR1), multidrug resistance-associated proteins MRPs (ABCC1), and breast cancer resistance protein BCRP (ABCG2) leading to the removal of drugs from the cell, thereby reducing their concentration below the effective level. They are expressed in a wide variety of organs, such as the liver, colon, kidney, brain, etc. [78]. MDR transporters perform critical roles, including the distribution, metabolism, and elimination of medicines, along with enabling cells to discharge anticancer or antibiotic drugs from the cell [66,79]. They have also been described as "hydrophobic vacuum cleaners" because of their capacity to remove drugs and lipids from the inner membrane leaflet [80]. Researchers have concentrated on finding inhibitors for these drug efflux pumps since they can increase the efficacy of anticancer and antibacterial medications [81].

1.4. NBD and TMD

Two solvent-exposed nucleotide-binding domains (NBD1 and NBD2) and two transmembrane domains (TMD1 and TMD2) make up the four functional domains of the typical ABC transporter (Figure 2A). The transmembrane domains are also known as MSD or membrane-spanning domains. Most of the eukaryotic ABC transporters are single polypeptides that contain all four domains, unlike bacterial ABC transporters, which can be composed of individual subunits or joined NBDs and TMDs. Eukaryotic ABC transporters can occasionally be built from similar or dissimilar half transporters [82]. ABC transporter structure can be divided into four groups named: full structure (NBD-TMD-NBD-TMD or TMD-NBD-TMD-NBD) [34], half structure (NBD-TMD or TMD-NBD), single structure (TMD or NBD)and ABC2 structure. ABC2 is the structure of non-transporter ABC proteins lacking TMDs. According to this system, ABC proteins in prokaryotes are either single or half structures, while those in eukaryotes can be either half or full [83].

NBD, also known as ATPase or ATP-binding sites, is a hydrophilic domain located in the cytoplasmic side of the ABC transporter and is composed of two subdomains referred to as RecA-type ATP-binding core and ABC-specific. RecA-type ATP-binding core is a central β-sheet flanked by four α-helices that participates in ATP binding and hydrolysis [42,84–87]. RecA-type ATP-binding core includes motifs or subdomains named Walker A and Walker B motifs, the Q-loop, the D-loop, and the H-switch [88,89] (Figure 2B). Walker A motif or phosphate-binding loop (P-loop) is a highly conserved sequence of amino acids, including glycine (G) and lysine (K)residues (GXXGXGKS/T), forming hydrogen bonds with the phosphate group of ATP. The threonine (T)/serine (S) residue is responsible for the interaction of nucleotide bases. Walker B motif ($\phi\phi\phi\phi$DE), where ϕ shows hydrophobic residues, contains aspartic acid (D) in order to coordinate magnesium (Mg^{2+}) required for ATP hydrolysis. This motif also has glutamate (E) that helps facilitate the ATP hydrolysis reaction as a general base by making water molecules more reactive. Walker A (P loop) and Walker B motifs are connected via a short loop named Q-loop which is crucial for coordinating nucleotide-binding sites (NBDs). This loop (ϕ(ϕ/Q)Q) includes glutamine residues that interact with γ-phosphate of the nucleotide [90]. Moreover, Q-loop plays a significant role in communicating with the TMDs. D-loop (SALD) connects two strands of the anti-parallel β-sheet within the RecA-like core and is involved in nucleotide binding and hydrolysis. As ATP hydrolysis happens, a conformational change occurs in a loop of amino acids known as the H-switch (switch region including conserved histidine residue), which is thought to be critical for connecting ATP hydrolysis to substrate transport. The ABC-specific subdomain is composed of three anti-parallel β-sheets known as ABC-β and an α-helical subdomain known as ABC-α. The signature motif (C-loop) is a short amino acid sequence with LSGGQ sequence [91] that is unique to ABC transporters and is situated between the Walker A and Walker B motifs, found at the end of ABC-α. The C-loop improves ATP binding stability and hydrolysis by interacting with ATP's γ-phosphate [42]. Walker-A motif (P loop), Walker-B motif, and C-loop (signature motif) are conserved sequence motifs within the NBDs that are essential for the ATP-binding and hydrolysis capability, along with interacting between NBDs and TMDs [88,92]. In the presence of two ATP molecules, NBDs can undergo the dimerization (sandwich dimer) required to generate energy for the process. An ABC transporter can hold two ATP molecules simultaneously between Walker A and Walker B motifs in one NBD and signature motif on the other NBD subunit [86,93]. Hydrolysis of ATP causes dissociation of the nucleotide-binding sites (NBS), releasing Pi and ADP.

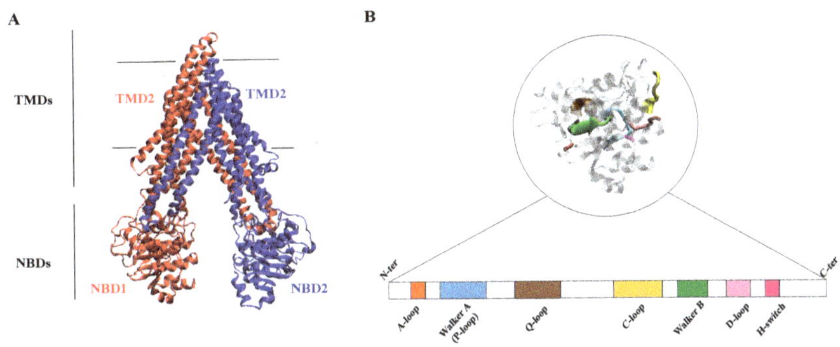

Figure 2. (**A**) The crystal structure of an ABC transporter including TMDs (TMD1 and TMD2) and NBDs (NBD1 and NBD2)(PDB ID:7OTI). (**B**) Schematic representation of NBD, which shows the positions of sequence motifs within NBDs.

TMD is the hydrophobic part of the ABC transporters, which spans the membrane leaflet, and substrates have their binding sites in TMD [94]. TMDs typically consist of six transmembrane α-helices, and each functional dimer TMD has around 12 of these transmembrane segments [95], except for some MRPs (Multidrug Resistance-Associated Proteins) with 17 α-helices including 3 TMDs [96]. The TMD dimers of ABC transporters serve as translocation pathways for substrates, facilitated by specialized transmembrane helices (TMHs) with specific substructures. When substrates bind to TMDs, it triggers a series of conformational changes that enable them to traverse the membrane. The hydrolysis of ATP by NBDs provides the essential energy for the process and drives these conformational modifications. The interface between TMDs and NBDs is established by small conserved helices known as coupling helices (CHs) that interact with the catalytic domain of NBDs [66] (Figure 1).

1.5. Alternating Access Mechanism (AAM)

The alternating access mechanism is used by all active membrane transporters, although they might vary in structure and mechanism [97–99]. In this mechanism (Figure 3), the transporter proteins undergo conformational changes to cross the substrate against its concentration gradient from one side of the membrane to another. During this process, the substrate binding site of the protein is exposed inside (inward-facing, IF) or outside (outward-facing, OF) the membrane alternately (IF $\leftrightarrow$ OF), passing through a number of potential intermediate phases [100–103], in such a way that ATP binding induces NBD dimerization and the development of the OF state. In contrast, NBD dissociation and the restoration of the IF state happens through ATP hydrolysis, resulting in the release of ADP and Pi [104,105]. Although the general structural state of the protein can remain IF or OF, various local conformational changes, such as mobility in amino acid side chains, loops, and helices, can block the substrate-binding site of these transporters [106].

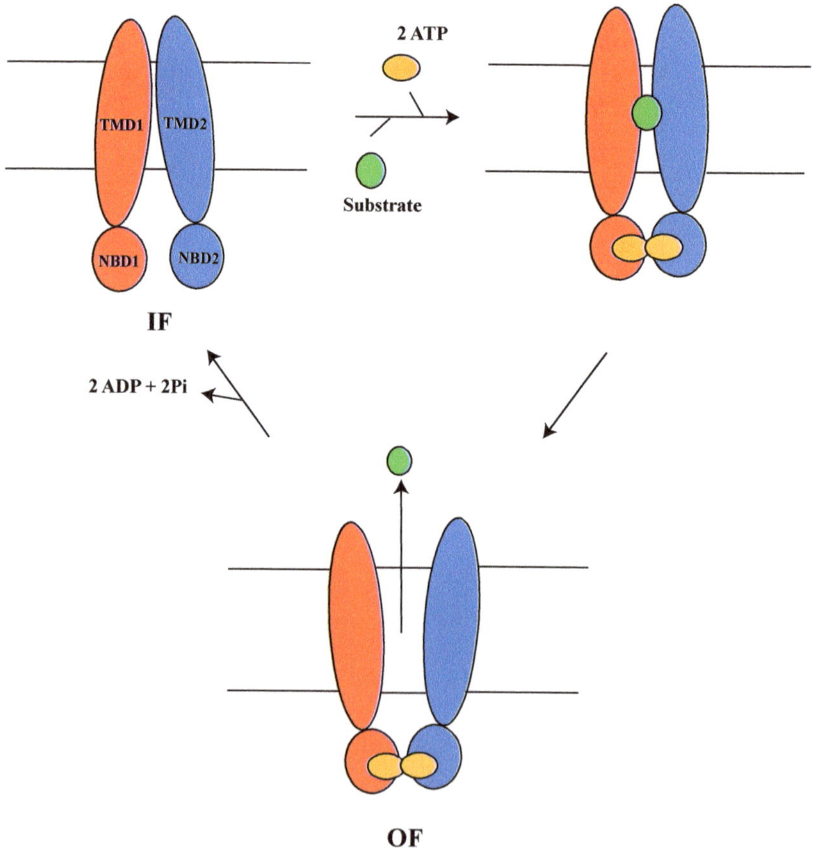

Figure 3. Schematic representation of the alternating access mechanism in ATP-binding (ABC) transporters. The substrate (green circle) initiates the transport cycle by binding to a substrate binding site formed by the TMDs. The NBDs then undergo a conformational change that permits ATP binding (orange ellipse) and the formation of a closed NBD dimer. A significant conformational change in the TMDs is triggered by the closed NBD dimer, and TMDs open toward the outside, allowing substrate translocation to begin. Dissociation of NBD dimer is triggered by ATP hydrolysis, and the transporter returns to an inward-facing conformation state by releasing phosphate and ADP.

2. Alternating Access Mechanism in Mammalian MDRs and Bacterial Homologs

This section will discuss four transporters, P-glycoprotein, Sav1866, MsbA, and MRP1, through reviewing computational and experimental papers. Pgp and MRP1 both function as multidrug resistance (MDR) proteins. Bacterial Pgp transporter homologs, such as Sav1866 and MsbA, are also discussed.

2.1. MRP1

MRPs (ABCC) are present in different tissues and are responsible for transporting various types of endo- and xenobiotics [107,108]. They not only function as drug efflux pumps for MDR but also have physiological roles such as detoxification, stress management, inflammation, and substance transportation [109]. Nine members of the ABCC family are known as MRPs. They are classified into long (ABCC1-3, ABCC6, and ABCC10) and short (ABCC4, ABCC5, ABCC11, and ABCC12) families [110].

MRP1 plays a role in protecting the body by pumping out endogenous substances and xenobiotics [111–113]. In normal cells, it functions as an efflux pump for the detoxification of xenobiotics generated by phase II enzymes and is present in several important tissues, such as the blood-brain barrier, lung, testis, kidney, intestinal epithelium, peripheral blood mononuclear cells, skeletal and cardiac muscle, and the placenta [114–116]. However, overexpression of MRP1 can occur in tumor cells, leading to reduced intracellular concentrations of antineoplastic drugs in cancer cells.

2.1.1. Structure of MRP1

The ABCC1 gene encodes the ATP binding cassette (ABC) transporter known as multidrug resistance protein 1 (MRP1). MRP1, similar to other ABC transporters, has TMDs and NBDs as structural elements. Each TMD has six transmembrane α-helices, and the NBDs are basolaterally localized [117]. MRP1 (1531 residues, 190 kDa [118]) also has a third TMD, called TMD0, with an extra-cytosolic NH2-terminus containing five predicted transmembrane segments [119] (Figure 4). TMD0 is found in the long group of ABCC transporters, including ABCC1-3, ABCC6, and ABCC10 [120]. The entire structure of MRP1, including the three TMDs and two NBDs, forms a stable network to enable domain-domain interaction, which facilitates the binding of substrate drugs in the cytoplasm and expels them to the extracellular milieu via ATP hydrolysis. TMD0 is not essential for the transport function or proper routing of MRP1 to the plasma membrane [57,121]. Lasso motif (L_0) as a conserved motif in all ABCC class of transporters [116] seems to play a role in the folding of the transporter and has a crucial role in MRP1 regulation [57].

MRP1 has been found to have a single ATPase site that can catalyze a reaction [122,123]. It has two nonequivalent NBDs. Once two ATP molecules connect to NBD's ATP sites, ATP hydrolysis first takes place in NBD2. Although, glutamate residue in walker B motif is responsible for attracting γ-phosphate of ATP in order to hydrolyze ATP through breaking phosphodiester bond between γ and β-phosphate [86,124], in MRP1, NBD1 has aspartate (D) instead of glutamate (E) residue, which aspartate may not be able to interact effectively with the water because of its short side chain. This distinction is exclusive to NBD1, while glutamate (E) can be found in this position in NBD2, thus, lending it a higher ability to hydrolyze the ATP. Moreover, NBD2 signature motif in MRP1 has a different sequence (LSVGQ) [57,122,125,126]. Additionally, NBD1 in MRP1 has a 13 amino acid deletion, which makes the interaction of this domain (NBD1) to TMD different and weaker compared to the interaction between NBD2 and TMD. The model is called "ball-and-socket", where the CH of TMD (ball) interacts with the surface of an NBD's cleft (socket). Therefore, there are fewer interactions between NBD1 and TMD due to the absence of 13 residues in the socket of NBD1 [57,127].

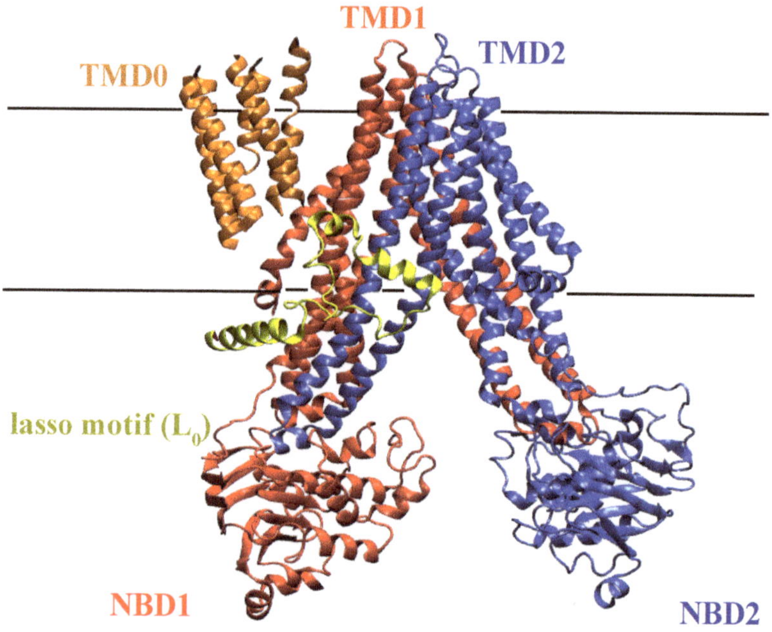

Figure 4. Crystal structure of the MRP1 (PDB:5UJA) including three transmembrane domains (TMD0, TMD1, TMD2), two cytosolic domains (NBD1 and NBD2), and Lasso motif (L_0) connecting TMD0 to the rest of the protein.

2.1.2. Substrate Binding Site of MRP1

When MRP1 functions as an efflux pump, the NBDs undergo a conformational change that causes a global conformational change, leading to a change in the shape of the substrate-binding pocket [128]. Cryo-EM studies have revealed the structures of MRP1 in two conformations: an apo form and a substrate-binding form [57,129]. These structures can explain the substrate specificity and polyspecificity of MRP1, as the TMDs are hydrophobic, and MRP1 substrates are typically amphipathic organic acids [130,131]. The substrate polyspecificity of MRP1 is achieved through the formation of a single bipartite binding site, which is flexible and predominantly composed of positively charged amino acids (P-pocket). MRP1's binding site also has a hydrophobic side or H-pocket, which is able to interact with hydrophobic substrates. This binding site allows MRP1 to recognize substrates with various chemical structures, especially organic acids [57,131].

The MRP1 protein provides cellular protection against oxidized heavy metal anions and can also regulate ion channel activity, including an increase in potassium ion channel activity with increased MRP1 expression [132]. MRP1 transports substances that are conjugated to glutathione (GSH), glucuronic acid, or sulfate, such as leukotrienes, alkylating agents, steroids, prostaglandin A2, bile salt derivatives, and folic acid [129,133,134]. Specifically, MRP1 can transport leukotriene C4 (LTC4), a product of arachidonic acid oxidation that is conjugated to antioxidant glutathione (GSH) and can cause bronchoconstriction and pulmonary edema [135]. MRP1 is highly expressed in white blood cells, lungs, and the trachea, where leukotriene synthesis occurs. This indicates that the transport of LTC4 may be an important physiological function of MRP1 [136,137]. In cancer cells, MRP1 can mediate multidrug resistance by promoting the efflux of glutathione-conjugated drugs [131]. MRP1 may also play a role in tumor invasion, metastasis, and disease outcome. MRP1 is involved in the blood-brain barrier and protects intracranial tissues from chemotherapeutic drugs [138]. Additionally, it is involved in mediating inflammation and removing toxic substances from the body [139].

2.1.3. AAM in MRP1

MRP1, present in all human tissues, is responsible for exporting endogenous and exogenous compounds, including arachidonic acid metabolites and drugs such as anticancer agents and antidepressants. Cryo-EM reconstructions of bovine MRP1 have provided insight into how the transporter is able to export a wide range of substrates and how these substrates stimulate ATP hydrolysis [57]. As mentioned above, the substrate-binding site is divided into a hydrophobic region (H-pocket) and a positively charged area (P-pocket). In the presence of ATP, the site changes to a low-affinity state, indicating that substrate release is independent of ATP hydrolysis [140]. The studies have shown that MRP1, in the absence of ATP or any substrate/drug, assumes an inward-facing conformation with the NBDs oriented away from each other and the translocation pathway near the cytoplasm [131]. When the TMDs approach each other, high substrate binding affinity is observed. Upon substrate binding, the NBDs move even closer to form a dimer, resulting in the outward-facing conformation of the ABC transporter [54,131]. Additionally, the extracellular ends of the TMD helices peel outward, causing the residues involved in generating the substrate-binding cavity to be pulled apart. As a result, the transporter's binding affinity for the substrate decreases significantly, leading to the efflux of the substrate/drug into the extracellular space [141].

The first phase of the transition from the protein's IF to occluded form is facilitated by the substrate, while the second step is carried out by ATP, which transfers the protein's occluded to the outward form of MRP1 [57,75]. A recent computational study investigated the transformation of MRP1 from IF to the occluded-facing form through co-transporting GSH and anticancer drugs such as mitoxantrone, epirubicin, and vincristine. GSH is required in most of the transportation by MRP1 [142]. They discovered that two TMHs (TM11 and TM17) of MRP1, GSH, and anticancer medications combine to form a structure known as a sandwich-like structure that causes the movement of transmembrane helices and NBDs; therefore, it causes the formation of the occluded-facing state. The formation of this sandwich-like structure is facilitated by some residues located on TM11, TM17, and TM6 that interact with antitumors and GSH in substrate binding sites of the MRP1. Trp1246 is the residue on TM17 in H-pocket, forming a bond with anticancer drugs; however, Lys332 of TM6 interacts with GSH in the P-pocket. A linker also forms between TM11 and TM17 through Phe594 of TM11. The simulations of the systems with anticancer drugs have been done in the absence and presence of GSH. The results indicated that MRP1 undergoes large conformational changes in the presence of both GSH and anticancer drugs. For instance, the distance between NBDs decreased, while ATP-binding sites between NBD1 and NBD2 were formed. However, this is not observed in the absence of even one of them. Some experimental evidence shows GSH is necessary to transport some of the drugs by MRP1 [132,143]. Zhao et al. (2020) also proposed unidirectional movement of NBD2 into NBD1 and believed that NBDs in MRP1 do not move toward each other like tweezers [93].

Johnson et al. (2018) investigated the role of ATP binding and hydrolysis in the release of substrates from the MRP1 by employing a combination of biochemical assays and molecular modeling. As a result, they found that releasing the substrate does not need ATP to be hydrolyzed; however, the hydrolysis of ATP is necessary to reuse the transporter for the next cycle [54].

A research study by Degorter et al. (2008) focused on the transport of drugs using the homology model of MRP1 with OF crystal structure of Sav1866 and conducted three different mutations on Y324 residue (Y324W, Y342F, Y324A), which is located in transmembrane helix number 6 of TMD1 (TM11). The results showed that systems with mutations to Alanine or tryptophan behaved similarly to WT systems, whereas a mutation to phenylalanine (Y342F) enhanced MRP1's activity to transfer the medicines [144]. Based on these results, Amram and colleagues started working on isolated TMDs [145]. They repeated the previous work with isolated TMDs in the POPC membrane. Once again, the Y342F system showed a higher clearance rate by remaining open among all the systems. According to research conducted by Weigl et al. (2018), the residue Phe583, located in TMD1 between

TMH10 and TMH11, is critical for its transport function. The F583A mutation in MRP1 inhibits ADP release, which traps the transporter in a substrate-releasing state and reduces its ligand affinity [146].

2.2. *P-glycoprotein*

P-glycoprotein (Pgp), which is also known as ABCB1 [43], is encoded by the MDR1 gene [147]. It is a transmembrane protein expressed in many tissues, including the cells lining the intestine, liver, and kidney [148,149]. It is a member of the ATP-binding cassette (ABC) transporter superfamily [150–152] and plays a crucial role in protecting the body from toxic substances by exporting them out of cells [153]. However, Pgp also causes multidrug resistance in cancer cells by pumping out anti-cancer drugs [154] including anthracyclines, taxanes, and vinca alkaloids [155,156], thereby reducing the effectiveness of these drugs, which makes it an important target for drug development [157]. Pgp functions as an efflux transporter by recognizing and binding a wide range of structurally diverse drugs and pumping them out of the cell [158,159]. The transport of drugs by Pgp is driven by the hydrolysis of ATP [62,160,161], which induces a conformational change in the protein and facilitates the movement of drugs across the membrane [160,162–164]. Dysregulation of Pgp has been implicated in various diseases, including cancer, Alzheimer's disease, and epilepsy [165,166]. In cancer, Pgp is known to play a key role in the development of drug resistance in cancer cells [155]. By pumping out chemotherapeutic drugs, Pgp makes tumors resistant to chemotherapy, thereby reducing their effectiveness in killing cancer cells [156]. Therefore, overexpression of Pgp in cancer cells can lead to multidrug resistance [165,167], which makes it difficult to treat cancer with a variety of different drugs [167]. It is also reported that in the central nervous system, Pgp plays a role in regulating the entry of drugs and toxins into the brain [168,169], and its dysregulation has been implicated in neurological disorders [170].

2.2.1. Structure of P-glycoprotein

Pgp is a 1280-residue [171], 170 kDa membrane protein [147,172–174] that exists as a heterodimer with pseudosymmetrical characteristics [175], and consists of two TMDs and two NBDs, one in each monomer [176]. Within Pgp, there exist two operational ATPase sites. Each site consists of the Walker A and Walker B motifs from one NBD, as well as the LSGGQ motif from the other NBD. At both of these sites, a glutamate residue, which is significantly conserved, functions as the catalytic base for ATP hydrolysis. If either of these catalytic glutamates undergoes mutation, the ATPase activity is severely diminished. Simultaneous mutations at both positions result in the trapping of Pgp in an ATP-occluded form, potentially with a closed-NBD dimer [62,177]. The NBDs possess multiple conserved sequences that play a crucial role in nucleotide binding and hydrolysis. In an NBD dimer, each of the two nucleotide-binding sites is formed by combining the Walker A motif, Walker B motif, A loop, H loop, and Q loop of one NBD with the D loop and signature motif of the other NBD. The Walker A motif is responsible for nucleotide phosphate binding, while the Walker B motif, along with the Q loop, coordinates Mg^{2+} and water at the catalytic site. The A loop contains an aromatic residue that interacts with the adenine ring of ATP, and the signature motif, D loop, and H loop contribute to the coordination of the ATP γ phosphate [128].

The general structure of this protein is TMD1-NBD1-TMD2-NBD2, with both the N-terminal and C-terminal of the protein located in the cytosol. In the inward-facing and outward-facing structures of human Pgp, the estimated distances between the C-terminal of NBD1 and NBD2 are approximately 30 Å and 11 Å, respectively [178]. The TMDs which form the substrate translocation pathway [179], are composed of six α-helices each and form a large central cavity that serves as the drug-binding site [172,180]. The NBDs contain the ATP-binding sites and are responsible for the energy-dependent transport of drugs across the membrane [181].

2.2.2. Substrate Binding Site of P-glycoprotein

Pgp substrates typically have weak amphipathic properties, are relatively lipophilic, and contain aromatic rings. They are also generally positively charged at normal pH levels [128,182,183]. The substrate binding pocket of Pgp is a large, highly complex, and polyspecific structure that plays a crucial role in substrate recognition and transport [174]. One of the most important features of the substrate binding pocket of Pgp is its ability to undergo conformational changes [184]. The pocket can adopt different conformations depending on the size and shape of the substrate. This flexibility is essential for accommodating structurally diverse substrates and maintaining a high affinity for binding them [184]. The substrate binding pocket of Pgp contains a network of hydrophobic and aromatic residues [185]. The hydrophobic residues include phenylalanine, leucine, and tryptophan, while the aromatic residues include tyrosine and tryptophan. These residues interact with the hydrophobic regions of the substrate and help to stabilize its binding within the pocket [172]. The degree of hydrophobicity observed in the drug-binding pocket of mouse Pgp is significantly greater than that observed in Sav1866 [186]. Therapeutic drugs, such as antiarrhythmic agents, calcium channel blockers, analgesics, chemotherapeutic drugs, HIV-protease inhibitors, antihistamines, antibiotics, and immunosuppressive agents, are among the Pgp substrates [128,187–189].

2.2.3. AAM in P-glycoprotein

The association of Pgp with multidrug resistance in cancer cells and the treatment failure of various drugs has led to an increased interest in understanding its mechanism of action [155,156]. The mechanism by which Pgp transports drugs across the cell membrane is called the alternating access mechanism [190,191]. The alternating access mechanism is a widely accepted model for the function of Pgp [174,190]. It suggests that the protein has two major conformations, each with a different orientation towards the inside or outside of the cell [40]. In the inward-facing conformation, the drug-binding pocket is accessible from the cytoplasmic side of the membrane. The drug molecule binds to the pocket and triggers a conformational change that moves the protein toward the outward-facing conformation. In the outward-facing conformation, the drug-binding pocket is accessible from the extracellular side of the membrane [172]. The switching between these two conformations is controlled by ATP hydrolysis, which provides the energy required to move the protein between the two conformations [62,160]. When ATP binds to Pgp, it causes a conformational change that results in the release of the drug molecule outside of the cell [171,192,193]. After ATP hydrolysis, Pgp undergoes a conformational reset and can bind to another drug molecule [162–164].

Electron microscopy and crystallographic studies have revealed the structure of the protein in both the inward- and outward-facing conformations [62,194–196], while Molecular dynamics (MD) simulations have been used to study the alternating access mechanisms [43,196,197]. MD is a powerful computational technique used to study the dynamics of biological macromolecules such as proteins [198,199], nucleic acids, and lipids [179,200,201]. It is an essential tool for studying the mechanisms of membrane transport proteins [106,184,202,203], including the alternating access mechanism in Pgp [43,196,197].

A recent study by Li et al. (2022) employed the two-state anisotropic network model to investigate the allosteric pathway of human Pgp in the substrate transport process, specifically from the inward-facing (IF) to the outward-facing (OF) state [204]. The authors found that the allosteric transitions proceed in a coupled manner and ultimately lead to conformational changes in the NBDs, resulting in the TMDs moving to the OF state via the allosteric propagation of intracellular helices IH1 and IH2. They indicate that the allosteric transition begins with the large-scale closing motion of the NBDs, accompanied by a significant twisting movement between them. This twisting motion becomes more pronounced near the transition state and is transmitted to TMDs via IH1 and IH2 intracellular helices. In conclusion, they reveal that this allosteric pathway is energetically favorable compared to alternative methods [204].

Another study by Vahedi et al. (2017) used computational analysis to investigate the significance of tyrosine residues in regulating Pgp-ATP hydrolysis via hydrogen bond formations with high-affinity modulators. The authors found that the mutation of 15 conserved residues in the drug-binding pocket of human Pgp to tyrosine resulted in no major effect on the total or cell surface expression of the mutant, indicating that tyrosine is critical in the drug-binding pocket and transport function of Pgp [205]. Verhalen et al. (2017) utilized double electron-electron resonance (DEER) and molecular dynamics (MD) simulations to investigate the ATP- and substrate-coupled conformational cycle of the mammalian ABC efflux transporter Pgp. They did this by introducing pairs of spin labels at selected residues to monitor the transition from an inward-facing (IF) to an outward-facing (OF) state. Their findings showed a two-stroke cycle in which ATP energy is harnessed in the NBDs, leading to conformational changes that reconfigure the TMD [43].

In addition to these studies, several other investigations have contributed to our understanding of the alternating access mechanism in Pgp. For example, a study by Futamata et al. (2020) used a FRET sensor to monitor the roles of ATP binding and ATP hydrolysis in the conformational changes of Pgp under a physiological membrane environment. The authors found that the ATP binding induces the conformational change to the outward-facing state, while ATP hydrolysis and subsequent release of γ-phosphate from both NBDs allow the outward-facing state to return to the original inward-facing state [178].

In determining the Molecular structure of human P-glycoprotein in the ATP-bound, outward-facing conformation, Kim et al. (2018) explained that the transition from inward-facing to outward-facing conformation involves a global movement of both halves of the molecule and extensive local rearrangements of the TM helices. They illustrated that the two "crossing" helices in each TMD (TM4 and 5 in TMD1, and TM10 (TM4) and TM11 (TM5) in TMD2) pivot inward, bringing the NBDs closer to each other. Additionally, the extracellular regions of TM7 (TM1) and 8 (TM2) move away from TM9 (TM3) to 12 (TM6), resulting in an outward-facing configuration [62] (Figure 5). The TM helices in TMD2 can be numbered from 1 to 6 or 7 to 12.

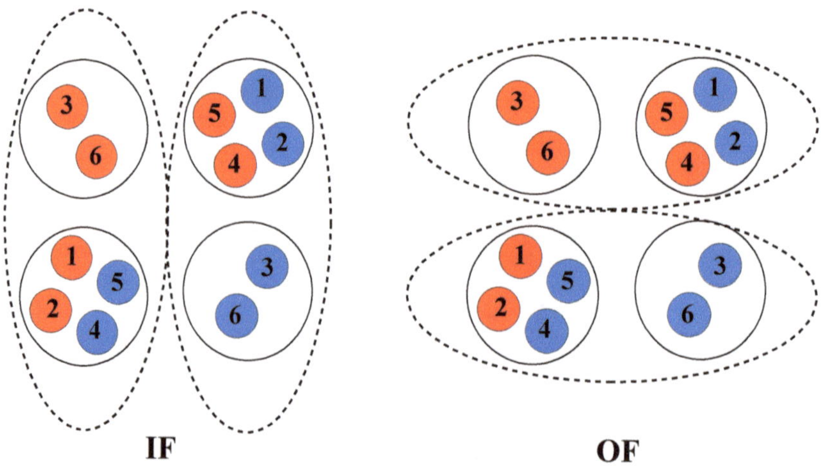

Figure 5. Schematic representation of transmembrane helices (TMs) arranged inward-facing (IF) and outward-facing (OF) orientation. Each TM helix is represented by a numbered circle in blue and red, indicating which subunit they belong to.

Loo et al. (2013) predict that a salt bridge formation between Glu256–Arg276 could enhance the folding of human Pgp by facilitating the packing of the TM segments. They explain the possibility as resulting from the linking of TM4 and TM5 segments by intracellular loop 2 (ICL2). The proposed function of the ICLs is to act as "ball-and-socket" joints that connect the TMDs to the NBDs and serve as transmission interfaces. ICL1 (TMD1-NBD1), ICL2 (TMD1-NBD2), ICL3 (TMD2-NBD2), and ICL4 (TMD1-NBD2) facilitate connections between domains [206]. In the crystal structure of *C. elegans* Pgp, ICL2 was predicted to be the most crucial link between the TMDs and NBDs in human Pgp [207], as 14 amino acids in this loop were projected to connect TMD1 to NBD2 via salt bridges, hydrogen bonds, or Van der Waals interactions. The structure of ICL2 is vital for Pgp biosynthesis since point mutations in this region hinder maturation [206].

Researchers in an experimental study investigated the role of the coupling helices (CH1 and CH2) in Pgp by conducting point mutations on these specific regions of the protein. They discovered that the mutations on CH2 have negative effects on Pgp activity; however, these specific mutations do not inhibit the function of Pgp when they are on CH1 [208]. According to the results of dynamic simulation analysis of human Pgp for generating the required conformational changes of this transporter, CH2 interactions with NBDs are higher and play a more important role compared to CH1[171].

Studies have demonstrated that lipid molecules (notably cholesterol) have a significant impact on regulating the structure and function of membrane proteins such as GPCRs [201] and membrane transporters, such as Pgp [209,210]. Pgp has been shown to have the ability to act as a lipid flippase, which means it can move cholesterol and phospholipids between the inner and outer layers of the cell membrane, redistributing and shuttling them [209]. By employing MD simulations, Thangapandian et al. (2020) investigated the role of cholesterol binding and translocation in Pgp [209]. From their simulations, they observed that the high-density binding region for cholesterol is located on the first transmembrane domain bundle of Pgp, specifically at the interface between TM1 and TM2, resulting in asymmetric cholesterol accumulation [209]. The study found that cholesterol binds more frequently to Pgp through its rough β face formed by the two protruding methyl groups, whereas the opposite smooth α face prefers packing alongside the membrane lipids. Additionally, the simulations captured one full and two partial cholesterol flipping events between the two leaflets of the bilayer mediated by the surface of Pgp. These events were observed in a region formed by helices TM1, TM2, and TM11, which featured two full and two partial CRAC/CARC motifs. During these events, Tyr49 and Tyr126 were identified as key residues that interacted with cholesterol. Experimentally, Clouser et al. (2020) aimed to investigate how cholesterol in a membrane impacts the conformational behavior of Pgp in lipid nanodiscs. Using hydrogen/deuterium exchange mass spectrometry, they found that cholesterol in the membrane causes asymmetric, long-range modifications in the distribution and exchange kinetics of conformations of the nucleotide-binding domains. They also showed that cholesterol presence enhances ATP hydrolysis and modifies lipid order and fluidity. These modifications link the effects of lipid composition on activity with specific changes in the Pgp conformational landscape [210].

These recent studies provide further insights into the alternating access mechanism in Pgp and identify the major domain sites, specifically the NBD and TMD, involved in the conformational changes that occur during substrate transport. Further research in this area will help to deepen our understanding of the mechanism of Pgp and may provide insights into new approaches for drug development.

2.3. *Sav1866*

Multidrug resistance (MDR) is a major obstacle in the treatment of bacterial infections, where pathogens become resistant to multiple antibiotics [211–213]. One of the key mechanisms of MDR is the over-expression of a protein called Sav1866 [214]. Sav1866 acts as a transporter, exporting a wide range of substrates, including drugs, chemotherapeutic agents, peptides, and lipids, out of bacterial cells and reducing their efficacy [215]. This qualifies it as an attractive target for understanding the mechanisms of drug resistance and developing strategies to overcome multidrug resistance (MDR) in bacteria and other microorganisms [165].

2.3.1. Structure of Sav1866

Sav1866 are found in various bacteria and archaea [216], including *Staphylococcus aureus* and *Mycobacterium tuberculosis* [217]. Sav1866 shares structural and functional similarities with eukaryotic ABC transporters, such as Pgp, and can serve as a model for understanding the mechanisms of drug resistance in both prokaryotic and eukaryotic organisms [218]. The structure of Sav1866 exists as a homodimer [67,179], consisting of two TMDs and two NBDs that are connected by a linker region [66,67]. The TMDs form a substrate translocation pathway, while the NBDs provide energy for substrate transport [179,219,220]. Following a recent classification of ABC transporters based on their distinct TMD folds, Sav1866 (Figure 6) belongs to the group IV classification based on their structural homology in the TMDs [42].

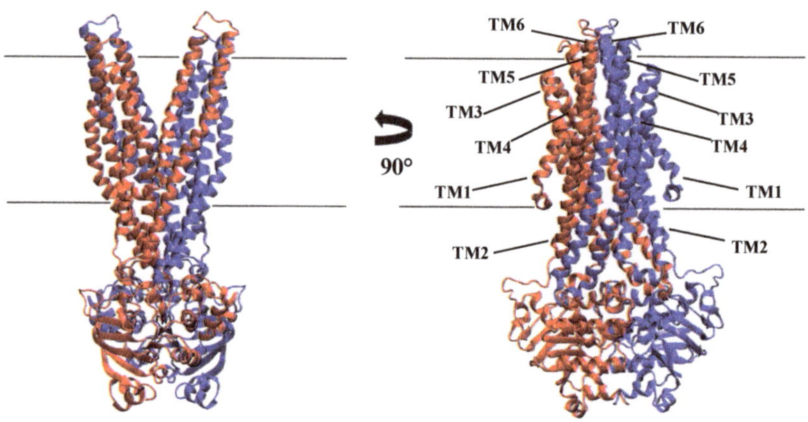

Figure 6. The cartoon representation of Sav1866 (PDB:2HYD) in OF state, showing both monomers in this structure with different colors. Each monomer has six transmembrane helices (TM1–TM6).

2.3.2. AAM in Sav1866

The alternating access mechanism allows for the selective transport of substrates while maintaining the integrity of the cell membrane [221]. The alternating access mechanism in Sav1866 involves two distinct conformations of the protein, referred to as the inward-facing conformation and the outward-facing conformation [165,179,186,222,223], with a third intermediate phase described as occluded conformation [179]. In the inward-facing conformation, the TMDs are closed, and the substrate-binding site is accessible from the intracellular side of the membrane. When ATP is bound and hydrolyzed by the NBDs, the protein undergoes a conformational change to the outward-facing conformation [224,225]. In this conformation, the TMDs open to the extracellular side of the membrane, and the substrate is released into the extracellular environment [216,219,223,226]. The opening of the TMDs involves the swapping of TMs 1 and 2 in the outward-facing dimer and TM 4 and 5 in the inward-facing dimer [39,103]. Therefore, the alternating access mechanism in Sav1866 alternates between an apo inward to a nucleotide-bound outward-facing conformation to transport their substrates across cellular membrane [179].

The conformational changes in Sav1866 that underlie the alternating access mechanism have been studied using a variety of techniques, including X-ray crystallography, electron microscopy, and molecular dynamics simulations [66,165,179,226]. These studies have provided valuable insights into the structural changes that occur during substrate transport and have helped to explain the alternating access mechanism in Sav1866. For example, the crystal structure of Sav1866 in the outward-facing state was determined by Dawson and Locher (2006) using X-ray crystallography [66]. It was revealed that Sav1866's structure comprises of TMDs that create a path for substrate translocation and two NBDs that supply the necessary energy for substrate transport. The substrate translocation pathway is closed in the outward-facing state, preventing the substrate from entering or leaving the binding site.

In the inward-facing state, the substrate translocation pathway opens, allowing the substrate to enter or leave the binding site. The switch between the outward-facing and inward-facing states is mediated by the movement of the TMDs and the coupling of this movement to the NBDs. The binding of ATP to the NBDs induces a conformational change in the TMDs, which results in the opening or closing of the substrate translocation pathway [66,67].

Molecular dynamics simulations have also been extensively used to study the alternating access mechanism in Sav1866 [165,179,224]. They have revealed that the protein undergoes a series of conformational changes involving the movement of helices and loops in the TMDs, as well as the repositioning of the NBDs, during substrate transport [179,219]. St-Pierre et al. (2012) employed MD simulations to investigate the stability of Sav1866 and study the ability of the protein to undergo conformational changes at physiological temperatures. They found that the protein transmembrane domain is not easily disturbed by large-scale motions of the NBDs [165].

Other studies have focused on the role and dependency of various lipids in the AAM of Sav1866 [179,227]. The composition of lipids has been known to alter the structural and functional dynamics of membrane proteins and membrane transporters [179]. For example, Immadiesity et al. (2019) used microsecond-level all-atom MD simulations to study the lipid-dependent characteristic of the alternating access mechanism in Sav1866. Their study focused on the conformational changes Sav1866 undergoes from an OF-IF transition in six different lipid environments. They show that the protein undergoes a large-scale conformational transition in the PE lipid bilayers, particularly in the POPE lipids, resulting in an IF-occluded conformation on the periplasmic side; however, this is not observed when the transporter is embedded in any of the PC environment [179]. The study identified a salt bridge (R81-D145) (Figure 7A) between two helices (TM2 and TM3) in the same monomer near a hinge region in one of the helices in POPE simulations, which was not observed in the crystal structure. The formation of this salt bridge was accompanied by the weakening of another salt bridge between TM6 and TM3 of the same monomer was present in the crystal structure: R296-D145 (Figure 7A), leading to the formation of the hinge region in TM3. The events were observed in one monomer in POPE simulations and were likely to promote periplasmic closure. The salt bridge brings the two bundles closer to each other while the hinge region allows for easier bending of one of the helices, which could ultimately lead to a closure of the periplasmic gate [179].

Aittoniemi et al. (2010) focused on the role of specific residues in the AAM of Sav1866 and showed that MgATP-binding residues and a network of charged residues at the dimer interface form a sequence of putative molecular switches that allows ATP hydrolysis only at one NBS [228]. The simulations found an asymmetry in the dynamic behavior of two identical Sav1866 monomers, with certain interactions only occurring in one monomer. These interactions involved the formation of a salt bridge (D423-K483) (Figure 7B), which has also been reported to occur in only one monomer of the homologous MgATP-bound NBD homodimer HlyB [229]. Notably, while the salt bridge was formed by residues within the same NBD monomer in HlyB, it is formed between the two NBDs in Sav1866. The formation of this salt bridge may be structurally coupled to conformational changes during ATP hydrolysis. The interactions could be dependent on the association of a Q-loop Gln residue (Q422) with MgATP, and when unbound from K483, D423 can interact with the coupling helix on the cytosolic loop (CL) of the opposite monomer, tightening NBD-TMD interactions. The formation of these interactions is preceded by the breaking of the D423 charge pairs with R474 of the short sequence motif in the NBD-TMD interface named X-loop [228]. Multiple computational studies on Sav1866 indicated that CHs forming interaction between TMDs and NBDs play a significant role in conformational changes of this transporter [224,228].

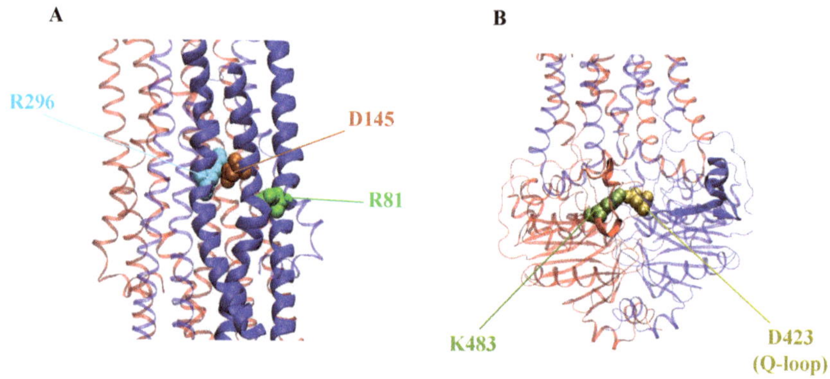

Figure 7. The cartoon representation of Sav1866 (PDB ID: 2HYD) shows the salt bridge interactions in different parts of the protein. Different colors indicate the different amino acids involved in the salt bridge interactions. (**A**) The salt bridge interaction between R81 and D145 of two helices (TM2 and TM3) in the same monomer. This figure also shows salt bridge interaction between R296 and D145 (TM6 and TM3 of the same monomer). (**B**) The salt bridge interaction between D423 and K483 is located in the Q-loop and region beside the signature sequence, respectively.

Overall, the AAM of Sav1866 is of significant interest because it is a model system for understanding the mechanisms of substrate transport by ABC transporters. The structural and functional studies of Sav1866 have provided valuable insights into the design of new antibiotics that can target the transporter specifically [217,230].

2.4. MsbA

In Gram-negative bacteria such as *Escherichia coli*, *Acinetobacter baumannii*, and *Salmonella typhimurium*, MsbA plays a crucial role as an ATP-binding cassette transporter [231]. MsbA, as a multidrug-resistant transporter (MDR), is crucial for the movement of drugs; it is also able to transfer other small molecules such as phospholipids and dyes [60,232]. Translocation of core-LPS is one of the functions mediated by MsbA [233]. Gram-negative bacteria are resistant to many antibiotics due to their outer cell membranes being permeabilized by lipopolysaccharide (LPS) [234], which consists of a hydrophobic lipid A anchor, a non-repeating core oligosaccharide, and a distal oligosaccharide [235]. MsbA transports lipid A and lipopolysaccharide across the inner membrane from the cytoplasmic to the periplasmic leaflet [236]. The core-LPS molecule is synthesized in the inner leaflet of the bacterial membrane, enters the MsbA transporter, and is flipped to the periplasm for the further process [69].

2.4.1. Structure of MsbA

MsbA (130 kDa, 584 residues [233,237]) is a homodimer that consists of two TMDs, each of which has six transmembrane (TM) helices and two cytosolic NBDs [69]. This protein is classified in group IV based on TMD fold arrangement (Figure 1). The TMDs of MsbA are responsible for accommodating various substrates and facilitating their translocation. Although the NBDs remain highly conserved, the TMDs vary in structure and sequence. MsbA's TMDs are made up of 12 transmembrane helices (TMHs), which form two wings that change their organization upon switching between the inward-facing (IF) and outward-facing (OF) conformations [71,238,239]. The core-LPS binding site, which accommodates the phosphorylated glucosamines(P-GlcN), acyl chains, and core sugars (core oligosaccharides) of the substrate, is formed by the TMHs and is accessible via a lateral portal formed by TMH4/6 of each monomer [68,69].

2.4.2. Substrate Binding Site of MsbA

The interior space of MsbA consists of a hydrophilic cavity and a hydrophobic pocket, which are exposed to the cytoplasmic and periplasmic sides, respectively. The P-GlcN and acyl chains of LPS interact with the hydrophobic inner cavity of MsbA; however, the inner core is located in a hydrophilic cavity. The interaction of the outer core is unclear [71,240].

According to a cryo-EM study [71], glucosamines have hydrophilic interactions with amino acids (Arg78, Arg148, Gln256, and Arg296 and Lys299) located at TM helices (TM2, 3, 5 and 6). They also mentioned that mutations in three conserved residues (Arg78, Arg148, and Lys299) to alanine affect lipopolysaccharide (LPS) binding. Discussions about MsbA-mediated LPS transport mechanisms are still ongoing. A study by Mi et al. (2017) proposed a model for MsbA flippase named "trap and flip" mediating LPS transport. The IF state of MsbA occurs in the absence of nucleotide or the presence of ADP in this model. The LPS substrate enters the binding site located in TMD and rearranges MsbA to allow ATP to bind to NBDs. The acyl chains of LPS are able to enter the periplasmic leaflet and trigger core oligosaccharides to move upward. The MsbA conformational changes and the translocation of LPS result in ATP hydrolysis and resume the IF state of MsbA [71].

2.4.3. AAM in MsbA

Across-membrane substrate transfer by MsbA is known to occur via an alternating access mechanism. The inward conformation undergoes a transition to the outward conformation upon binding of two ATPs and the substrate. A computational study by Moradi et al. (2013) sheds light on the study of large-scale conformational changes in ABC transporters such as MsbA using nonequilibrium MD simulations. This study highlights the critical role of TMD-NBD interactions in alternating access mechanisms that govern the function of MsbA. They used multiple collective variables to induce the orientation of NBDs and TMDs and induce the transition OF to IF state [241]. A combination of nonequilibrium simulations and free energy calculations for investigating large-scale conformational changes in proteins is proposed in another computational study by the same authors. Since the IF to OF transition occurs over extremely long time scales, MD simulations cannot be used to directly investigate this phenomenon. As MsbA crystal structure exists in three states: OF, IF-open, and IF-closed (Figure 8), it can be used to investigate the transition by employing collective variable (CVs) and biasing protocols to bias the system towards the specific states. They mentioned an asymmetrical arrangement of MsbA's TMDs based on a significant salt bridge between residues D252 (TM5) and K299 (TM6) in the IF state. They could find this salt bridge once in the IF form of MsbA instead of twice or none [242]. This finding agrees with experimental evidence indicating the asymmetrical behavior of TMDs in MsbA [243].

In addition to accommodating the substrate, the TMDs must communicate with the NBDs for substrate release [244]. This communication is facilitated by intracellular loops called coupling helices (CH1/2), located at grooves on the NBD surface [245–247]. A study has been done on the MsbA transporter using Molecular dynamics simulations and enzyme assays to investigate the role of the coupling helices (CHs) at the interface between the NBD and the TMD. They used MsbA and mutated either CH1 or CH2 residues to alanine (A). The experimental and computational results indicated that ATP affinity is decreased for both systems, including mutations on either CHs. However, the ATPase activity of only mutated CH2 is reduced as conserved His357 of H-loop in ATP binding pocket (ABP) had fewer interactions with the γ-phosphate of ATP. Hence, both CHs contribute to ATP binding, but CH2 plays more significant roles in ATPase function [247].

TMH3/4 of each TMD form a tetrahelix bundle at the intracellular extensions of the coupling helix (CH), which is essential for ATPase activity and conformational changes related to ATP binding in the TMD region [248].

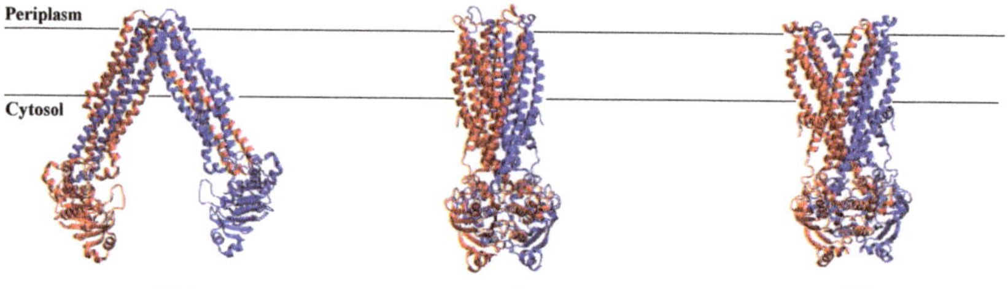

Figure 8. Cartoon representation of MsbA crystal structures in inward-facing open (IF(o)), closed (IF(c)), and outward-facing (OF) conformations. PDB IDs used to generate these structures are 8DMO, 7BCW, and 8DMM, respectively.

In MsbA, both of NBDs have all conserved motifs (Walker A/B, Signature Motif, A/D/Q-loop, H-Switch) as their ATP-binding pocket (ABP), and they can be dimerized by turning from the top down (head-to-tail manner) and sandwiching two ATP molecules between them [70,247]. Hydrolysis of ATP takes place in MsbA via both NBDs, as both contain all required motifs; however, whether this happens simultaneously is unclear.

NBD of MsbA in *Escherichia coli* was the subject of a study that integrated site-directed mutagenesis and biochemical assays. The results of this study indicate that the mutations in particular amino acid residues of MsbA's nucleotide-binding domains (e.g., Leucine 504 of Walker B motif) significantly alter the protein's ability to bind and hydrolyze ATP, as well as affect the rearrangement of the transporter [249].

Under circumstances of ATP depletion, NBDs in MsbA can also perform the reverse adenylate kinase reaction (AK reaction), which involves the conversion of ADP to ATP and AMP (2ADP ↔ ATP + AMP) [250–252]. There is an experimental study on MsbA proposing a novel model called "coupled ATPase-adenylate kinase mechanism" [253]. This study suggests that ATP hydrolysis in some ABC transporters, such as MsbA, is coupled with the activity of adenylate kinase. The AK reaction employs the conserved NBD motifs that are also required for ATP hydrolysis. Furthermore, solid-state NMR demonstrates that the Q-loop is involved in transient nucleotide binding during the AK reaction [253,254].

3. Conclusions

In this review paper, we have described ATP binding cassette transporter (ABC) classifications and focused on the specific type of ABC transporters named multidrug resistance (MDR) proteins, including mammalian Pgp and MRP1, as well as bacterial homologs of Pgp (Sav1866 and MsbA). We mentioned that the above transporters belong to type IV transporters based on their TMD fold arrangement. The functional core of this class of proteins has two TMDs (each one has six TM helices (TMHs)), and two NBDs, although MRP1 has an extra TMDs named TMD0 (including five TMHs), and its function for the transportation of substrates is not clear yet. In the IF state (Figure 5), each wing of these proteins consists of six helices, four of which come from TMD1 (H1, H2, H3, H6) and two of which come from TMD2 (H4 and H5), whereas in the OF state, the arrangement is changed, with TMHs from domain one including TMH1,2 plus TMH from domain two composed of TMH3, 4, 5, 6. However, the structure of NBDs in prokaryotic ABC proteins such as Sav1866, MsbA, and mammalian Pgp are identical, while NBDs in MRP1 are distinct. The presence of two ATP molecules is required to form an interface between two binding sites of NBD domains and in all of these transporters. ATP hydrolysis takes place following substrate transport and is necessary for reusing the transporters for the next cycle of substrate transportation. NBD2 is the only NBD domain in MRP1 that is able to hydrolyze the ATP molecule; however, both NBDs of Pgp, Sav1866, and MsbA are able to carry out this reaction. Moreover, NBD dimer dissociation can be accomplished with only one ATP hydrolysis. For the full transporters to properly couple dynamically, the coupling helices (CHs) must play a crucial role in the coordinated motion of complete transporters. These findings suggest that CHs mediate interactions at interfaces between NBDs and TMDs, facilitating the propagation of conformational changes from NBDs to TMDs. Furthermore, MsbA is the only one of these four transporters to exhibit the adenylate kinase reaction (AK reaction) associated with NBD.

Understanding the involvement of certain protein domains in inducing conformational changes, such as transmembrane helices (TMHs), is an important area of research with considerable prospects for the development of more effective therapeutic approaches. For instance, mammalian Pgp and MRP1 are well-known contributors to multidrug resistance in cancer chemotherapy. Knowing the structural features and functional mechanisms of these transporters can help develop targeted therapies to overcome drug resistance. We can improve patient outcomes and enhance chemotherapy effectiveness by inhibiting these transporters. Additionally, the structural and mechanistic similarities of Sav1866 and MsbA proteins to human Pgp and MRP1 provide valuable information and make them extremely helpful. These bacterial transporters play a vital role in bacterial drug resistance, and understanding their structure-function relationships can inform the development of novel strategies to tackle multidrug-resistant bacterial infections. Multiple conformations of these transporters exist throughout their transport cycle, and by focusing on particular conformations, researchers can investigate various functional aspects and create a variety of treatment approaches. Namely, inhibiting substrate binding can prevent the efflux of drugs and enhance their efficacy. Inhibiting ATP hydrolysis, on the other hand, can disrupt the energy supply necessary for transporter function. Developing therapies to inhibit the dissociation of NBDs can also impact the transport cycle and disrupt the drug resistance mechanisms.

We expect that the information provided in this review will provide insight into all aspects of mentioned proteins. Although much has been learned about these transporters, there is still much more to discover. More research into individual transmembrane helices (TMHs) and their communication with the NBDs and substrates is needed to better understand the role of specific protein domains in triggering conformational changes.

Author Contributions: Conceptualization, S.A.B. and M.M.; methodology, S.A.B. and M.M.; software, S.A.B. and M.M.; validation, S.A.B., U.H.I., E.K. and M.M.; formal analysis, S.A.B., U.H.I. and E.K.; investigation, S.A.B., U.H.I., E.K. and M.M.; resources, S.A.B., U.H.I., E.K. and M.M.; data curation, S.A.B., U.H.I., E.K. and M.M.; writing—original draft preparation, S.A.B., U.H.I., E.K. and M.M.; writing—review and editing, S.A.B. and M.M.; visualization, S.A.B., U.H.I., E.K. and M.M.; supervision, M.M.; project administration, S.A.B. and M.M.; funding acquisition, M.M. All authors have read and agreed to the published version of the manuscript.

Funding: This work was supported by the National Institute of General Medical Sciences of the National Institutes of Health under award number R35GM147423. This work was also supported by the Arkansas Biosciences Institute.

Institutional Review Board Statement: Not applicable.

Data Availability Statement: No new data are reported in this review article.

Conflicts of Interest: The authors declare no conflict of interest.

Abbreviations

The following abbreviations are used in this manuscript:

CRAC	Cholesterol Recognition/Interaction Amino Acid Consensus (CRAC)
ABC	Adenosine triphosphate binding cassette
GSH	Conjugated to antioxidant glutathione
Cryo-EM	Cryogenic Electron Microscopy
X-ray	X-ray Diffraction
MD(S)	Molecular dynamics (simulation)
MRP1	Multidrug resistance protein 1
AAM	Alternating Access Mechanism

MFS	Major facilitator superfamily
MATE	Multi-Antimicrobial Extrusion
AK	Adenylate kinase reaction
MDR	Multi-drug resistance
TMD	Transmembrane domain
MSD	membrane spanning domain
NBD	Nuclotide-binding domain
SMR	Small Multidrug Resistance
RND	Resistance-Nodulation-Division
ATP	Adenosine triphosphate
CV	Collective variable
TMH	Transmembrane helix
IH	Intracellular helix
ICL	Intracellular loop
CLR	Cholesterol
Pgp	P-glycoprotein
TM	Transmembrane
IF	Inward facing
OF	Outward facing
CH	Coupling helix
EH	Elbow helix
Pi	Inorganic phosphate
C	C terminus
N	N terminus

References

1. Muriithi, W.; Macharia, L.W.; Heming, C.P.; Echevarria, J.L.; Nyachieo, A.; Niemeyer Filho, P.; Neto, V.M. ABC transporters and the hallmarks of cancer: Roles in cancer aggressiveness beyond multidrug resistance. *Cancer Biol. Med.* **2020**, *17*, 253. [CrossRef] [PubMed]
2. He, J.; Fortunati, E.; Liu, D.X.; Li, Y. Pleiotropic roles of ABC transporters in breast cancer. *Int. J. Mol. Sci.* **2021**, *22*, 3199. [CrossRef] [PubMed]
3. Wang, J.Q.; Wu, Z.X.; Yang, Y.; Teng, Q.X.; Li, Y.D.; Lei, Z.N.; Jani, K.A.; Kaushal, N.; Chen, Z.S. ATP-binding cassette (ABC) transporters in cancer: A review of recent updates. *J. Evid.-Based Med.* **2021**, *14*, 232–256. [CrossRef] [PubMed]
4. Xiao, H.; Zheng, Y.; Ma, L.; Tian, L.; Sun, Q. Clinically-relevant ABC transporter for anti-cancer drug resistance. *Front. Pharmacol.* **2021**, *12*, 648407. [CrossRef]
5. Catalano, A.; Iacopetta, D.; Ceramella, J.; Scumaci, D.; Giuzio, F.; Saturnino, C.; Aquaro, S.; Rosano, C.; Sinicropi, M.S. Multidrug resistance (MDR): A widespread phenomenon in pharmacological therapies. *Molecules* **2022**, *27*, 616. [CrossRef]
6. Akhtar, A.A.; Turner, D.P. The role of bacterial ATP-binding cassette (ABC) transporters in pathogenesis and virulence: Therapeutic and vaccine potential. *Microb. Pathog.* **2022**, *171*, 105734. [CrossRef]
7. Pote, M.S.; Gacche, R.N. ATP-binding cassette efflux transporters and MDR in cancer. *Drug Discov. Today* **2023**, *28*, 103537. [CrossRef]
8. Dhakne, P.; Pillai, M.; Mishra, S.; Chatterjee, B.; Tekade, R.K.; Sengupta, P. Refinement of safety and efficacy of anti-cancer chemotherapeutics by tailoring their site-specific intracellular bioavailability through transporter modulation. *Biochim. Biophys. Acta (BBA)—Rev. Cancer* **2023**, *1878*, 188906. [CrossRef]
9. Duan, C.; Yu, M.; Xu, J.; Li, B.Y.; Zhao, Y.; Kankala, R.K. Overcoming Cancer Multi-drug Resistance (MDR): Reasons, mechanisms, nanotherapeutic solutions, and challenges. *Biomed. Pharmacother.* **2023**, *162*, 114643. [CrossRef]
10. Pucelik, B.; Dąbrowski, J.M. Photodynamic inactivation (PDI) as a promising alternative to current pharmaceuticals for the treatment of resistant microorganisms. *Adv. Inorg. Chem.* **2022**, *79*, 65.
11. Chitsaz, M.; Brown, M.H. The role played by drug efflux pumps in bacterial multidrug resistance. *Essays Biochem.* **2017**, *61*, 127–139.
12. Huang, L.; Wu, C.; Gao, H.; Xu, C.; Dai, M.; Huang, L.; Hao, H.; Wang, X.; Cheng, G. Bacterial multidrug efflux pumps at the frontline of antimicrobial resistance: An overview. *Antibiotics* **2022**, *11*, 520. [CrossRef]
13. da Silva, P.E.A.; Machado, D.; Ramos, D.; Couto, I.; Von Groll, A.; Viveiros, M. Efflux pumps in mycobacteria: Antimicrobial resistance, physiological functions, and role in pathogenicity. In *Efflux-Mediated Antimicrobial Resistance in Bacteria: Mechanisms, Regulation and Clinical Implications*; Adis: Cham, Switzerland, 2016; pp. 527–559.
14. Li, X.Z.; Nikaido, H. Efflux-mediated drug resistance in bacteria: An update. *Drugs* **2009**, *69*, 1555–1623. [CrossRef]
15. Chetri, S. The culmination of multidrug-resistant efflux pumps vs. meager antibiotic arsenal era: Urgent need for an improved new generation of EPIs. *Front. Microbiol.* **2023**, *14*, 1149418. [CrossRef]

16. Kumar, S.; He, G.; Kakarla, P.; Shrestha, U.; KC, R.; Ranaweera, I.; Mark Willmon, T.; R. Barr, S.; J. Hernandez, A.; F. Varela, M. Bacterial multidrug efflux pumps of the major facilitator superfamily as targets for modulation. *Infect. Disord. Drug Targets (Former. Curr. Drug Targets Infect. Disord.)* **2016**, *16*, 28–43.
17. Pasqua, M.; Grossi, M.; Zennaro, A.; Fanelli, G.; Micheli, G.; Barras, F.; Colonna, B.; Prosseda, G. The varied role of efflux pumps of the MFS family in the interplay of bacteria with animal and plant cells. *Microorganisms* **2019**, *7*, 285. [CrossRef]
18. Varela, M.F.; Stephen, J.; Bharti, D.; Lekshmi, M.; Kumar, S. Inhibition of Multidrug Efflux Pumps Belonging to the Major Facilitator Superfamily in Bacterial Pathogens. *Biomedicines* **2023**, *11*, 1448. [CrossRef]
19. Drew, D.; North, R.A.; Nagarathinam, K.; Tanabe, M. Structures and general transport mechanisms by the major facilitator superfamily (MFS). *Chem. Rev.* **2021**, *121*, 5289–5335. [CrossRef]
20. Kusakizako, T.; Miyauchi, H.; Ishitani, R.; Nureki, O. Structural biology of the multidrug and toxic compound extrusion superfamily transporters. *Biochim. Biophys. Acta (BBA)—Biomembr.* **2020**, *1862*, 183154. [CrossRef]
21. Zhang, X.; He, X.; Baker, J.; Tama, F.; Chang, G.; Wright, S.H. Twelve transmembrane helices form the functional core of mammalian MATE1 (multidrug and toxin extruder 1) protein. *J. Biol. Chem.* **2012**, *287*, 27971–27982. [CrossRef]
22. Jagessar, K.L.; Claxton, D.P.; Stein, R.A.; Mchaourab, H.S. Sequence and structural determinants of ligand-dependent alternating access of a MATE transporter. *Proc. Natl. Acad. Sci. USA* **2020**, *117*, 4732–4740. [CrossRef] [PubMed]
23. Omote, H.; Hiasa, M.; Matsumoto, T.; Otsuka, M.; Moriyama, Y. The MATE proteins as fundamental transporters of metabolic and xenobiotic organic cations. *Trends Pharmacol. Sci.* **2006**, *27*, 587–593. [CrossRef] [PubMed]
24. Lu, M.; Radchenko, M.; Symersky, J.; Nie, R.; Guo, Y. Structural insights into H + -coupled multidrug extrusion by a MATE transporter. *Nat. Struct. Mol. Biol.* **2013**, *20*, 1310–1317. [CrossRef] [PubMed]
25. Du, D.; Wang-Kan, X.; Neuberger, A.; Van Veen, H.W.; Pos, K.M.; Piddock, L.J.; Luisi, B.F. Multidrug efflux pumps: Structure, function and regulation. *Nat. Rev. Microbiol.* **2018**, *16*, 523–539. [CrossRef]
26. Spengler, G.; Kincses, A.; Gajdács, M.; Amaral, L. New roads leading to old destinations: Efflux pumps as targets to reverse multidrug resistance in bacteria. *Molecules* **2017**, *22*, 468. [CrossRef]
27. Handzlik, J.; Matys, A.; Kieć-Kononowicz, K. Recent advances in multi-drug resistance (MDR) efflux pump inhibitors of Gram-positive bacteria *S. aureus*. *Antibiotics* **2013**, *2*, 28–45. [CrossRef]
28. Nikaido, H.; Takatsuka, Y. Mechanisms of RND multidrug efflux pumps. *Biochim. Biophys. Acta (BBA)—Proteins Proteom.* **2009**, *1794*, 769–781. [CrossRef]
29. Daury, L.; Orange, F.; Taveau, J.C.; Verchere, A.; Monlezun, L.; Gounou, C.; Marreddy, R.K.; Picard, M.; Broutin, I.; Pos, K.M.; et al. Tripartite assembly of RND multidrug efflux pumps. *Nat. Commun.* **2016**, *7*, 10731. [CrossRef] [PubMed]
30. Dreier, J.; Ruggerone, P. Interaction of antibacterial compounds with RND efflux pumps in *Pseudomonas aeruginosa*. *Front. Microbiol.* **2015**, *6*, 660. [CrossRef] [PubMed]
31. Klenotic, P.A.; Moseng, M.A.; Morgan, C.E.; Yu, E.W. Structural and functional diversity of resistance–nodulation–cell division transporters. *Chem. Rev.* **2020**, *121*, 5378–5416. [CrossRef]
32. Callaghan, R.; Gelissen, I.C.; George, A.M.; Hartz, A.M. Mamma Mia, P-glycoprotein binds again. *FEBS Lett.* **2020**, *594*, 4076–4084. [CrossRef] [PubMed]
33. Srikant, S.; Gaudet, R. Mechanics and pharmacology of substrate selection and transport by eukaryotic ABC exporters. *Nat. Struct. Mol. Biol.* **2019**, *26*, 792–801. [CrossRef] [PubMed]
34. Holland, I.B.; Cole, S.P.; Kuchler, K.; Higgins, C.F. *ABC Proteins: From Bacteria to Man*; Elsevier: Amsterdam, The Netherlands, 2003.
35. Huang, J.; Ecker, G.F. A Structure-Based View on ABC-Transporter Linked to Multidrug Resistance. *Molecules* **2023**, *28*, 495. [CrossRef] [PubMed]
36. Hohl, M.; Briand, C.; Grütter, M.G.; Seeger, M.A. Crystal structure of a heterodimeric ABC transporter in its inward-facing conformation. *Nat. Struct. Mol. Biol.* **2012**, *19*, 395–402. [CrossRef] [PubMed]
37. Procko, E.; O'Mara, M.L.; Bennett, W.D.; Tieleman, D.P.; Gaudet, R. The mechanism of ABC transporters: General lessons from structural and functional studies of an antigenic peptide transporter. *FASEB J.* **2009**, *23*, 1287–1302. [CrossRef] [PubMed]
38. Symmons, M.F.; Bokma, E.; Koronakis, E.; Hughes, C.; Koronakis, V. The assembled structure of a complete tripartite bacterial multidrug efflux pump. *Proc. Natl. Acad. Sci. USA* **2009**, *106*, 7173–7178. [CrossRef] [PubMed]
39. Ford, R.C.; Beis, K. Learning the ABCs one at a time: Structure and mechanism of ABC transporters. *Biochem. Soc. Trans.* **2019**, *47*, 23–36. [CrossRef] [PubMed]
40. George, A.M.; Jones, P.M. Perspectives on the structure–function of ABC transporters: The Switch and Constant Contact Models. *Prog. Biophys. Mol. Biol.* **2012**, *109*, 95–107. [CrossRef] [PubMed]
41. Davidson, A.L.; Dassa, E.; Orelle, C.; Chen, J. Structure, function, and evolution of bacterial ATP-binding cassette systems. *Microbiol. Mol. Biol. Rev.* **2008**, *72*, 317–364. [CrossRef]
42. Thomas, C.; Aller, S.G.; Beis, K.; Carpenter, E.P.; Chang, G.; Chen, L.; Dassa, E.; Dean, M.; Duong Van Hoa, F.; Ekiert, D.; et al. Structural and functional diversity calls for a new classification of ABC transporters. *FEBS Lett.* **2020**, *594*, 3767–3775. [CrossRef] [PubMed]
43. Verhalen, B.; Dastvan, R.; Thangapandian, S.; Peskova, Y.; Koteiche, H.A.; Nakamoto, R.K.; Tajkhorshid, E.; Mchaourab, H.S. Energy transduction and alternating access of the mammalian ABC transporter P-glycoprotein. *Nature* **2017**, *543*, 738–741. [CrossRef] [PubMed]

44. Sharom, F.J. The P-glycoprotein multidrug transporter. *Essays Biochem.* **2011**, *50*, 161–178. [CrossRef] [PubMed]
45. Sarkadi, B.; Homolya, L.; Szakács, G.; Váradi, A. Human multidrug resistance ABCB and ABCG transporters: Participation in a chemoimmunity defense system. *Physiol. Rev.* **2006**, *86*, 1179–1236. [CrossRef] [PubMed]
46. Moitra, K.; Dean, M. Evolution of ABC transporters by gene duplication and their role in human disease. *Biol. Chem.* **2011**, *392*, 29–37. [CrossRef] [PubMed]
47. Saurin, W.; Hofnung, M.; Dassa, E. Getting in or out: Early segregation between importers and exporters in the evolution of ATP-binding cassette (ABC) transporters. *J. Mol. Evol.* **1999**, *48*, 22–41. [CrossRef]
48. Bouige, P.; Laurent, D.; Piloyan, L.; Dassa, E. Phylogenetic and functional classification of ATP-binding cassette (ABC) systems. *Curr. Protein Pept. Sci.* **2002**, *3*, 541–559. [CrossRef] [PubMed]
49. Qian, H.; Zhao, X.; Cao, P.; Lei, J.; Yan, N.; Gong, X. Structure of the human lipid exporter ABCA1. *Cell* **2017**, *169*, 1228–1239. [CrossRef] [PubMed]
50. Taylor, N.M.; Manolaridis, I.; Jackson, S.M.; Kowal, J.; Stahlberg, H.; Locher, K.P. Structure of the human multidrug transporter ABCG2. *Nature* **2017**, *546*, 504–509. [CrossRef] [PubMed]
51. Yang, H.B.; Hou, W.T.; Cheng, M.T.; Jiang, Y.L.; Chen, Y.; Zhou, C.Z. Structure of a MacAB-like efflux pump from Streptococcus pneumoniae. *Nat. Commun.* **2018**, *9*, 196. [CrossRef]
52. Bi, Y.; Mann, E.; Whitfield, C.; Zimmer, J. Architecture of a channel-forming O-antigen polysaccharide ABC transporter. *Nature* **2018**, *553*, 361–365. [CrossRef] [PubMed]
53. Chen, L.; Hou, W.T.; Fan, T.; Liu, B.; Pan, T.; Li, Y.H.; Jiang, Y.L.; Wen, W.; Chen, Z.P.; Sun, L.; et al. Cryo-electron microscopy structure and transport mechanism of a wall teichoic acid ABC transporter. *MBio* **2020**, *11*, e02749-19. [CrossRef] [PubMed]
54. Johnson, Z.L.; Chen, J. ATP binding enables substrate release from multidrug resistance protein 1. *Cell* **2018**, *172*, 81–89. [CrossRef] [PubMed]
55. Wang, L.; Johnson, Z.L.; Wasserman, M.R.; Levring, J.; Chen, J.; Liu, S. Characterization of the kinetic cycle of an ABC transporter by single-molecule and cryo-EM analyses. *eLife* **2020**, *9*, e56451. [CrossRef] [PubMed]
56. Pietz, H.L.; Abbas, A.; Johnson, Z.L.; Oldham, M.L.; Suga, H.; Chen, J. A macrocyclic peptide inhibitor traps MRP1 in a catalytically incompetent conformation. *Proc. Natl. Acad. Sci. USA* **2023**, *120*, e2220012120. [CrossRef] [PubMed]
57. Johnson, Z.L.; Chen, J. Structural basis of substrate recognition by the multidrug resistance protein MRP1. *Cell* **2017**, *168*, 1075–1085. [CrossRef]
58. Khandelwal, N.K.; Millan, C.R.; Zangari, S.I.; Avila, S.; Williams, D.; Thaker, T.M.; Tomasiak, T.M. The structural basis for regulation of the glutathione transporter Ycf1 by regulatory domain phosphorylation. *Nat. Commun.* **2022**, *13*, 1278. [CrossRef]
59. Esser, L.; Zhou, F.; Pluchino, K.M.; Shiloach, J.; Ma, J.; Tang, W.k.; Gutierrez, C.; Zhang, A.; Shukla, S.; Madigan, J.P.; et al. Structures of the Multidrug Transporter P-glycoprotein Reveal Asymmetric ATP Binding and the Mechanism of Polyspecificity. *J. Biol. Chem.* **2017**, *292*, 446–461. [CrossRef] [PubMed]
60. Eckford, P.D.; Sharom, F.J. The reconstituted Escherichia coli MsbA protein displays lipid flippase activity. *Biochem. J.* **2010**, *429*, 195–203. [CrossRef] [PubMed]
61. Szewczyk, P.; Tao, H.; McGrath, A.P.; Villaluz, M.; Rees, S.D.; Lee, S.C.; Doshi, R.; Urbatsch, I.L.; Zhang, Q.; Chang, G. Snapshots of ligand entry, malleable binding and induced helical movement in P-glycoprotein. *Acta Crystallogr. Sect. D Biol. Crystallogr.* **2015**, *71*, 732–741. [CrossRef] [PubMed]
62. Kim, Y.; Chen, J. Molecular structure of human P-glycoprotein in the ATP-bound, outward-facing conformation. *Science* **2018**, *359*, 915–919. [CrossRef] [PubMed]
63. Nicklisch, S.C.; Rees, S.D.; McGrath, A.P.; Gökirmak, T.; Bonito, L.T.; Vermeer, L.M.; Cregger, C.; Loewen, G.; Sandin, S.; Chang, G.; et al. Global marine pollutants inhibit P-glycoprotein: Environmental levels, inhibitory effects, and cocrystal structure. *Sci. Adv.* **2016**, *2*, e1600001. [CrossRef] [PubMed]
64. Li, J.; Jaimes, K.F.; Aller, S.G. Refined structures of mouse P-glycoprotein. *Protein Sci.* **2014**, *23*, 34–46. [CrossRef] [PubMed]
65. Le, C.A.; Harvey, D.S.; Aller, S.G. Structural definition of polyspecific compensatory ligand recognition by P-glycoprotein. *IUCrJ* **2020**, *7*, 663–672. [CrossRef]
66. Dawson, R.J.P.; Locher, K.P. Structure of a bacterial multidrug ABC transporter. *Nature* **2006**, *443*, 180–185. [CrossRef] [PubMed]
67. Dawson, R.J.P.; Locher, K.P. Structure of the multidrug ABC transporter Sav1866 from Staphylococcus aureus in complex with AMP-PNP. *FEBS Lett.* **2007**, *581*, 935–938. [CrossRef] [PubMed]
68. Ho, H.; Miu, A.; Alexander, M.K.; Garcia, N.K.; Oh, A.; Zilberleyb, I.; Reichelt, M.; Austin, C.D.; Tam, C.; Shriver, S.; et al. Structural basis for dual-mode inhibition of the ABC transporter MsbA. *Nature* **2018**, *557*, 196–201. [CrossRef] [PubMed]
69. Padayatti, P.S.; Lee, S.C.; Stanfield, R.L.; Wen, P.C.; Tajkhorshid, E.; Wilson, I.A.; Zhang, Q. Structural insights into the lipid A transport pathway in MsbA. *Structure* **2019**, *27*, 1114–1123. [CrossRef] [PubMed]
70. Ward, A.; Reyes, C.L.; Yu, J.; Roth, C.B.; Chang, G. Flexibility in the ABC transporter MsbA: Alternating access with a twist. *Proc. Natl. Acad. Sci. USA* **2007**, *104*, 19005–19010. [CrossRef]
71. Mi, W.; Li, Y.; Yoon, S.H.; Ernst, R.K.; Walz, T.; Liao, M. Structural basis of MsbA-mediated lipopolysaccharide transport. *Nature* **2017**, *549*, 233–237. [CrossRef] [PubMed]
72. Angiulli, G.; Dhupar, H.S.; Suzuki, H.; Wason, I.S.; Duong Van Hoa, F.; Walz, T. New approach for membrane protein reconstitution into peptidiscs and basis for their adaptability to different proteins. *eLife* **2020**, *9*, e53530. [CrossRef] [PubMed]

73. Li, W.; Zhang, H.; Assaraf, Y.G.; Zhao, K.; Xu, X.; Xie, J.; Yang, D.H.; Chen, Z.S. Overcoming ABC transporter-mediated multidrug resistance: Molecular mechanisms and novel therapeutic drug strategies. *Drug Resist. Updat.* **2016**, *27*, 14–29. [CrossRef] [PubMed]
74. Greene, N.P.; Kaplan, E.; Crow, A.; Koronakis, V. Antibiotic resistance mediated by the MacB ABC transporter family: A structural and functional perspective. *Front. Microbiol.* **2018**, *9*, 950. [CrossRef] [PubMed]
75. Robey, R.W.; Pluchino, K.M.; Hall, M.D.; Fojo, A.T.; Bates, S.E.; Gottesman, M.M. Revisiting the role of ABC transporters in multidrug-resistant cancer. *Nat. Rev. Cancer* **2018**, *18*, 452–464. [CrossRef]
76. Jernigan, J.A.; Hatfield, K.M.; Wolford, H.; Nelson, R.E.; Olubajo, B.; Reddy, S.C.; McCarthy, N.; Paul, P.; McDonald, L.C.; Kallen, A.; et al. Multidrug-resistant bacterial infections in US hospitalized patients, 2012–2017. *N. Engl. J. Med.* **2020**, *382*, 1309–1319. [CrossRef]
77. Kumar, S.; Anwer, R.; Azzi, A. Virulence potential and treatment options of multidrug-resistant (MDR) Acinetobacter baumannii. *Microorganisms* **2021**, *9*, 2104. [CrossRef] [PubMed]
78. Liu, X. ABC family transporters. In *Drug Transporters in Drug Disposition, Effects and Toxicity*; Springer: Singapore, 2019; pp. 13–100.
79. Zahreddine, H.; Borden, K. Mechanisms and insights into drug resistance in cancer. *Front. Pharmacol.* **2013**, *4*, 28. [CrossRef] [PubMed]
80. Chang, G.; Roth, C.B. Structure of MsbA from *E. coli*: A homolog of the multidrug resistance ATP binding cassette (ABC) transporters. *Science* **2001**, *293*, 1793–1800. [CrossRef] [PubMed]
81. Dewanjee, S.; Dua, T.K.; Bhattacharjee, N.; Das, A.; Gangopadhyay, M.; Khanra, R.; Joardar, S.; Riaz, M.; De Feo, V.; Zia-Ul-Haq, M. Natural products as alternative choices for P-glycoprotein (P-gp) inhibition. *Molecules* **2017**, *22*, 871. [CrossRef] [PubMed]
82. Wilkens, S. Structure and mechanism of ABC transporters. *F1000prime Rep.* **2015**, *7*, 14. [CrossRef] [PubMed]
83. Xiong, J.; Feng, J.; Yuan, D.; Zhou, J.; Miao, W. Tracing the structural evolution of eukaryotic ATP binding cassette transporter superfamily. *Sci. Rep.* **2015**, *5*, 16724. [CrossRef] [PubMed]
84. Hung, L.W.; Wang, I.X.; Nikaido, K.; Liu, P.Q.; Ames, G.F.L.; Kim, S.H. Crystal structure of the ATP-binding subunit of an ABC transporter. *Nature* **1998**, *396*, 703–707. [CrossRef] [PubMed]
85. Hopfner, K.P.; Karcher, A.; Shin, D.S.; Craig, L.; Arthur, L.M.; Carney, J.P.; Tainer, J.A. Structural biology of Rad50 ATPase: ATP-driven conformational control in DNA double-strand break repair and the ABC-ATPase superfamily. *Cell* **2000**, *101*, 789–800. [CrossRef] [PubMed]
86. Smith, P.C.; Karpowich, N.; Millen, L.; Moody, J.E.; Rosen, J.; Thomas, P.J.; Hunt, J.F. ATP binding to the motor domain from an ABC transporter drives formation of a nucleotide sandwich dimer. *Mol. Cell* **2002**, *10*, 139–149. [CrossRef] [PubMed]
87. Zolnerciks, J.K.; Andress, E.J.; Nicolaou, M.; Linton, K.J. Structure of ABC transporters. *Essays Biochem.* **2011**, *50*, 43. [PubMed]
88. Seyffer, F.; Tampé, R. ABC transporters in adaptive immunity. *Biochim. Biophys. Acta (BBA)—Gen. Subj.* **2015**, *1850*, 449–460. [CrossRef]
89. Vetter, I.R.; Wittinghofer, A. Nucleoside triphosphate-binding proteins: Different scaffolds to achieve phosphoryl transfer. *Q. Rev. Biophys.* **1999**, *32*, 1–56. [CrossRef] [PubMed]
90. Jones, P.M.; George, A.M. Subunit interactions in ABC transporters: Towards a functional architecture. *FEMS Microbiol. Lett.* **1999**, *179*, 187–202. [CrossRef] [PubMed]
91. Higgins, C.F.; Hiles, I.D.; Salmond, G.P.; Gill, D.R.; Downie, J.A.; Evans, I.J.; Holland, I.B.; Gray, L.; Buckel, S.D.; Bell, A.W.; et al. A family of related ATP-binding subunits coupled to many distinct biological processes in bacteria. *Nature* **1986**, *323*, 448–450. [CrossRef] [PubMed]
92. Locher, K.P. Mechanistic diversity in ATP-binding cassette (ABC) transporters. *Nat. Struct. Mol. Biol.* **2016**, *23*, 487–493. [CrossRef] [PubMed]
93. Chen, J.; Lu, G.; Lin, J.; Davidson, A.L.; Quiocho, F.A. A tweezers-like motion of the ATP-binding cassette dimer in an ABC transport cycle. *Mol. Cell* **2003**, *12*, 651–661. [CrossRef] [PubMed]
94. Kunjachan, S.; Rychlik, B.; Storm, G.; Kiessling, F.; Lammers, T. Multidrug resistance: Physiological principles and nanomedical solutions. *Adv. Drug Deliv. Rev.* **2013**, *65*, 1852–1865. [CrossRef] [PubMed]
95. Biemans-Oldehinkel, E.; Doeven, M.K.; Poolman, B. ABC transporter architecture and regulatory roles of accessory domains. *FEBS Lett.* **2006**, *580*, 1023–1035. [CrossRef] [PubMed]
96. Borst, P.; Evers, R.; Kool, M.; Wijnholds, J. A family of drug transporters: The multidrug resistance-associated proteins. *J. Natl. Cancer Inst.* **2000**, *92*, 1295–1302. [CrossRef] [PubMed]
97. Jardetzky, O. Simple allosteric model for membrane pumps. *Nature* **1966**, *211*, 969–970. [CrossRef] [PubMed]
98. Mitchell, P. A general theory of membrane transport from studies of bacteria. *Nature* **1957**, *180*, 134–136. [CrossRef] [PubMed]
99. Tanford, C. Simple model for the chemical potential change of a transported ion in active transport. *Proc. Natl. Acad. Sci. USA* **1982**, *79*, 2882–2884. [CrossRef]
100. Patlak, C.S. Contributions to the theory of active transport: II. The gate type non-carrier mechanism and generalizations concerning tracer flow, efficiency, and measurement of energy expenditure. *Bull. Math. Biophys.* **1957**, *19*, 209–235. [CrossRef]
101. Yan, N. Structural biology of the major facilitator superfamily transporters. *Annu. Rev. Biophys.* **2015**, *44*, 257–283. [CrossRef]
102. Drew, D.; Boudker, O. Shared molecular mechanisms of membrane transporters. *Annu. Rev. Biochem.* **2016**, *85*, 543–572. [CrossRef]

103. Kroll, T.; Prescher, M.; Smits, S.H.J.; Schmitt, L. Structure and Function of Hepatobiliary ATP Binding Cassette Transporters. *Chem. Rev.* **2021**, *121*, 5240–5288. [CrossRef]
104. Locher, K.P. Structure and mechanism of ATP-binding cassette transporters. *Philos. Trans. R. Soc. B Biol. Sci.* **2009**, *364*, 239–245. [CrossRef]
105. Oldham, M.L.; Davidson, A.L.; Chen, J. Structural insights into ABC transporter mechanism. *Curr. Opin. Struct. Biol.* **2008**, *18*, 726–733. [CrossRef]
106. Enkavi, G.; Li, J.; Wen, P.; Thangapandian, S.; Moradi, M.; Jiang, T.; Han, W.; Tajkhorshid, E. Chapter Four—A Microscopic View of the Mechanisms of Active Transport Across the Cellular Membrane. *Annu. Rep. Comput. Chem.* **2014**, *10*, 77–125. [CrossRef]
107. Deeley, R.G.; Westlake, C.; Cole, S.P. Transmembrane transport of endo-and xenobiotics by mammalian ATP-binding cassette multidrug resistance proteins. *Physiol. Rev.* **2006**, *86*, 849–899. [CrossRef]
108. Zhang, Y.K.; Wang, Y.J.; Gupta, P.; Chen, Z.S. Multidrug resistance proteins (MRPs) and cancer therapy. *AAPS J.* **2015**, *17*, 802–812. [CrossRef]
109. Ivanyuk, A.; Livio, F.; Biollaz, J.; Buclin, T. Renal drug transporters and drug interactions. *Clin. Pharmacokinet.* **2017**, *56*, 825–892. [CrossRef]
110. Wang, J.Q.; Cui, Q.; Lei, Z.N.; Teng, Q.X.; Ji, N.; Lin, L.; Liu, Z.; Chen, Z.S. Insights on the structure–function relationship of human multidrug resistance protein 7 (MRP7/ABCC10) from molecular dynamics simulations and docking studies. *MedComm* **2021**, *2*, 221–235. [CrossRef]
111. Roy, U.; Barber, P.; Tse-Dinh, Y.C.; Batrakova, E.V.; Mondal, D.; Nair, M. Role of MRP transporters in regulating antimicrobial drug inefficacy and oxidative stress-induced pathogenesis during HIV-1 and TB infections. *Front. Microbiol.* **2015**, *6*, 948. [CrossRef]
112. Natarajan, K.; Xie, Y.; Baer, M.R.; Ross, D.D. Role of breast cancer resistance protein (BCRP/ABCG2) in cancer drug resistance. *Biochem. Pharmacol.* **2012**, *83*, 1084–1103. [CrossRef]
113. Shukalek, C.B.; Swanlund, D.P.; Rousseau, R.K.; Weigl, K.E.; Marensi, V.; Cole, S.P.; Leslie, E.M. Arsenic triglutathione [As (GS) 3] transport by multidrug resistance protein 1 (MRP1/ABCC1) is selectively modified by phosphorylation of Tyr920/Ser921 and glycosylation of Asn19/Asn23. *Mol. Pharmacol.* **2016**, *90*, 127–139. [CrossRef]
114. Yin, J.; Zhang, J. Multidrug resistance-associated protein 1 (MRP1/ABCC1) polymorphism: From discovery to clinical application. *Zhong Nan Xue Xue Bao Xue Ban J. Cent. South Univ. Med. Sci.* **2011**, *36*, 927.
115. Grube, M.; Hagen, P.; Jedlitschky, G. Neurosteroid transport in the brain: Role of ABC and SLC transporters. *Front. Pharmacol.* **2018**, *9*, 354. [CrossRef] [PubMed]
116. Liu, H.; Wu, X.; Huang, J.; Peng, J.; Guo, L. miR-7 modulates chemoresistance of small cell lung cancer by repressing MRP 1/ABCC 1. *Int. J. Exp. Pathol.* **2015**, *96*, 240–247. [CrossRef] [PubMed]
117. Miglionico, R.; Gerbino, A.; Ostuni, A.; Armentano, M.F.; Monné, M.; Carmosino, M.; Bisaccia, F. New insights into the roles of the N-terminal region of the ABCC6 transporter. *J. Bioenerg. Biomembr.* **2016**, *48*, 259–267. [CrossRef] [PubMed]
118. Cole, S.P.; Sparks, K.E.; Fraser, K.; Loe, D.W.; Grant, C.E.; Wilson, G.M.; Deeley, R.G. Pharmacological characterization of multidrug resistant MRP-transfected human tumor cells. *Cancer Res.* **1994**, *54*, 5902–5910.
119. Slot, A.J.; Molinski, S.V.; Cole, S.P. Mammalian multidrug-resistance proteins (MRPs). *Essays Biochem.* **2011**, *50*, 179–207.
120. Cui, Q.; Yang, Y.; Ji, N.; Wang, J.Q.; Ren, L.; Yang, D.H.; Chen, Z.S. Gaseous signaling molecules and their application in resistant cancer treatment: From invisible to visible. *Future Med. Chem.* **2019**, *11*, 323–336. [CrossRef]
121. Bakos, E.; Evers, R.; Szakács, G.; Tusnády, G.E.; Welker, E.; Szabó, K.; de Haas, M.; van Deemter, L.; Borst, P.; Váradi, A.; et al. Functional multidrug resistance protein (MRP1) lacking the N-terminal transmembrane domain. *J. Biol. Chem.* **1998**, *273*, 32167–32175. [CrossRef]
122. Hou, Y.x.; Cui, L.; Riordan, J.R.; Chang, X.b. Allosteric interactions between the two non-equivalent nucleotide binding domains of multidrug resistance protein MRP1. *J. Biol. Chem.* **2000**, *275*, 20280–20287. [CrossRef]
123. Gao, M.; Cui, H.R.; Loe, D.W.; Grant, C.E.; Almquist, K.C.; Cole, S.P.; Deeley, R.G. Comparison of the functional characteristics of the nucleotide binding domains of multidrug resistance protein 1. *J. Biol. Chem.* **2000**, *275*, 13098–13108. [CrossRef]
124. Moody, J.E.; Millen, L.; Binns, D.; Hunt, J.F.; Thomas, P.J. Cooperative, ATP-dependent association of the nucleotide binding cassettes during the catalytic cycle of ATP-binding cassette transporters. *J. Biol. Chem.* **2002**, *277*, 21111–21114. [CrossRef]
125. Nagata, K.; Nishitani, M.; Matsuo, M.; Kioka, N.; Amachi, T.; Ueda, K. Nonequivalent nucleotide trapping in the two nucleotide binding folds of the human multidrug resistance protein MRP1. *J. Biol. Chem.* **2000**, *275*, 17626–17630. [CrossRef]
126. Payen, L.; Gao, M.; Westlake, C.; Theis, A.; Cole, S.P.; Deeley, R.G. Functional interactions between nucleotide binding domains and leukotriene C4 binding sites of multidrug resistance protein 1 (ABCC1). *Mol. Pharmacol.* **2005**, *67*, 1944–1953. [CrossRef]
127. Tóth, Á.; Janaszkiewicz, A.; Crespi, V.; Di Meo, F. On the interplay between lipids and asymmetric dynamics of an NBS degenerate ABC transporter. *Commun. Biol.* **2023**, *6*, 149. [CrossRef]
128. Arana, M.R.; Altenberg, G.A. ATP-binding Cassette Exporters: Structure and Mechanism with a Focus on P-glycoprotein and MRP1. *Curr. Med. Chem.* **2019**, *26*, 1062–1078. [CrossRef]
129. Conseil, G.; Arama-Chayoth, M.; Tsfadia, Y.; Cole, S.P. Structure-guided probing of the leukotriene C4 binding site in human multidrug resistance protein 1 (MRP1; ABCC1). *FASEB J.* **2019**, *33*, 10692–10704. [CrossRef]
130. Raichaudhuri, A. Arabidopsis thaliana MRP1 (AtABCC1) nucleotide binding domain contributes to arsenic stress tolerance with serine triad phosphorylation. *Plant Physiol. Biochem.* **2016**, *108*, 109–120.

131. Cole, S.P. Multidrug resistance protein 1 (MRP1, ABCC1), a "multitasking" ATP-binding cassette (ABC) transporter. *J. Biol. Chem.* **2014**, *289*, 30880–30888. [CrossRef]
132. He, S.M.; Li, R.; R Kanwar, J.; Zhou, S.F. Structural and functional properties of human multidrug resistance protein 1 (MRP1/ABCC1). *Curr. Med. Chem.* **2011**, *18*, 439–481. [CrossRef]
133. Ghanem, C.I.; Manautou, J.E. Modulation of hepatic MRP3/ABCC3 by xenobiotics and pathophysiological conditions: Role in drug pharmacokinetics. *Curr. Med. Chem.* **2019**, *26*, 1185–1223.
134. Zhang, Y.; Yang, S.H.; Guo, X.L. New insights into Vinca alkaloids resistance mechanism and circumvention in lung cancer. *Biomed. Pharmacother.* **2017**, *96*, 659–666. [CrossRef] [PubMed]
135. Smith, E.E.; Conseil, G.; Cole, S.P. Conserved amino acids in the region connecting membrane spanning domain 1 to nucleotide binding domain 1 are essential for expression of the MRP1 (ABCC1) transporter. *PLoS ONE* **2021**, *16*, e0246727. [CrossRef]
136. Cort, A.; Ozben, T.; Saso, L.; De Luca, C.; Korkina, L. Redox control of multidrug resistance and its possible modulation by antioxidants. *Oxidative Med. Cell. Longev.* **2016**, *2016*, 4251912. [CrossRef] [PubMed]
137. Du, Y.; Chen, B. Detection approaches for multidrug resistance genes of leukemia. *Drug Des. Dev. Ther.* **2017**, *11*, 1255–1261. [CrossRef] [PubMed]
138. Shang, J.; Chen, W.M.; Wang, Z.H.; Wei, T.N.; Chen, Z.Z.; Wu, W.B. CircPAN3 mediates drug resistance in acute myeloid leukemia through the miR-153-5p/miR-183-5p–XIAP axis. *Exp. Hematol.* **2019**, *70*, 42–54. [CrossRef]
139. Feldman, R.E.; Lam, A.C.; Sadow, P.M.; Bleier, B.S. P-glycoprotein is a marker of tissue eosinophilia and radiographic inflammation in chronic rhinosinusitis without nasal polyps. In *International Forum of Allergy & Rhinology*; Wiley Online Library: Hoboken, NJ, USA, 2013; Volume 3, pp. 684–687.
140. Westlake, C.J.; Qian, Y.M.; Gao, M.; Vasa, M.; Cole, S.P.; Deeley, R.G. Identification of the structural and functional boundaries of the multidrug resistance protein 1 cytoplasmic loop 3. *Biochemistry* **2003**, *42*, 14099–14113. [CrossRef]
141. Hanssen, K.M.; Wheatley, M.S.; Yu, D.M.; Conseil, G.; Norris, M.D.; Haber, M.; Cole, S.P.; Fletcher, J.I. GSH facilitates the binding and inhibitory activity of novel multidrug resistance protein 1 (MRP1) modulators. *FEBS J.* **2022**, *289*, 3854–3875. [CrossRef]
142. Zhao, Z.J.; Gao, X.Y.; Zeng, J.C.; Zhang, S.L.; Meng, X.M.; Shen, Y.J.; Sheng, X.H. Theoretical insights into the cotransport mechanism of GSH with anticancer drugs by MRP1. *J. Phys. Chem. B* **2020**, *124*, 9803–9811. [CrossRef]
143. Cole, S.P. Targeting multidrug resistance protein 1 (MRP1, ABCC1): Past, present, and future. *Annu. Rev. Pharmacol. Toxicol.* **2014**, *54*, 95–117. [CrossRef]
144. DeGorter, M.K.; Conseil, G.; Deeley, R.G.; Campbell, R.L.; Cole, S.P. Molecular modeling of the human multidrug resistance protein 1 (MRP1/ABCC1). *Biochem. Biophys. Res. Commun.* **2008**, *365*, 29–34. [CrossRef]
145. Amram, S.; Ganoth, A.; Tichon, O.; Peer, D.; Nachliel, E.; Gutman, M.; Tsfadia, Y. Structural characterization of the drug translocation path of MRP1/ABCC1. *Israel J. Chem.* **2014**, *54*, 1382–1393. [CrossRef]
146. Weigl, K.E.; Conseil, G.; Rothnie, A.J.; Arama, M.; Tsfadia, Y.; Cole, S.P. An outward-facing aromatic amino acid is crucial for signaling between the membrane-spanning and nucleotide-binding domains of multidrug resistance protein 1 (MRP1; ABCC1). *Mol. Pharmacol.* **2018**, *94*, 1069–1078. [CrossRef]
147. Jagodinsky, J.C.; Akgun, U. Characterizing the binding interactions between P-glycoprotein and eight known cardiovascular transport substrates. *Pharmacol. Res. Perspect.* **2015**, *3*, e00114. [CrossRef]
148. Ferreira, R.J.; Ferreira, M.J.U.; dos Santos, D.J.V.A. Insights on P-Glycoprotein's Efflux Mechanism Obtained by Molecular Dynamics Simulations. *J. Chem. Theory Comput.* **2012**, *8*, 1853–1864. [CrossRef]
149. Dastvan, R.; Mishra, S.; Peskova, Y.B.; Nakamoto, R.K.; Mchaourab, H.S. Mechanism of allosteric modulation of P-glycoprotein by transport substrates and inhibitors. *Science* **2019**, *364*, 689–692. [CrossRef]
150. Domicevica, L.; Biggin, P. Homology modelling of human P-glycoprotein. *Biochem. Soc. Trans.* **2015**, *43*, 952–958. [CrossRef]
151. Prajapati, R.; Sangamwar, A.T. Translocation mechanism of P-glycoprotein and conformational changes occurring at drug-binding site: Insights from multi-targeted molecular dynamics. *Biochim. Biophys. Acta (BBA)—Biomembr.* **2014**, *1838*, 2882–2898. [CrossRef]
152. Manoharan, J.P.; Karunakaran, K.N.; Vidyalakshmi, S.; Dhananjayan, K. Computational binding affinity and molecular dynamic characterization of annonaceous acetogenins at nucleotide binding domain (NBD) of multi-drug resistance ATP-binding cassette sub-family B member 1 (ABCB1). *J. Biomol. Struct. Dyn.* **2023**, *41*, 821–832. [CrossRef]
153. Esposito, C.; Wang, S.; Lange, U.E.W.; Oellien, F.; Riniker, S. Combining Machine Learning and Molecular Dynamics to Predict P-Glycoprotein Substrates. *J. Chem. Inf. Model.* **2020**, *60*, 4730–4749. [CrossRef]
154. Prajapati, R.; Singh, U.; Patil, A.; Khomane, K.S.; Bagul, P.; Bansal, A.K.; Sangamwar, A.T. In silico model for P-glycoprotein substrate prediction: Insights from molecular dynamics and in vitro studies. *J. Comput.-Aided Mol. Des.* **2013**, *27*, 347–363. [CrossRef]
155. Karthika, C.; Sureshkumar, R.; Zehravi, M.; Akter, R.; Ali, F.; Ramproshad, S.; Mondal, B.; Tagde, P.; Ahmed, Z.; Khan, F.S.; et al. Multidrug Resistance of Cancer Cells and the Vital Role of P-Glycoprotein. *Life* **2022**, *12*, 897. [CrossRef] [PubMed]
156. Waghray, D.; Zhang, Q. Inhibit or Evade Multidrug Resistance P-Glycoprotein in Cancer Treatment. *J. Med. Chem.* **2018**, *61*, 5108–5121. [CrossRef] [PubMed]
157. Nguyen, T.T.L.; Duong, V.A.; Maeng, H.J. Pharmaceutical Formulations with P-Glycoprotein Inhibitory Effect as Promising Approaches for Enhancing Oral Drug Absorption and Bioavailability. *Pharmaceutics* **2021**, *13*, 1103. [CrossRef] [PubMed]
158. Chang, S.Y.; Liu, F.F.; Dong, X.Y.; Sun, Y. Molecular insight into conformational transmission of human P-glycoprotein. *J. Chem. Phys.* **2013**, *139*, 225102. [CrossRef]

159. Wen, P.C.; Verhalen, B.; Wilkens, S.; Mchaourab, H.S.; Tajkhorshid, E. On the Origin of Large Flexibility of P-glycoprotein in the Inward-facing State. *J. Biol. Chem.* **2013**, *288*, 19211–19220. [CrossRef]
160. Hulyer, A.R.C.; Briggs, D.A.; O'Mara, M.L.; Kerr, I.D.; Harmer, J.R.; Callaghan, R. Cross-linking, DEER-spectroscopy and molecular dynamics confirm the inward facing state of P-glycoprotein in a lipid membrane. *J. Struct. Biol.* **2020**, *211*, 107513. [CrossRef]
161. Zhang, Y.; Gong, W.; Wang, Y.; Liu, Y.; Li, C. Exploring movement and energy in human P-glycoprotein conformational rearrangement. *J. Biomol. Struct. Dyn.* **2019**, *37*, 1104–1119. [CrossRef]
162. Al-Shawi, M.K.; Polar, M.K.; Omote, H.; Figler, R.A. Transition state analysis of the coupling of drug transport to ATP hydrolysis by P-glycoprotein. *J. Biol. Chem.* **2003**, *278*, 52629–52640. [CrossRef]
163. Frank, G.A.; Shukla, S.; Rao, P.; Borgnia, M.J.; Bartesaghi, A.; Merk, A.; Mobin, A.; Esser, L.; Earl, L.A.; Gottesman, M.M.; et al. Cryo-EM analysis of the conformational landscape of human P-glycoprotein (ABCB1) during its catalytic cycle. *Mol. Pharmacol.* **2016**, *90*, 35–41.
164. Rosenberg, M.F.; Kamis, A.B.; Callaghan, R.; Higgins, C.F.; Ford, R.C. Three-dimensional structures of the mammalian multidrug resistance P-glycoprotein demonstrate major conformational changes in the transmembrane domains upon nucleotide binding. *J. Biol. Chem.* **2003**, *278*, 8294–8299.
165. St-Pierre, J.F.; Bunker, A.; Róg, T.; Karttunen, M.; Mousseau, N. Molecular Dynamics Simulations of the Bacterial ABC Transporter SAV1866 in the Closed Form. *J. Phys. Chem. B* **2012**, *116*, 2934–2942. [CrossRef]
166. Taggi, V.; Riera Romo, M.; Piquette-Miller, M.; zu Schwabedissen, H.E.; Neuhoff, S. Transporter Regulation in Critical Protective Barriers: Focus on Brain and Placenta. *Pharmaceutics* **2022**, *14*, 1376. [CrossRef]
167. Nanayakkara, A.K.; Follit, C.A.; Chen, G.; Williams, N.S.; Vogel, P.D.; Wise, J.G. Targeted inhibitors of P-glycoprotein increase chemotherapeutic-induced mortality of multidrug resistant tumor cells. *Sci. Rep.* **2018**, *8*, 967. [CrossRef]
168. Zibell, G.; Unkrüer, B.; Pekcec, A.; Hartz, A.M.; Bauer, B.; Miller, D.S.; Potschka, H. Prevention of seizure-induced up-regulation of endothelial P-glycoprotein by COX-2 inhibition. *Neuropharmacology* **2009**, *56*, 849–855. [CrossRef]
169. Bauer, B.; Hartz, A.M.; Pekcec, A.; Toellner, K.; Miller, D.S.; Potschka, H. Seizure-induced up-regulation of P-glycoprotein at the blood-brain barrier through glutamate and cyclooxygenase-2 signaling. *Mol. Pharmacol.* **2008**, *73*, 1444–1453. [CrossRef]
170. Chai, A.B.; Callaghan, R.; Gelissen, I.C. Regulation of P-Glycoprotein in the Brain. *Int. J. Mol. Sci.* **2022**, *23*, 14667. [CrossRef]
171. Wise, J.G. Catalytic transitions in the human MDR1 P-glycoprotein drug binding sites. *Biochemistry* **2012**, *51*, 5125–5141. [CrossRef]
172. Aller, S.G.; Yu, J.; Ward, A.; Weng, Y.; Chittaboina, S.; Zhuo, R.; Harrell, P.M.; Trinh, Y.T.; Zhang, Q.; Urbatsch, I.L.; et al. Structure of P-Glycoprotein Reveals a Molecular Basis for Poly-Specific Drug Binding. *Science* **2009**, *323*, 1718–1722. [CrossRef]
173. Xing, J.; Huang, S.; Heng, Y.; Mei, H.; Pan, X. Computational Insights into Allosteric Conformational Modulation of P-Glycoprotein by Substrate and Inhibitor Binding. *Molecules* **2020**, *25*, 6006. [CrossRef]
174. Ward, A.B.; Szewczyk, P.; Grimard, V.; Lee, C.W.; Martinez, L.; Doshi, R.; Caya, A.; Villaluz, M.; Pardon, E.; Cregger, C.; et al. Structures of P-glycoprotein reveal its conformational flexibility and an epitope on the nucleotide-binding domain. *Proc. Natl. Acad. Sci. USA* **2013**, *110*, 13386–13391. [CrossRef]
175. Klepsch, F.; Ecker, G.F. Impact of the recent mouse P-glycoprotein structure for structure-based ligand design. *Mol. Inform.* **2010**, *29*, 276–286. [CrossRef] [PubMed]
176. Rosenberg, M.F.; Velarde, G.; Ford, R.C.; Martin, C.; Berridge, G.; Kerr, I.D.; Callaghan, R.; Schmidlin, A.; Wooding, C.; Linton, K.J.; et al. Repacking of the transmembrane domains of P-glycoprotein during the transport ATPase cycle. *EMBO J.* **2001**, *20*, 5615–5625. [CrossRef] [PubMed]
177. Carrier, I.; Gros, P. Investigating the role of the invariant carboxylate residues E552 and E1197 in the catalytic activity of Abcb1a (mouse Mdr3). *FEBS J.* **2008**, *275*, 3312–3324. [CrossRef]
178. Futamata, R.; Ogasawara, F.; Ichikawa, T.; Kodan, A.; Kimura, Y.; Kioka, N.; Ueda, K. In vivo FRET analyses reveal a role of ATP hydrolysis–associated conformational changes in human P-glycoprotein. *J. Biol. Chem.* **2020**, *295*, 5002–5011. [CrossRef] [PubMed]
179. Immadisetty, K.; Hettige, J.; Moradi, M. Lipid-Dependent Alternating Access Mechanism of a Bacterial Multidrug ABC Exporter. *ACS Cent. Sci.* **2019**, *5*, 43–56. [CrossRef]
180. Kodan, A.; Yamaguchi, T.; Nakatsu, T.; Sakiyama, K.; Hipolito, C.J.; Fujioka, A.; Hirokane, R.; Ikeguchi, K.; Watanabe, B.; Hiratake, J.; et al. Structural basis for gating mechanisms of a eukaryotic P-glycoprotein homolog. *Proc. Natl. Acad. Sci. USA* **2014**, *111*, 4049–4054. [CrossRef]
181. Bársony, O.; Szalóki, G.; Türk, D.; Tarapcsák, S.; Gutay-Tóth, Z.; Bacsó, Z.; Holb, I.J.; Székvölgyi, L.; Szabó, G.; Csanády, L.; et al. A single active catalytic site is sufficient to promote transport in P-glycoprotein. *Sci. Rep.* **2016**, *6*, 24810. [CrossRef]
182. Sharom, F.J.; Lugo, M.R.; Eckford, P.D. New insights into the drug binding, transport and lipid flippase activities of the p-glycoprotein multidrug transporter. *J. Bioenerg. Biomembr.* **2005**, *37*, 481–487. [CrossRef]
183. Raub, T.J. P-glycoprotein recognition of substrates and circumvention through rational drug design. *Mol. Pharm.* **2006**, *3*, 3–25. [CrossRef]
184. Mora Lagares, L.; Pérez-Castillo, Y.; Minovski, N.; Novič, M. Structure–Function Relationships in the Human P-Glycoprotein (ABCB1): Insights from Molecular Dynamics Simulations. *Int. J. Mol. Sci.* **2022**, *23*, 362. [CrossRef]

185. Swartz, D.J.; Weber, J.; Urbatsch, I.L. P-glycoprotein is fully active after multiple tryptophan substitutions. *Biochim. Biophys. Acta (BBA)—Biomembr.* **2013**, *1828*, 1159–1168. [CrossRef]
186. Pan, L.; Aller, S.G. Equilibrated Atomic Models of Outward-Facing P-glycoprotein and Effect of ATP Binding on Structural Dynamics. *Sci. Rep.* **2015**, *5*, 7880. [CrossRef]
187. Gil-Martins, E.; Barbosa, D.J.; Silva, V.; Remião, F.; Silva, R. Dysfunction of ABC transporters at the blood-brain barrier: Role in neurological disorders. *Pharmacol. Ther.* **2020**, *213*, 107554. [CrossRef]
188. Ledwitch, K.V.; Roberts, A.G. Cardiovascular ion channel inhibitor drug-drug interactions with P-glycoprotein. *AAPS J.* **2017**, *19*, 409–420. [CrossRef]
189. Miller, D.S. Regulation of ABC transporters blood–brain barrier: The good, the bad, and the ugly. *Adv. Cancer Res.* **2015**, *125*, 43–70.
190. Al-Shawi, M.K.; Omote, H. The Remarkable Transport Mechanism of P-Glycoprotein: A Multidrug Transporter. *J. Bioenerg. Biomembr.* **2005**, *37*, 489–496. [CrossRef]
191. Seelig, A. P-Glycoprotein: One Mechanism, many tasks and the Consequences for Pharmacotherapy of Cancers. *Front. Oncol.* **2020**, *10*, 576559. [CrossRef]
192. van Wonderen, J.H.; McMahon, R.M.; O'Mara, M.L.; McDevitt, C.A.; Thomson, A.J.; Kerr, I.D.; MacMillan, F.; Callaghan, R. The central cavity of ABCB 1 undergoes alternating access during ATP hydrolysis. *FEBS J.* **2014**, *281*, 2190–2201. [CrossRef]
193. Martin, C.; Berridge, G.; Mistry, P.; Higgins, C.; Charlton, P.; Callaghan, R. Drug binding sites on P-glycoprotein are altered by ATP binding prior to nucleotide hydrolysis. *Biochemistry* **2000**, *39*, 11901–11906. [CrossRef]
194. Urgaonkar, S.; Nosol, K.; Said, A.M.; Nasief, N.N.; Bu, Y.; Locher, K.P.; Lau, J.Y.N.; Smolinski, M.P. Discovery and Characterization of Potent Dual P-Glycoprotein and CYP3A4 Inhibitors: Design, Synthesis, Cryo-EM Analysis, and Biological Evaluations. *J. Med. Chem.* **2022**, *65*, 191–216. [CrossRef]
195. Thonghin, N.; Collins, R.F.; Barbieri, A.; Shafi, T.; Siebert, A.; Ford, R.C. Novel features in the structure of P-glycoprotein (ABCB1) in the post-hydrolytic state as determined at 7.9,Å resolution. *BMC Struct. Biol.* **2018**, *18*, 17. [CrossRef] [PubMed]
196. Kodan, A.; Yamaguchi, T.; Nakatsu, T.; Matsuoka, K.; Kimura, Y.; Ueda, K.; Kato, H. Inward- and outward-facing X-ray crystal structures of homodimeric P-glycoprotein CmABCB1. *Nat. Commun.* **2019**, *10*, 88. [CrossRef] [PubMed]
197. Kapoor, K.; Pant, S.; Tajkhorshid, E. Active participation of membrane lipids in inhibition of multidrug transporter P-glycoprotein. *Chem. Sci.* **2021**, *12*, 6293–6306. [CrossRef] [PubMed]
198. Govind Kumar, V.; Ogden, D.S.; Isu, U.H.; Polasa, A.; Losey, J.; Moradi, M. Prefusion spike protein conformational changes are slower in SARS-CoV-2 than in SARS-CoV-1. *J. Biol. Chem.* **2022**, *298*, 101814. [CrossRef] [PubMed]
199. Polasa, A.; Moradi, M. Deciphering the Inter-domain Decoupling in the Gram-negative Bacterial Membrane Insertase. *bioRxiv* **2022**. [CrossRef]
200. Domicevica, L.; Koldsø, H.; Biggin, P.C. Multiscale molecular dynamics simulations of lipid interactions with P-glycoprotein in a complex membrane. *J. Mol. Graph. Model.* **2018**, *80*, 147–156. [CrossRef]
201. Isu, U.H.; Badiee, S.A.; Khodadadi, E.; Moradi, M. Cholesterol in Class C GPCRs: Role, Relevance, and Localization. *Membranes* **2023**, *13*, 301. [CrossRef]
202. Wang, L.; O'Mara, M.L. Effect of the Force Field on Molecular Dynamics Simulations of the Multidrug Efflux Protein P-Glycoprotein. *J. Chem. Theory Comput.* **2021**, *17*, 6491–6508. [CrossRef]
203. Behmard, E.; Barzegari, E.; Najafipour, S.; Kouhpayeh, A.; Ghasemi, Y.; Asadi-Pooya, A.A. Efflux dynamics of the antiseizure drug, levetiracetam, through the P-glycoprotein channel revealed by advanced comparative molecular simulations. *Sci. Rep.* **2022**, *12*, 13674. [CrossRef]
204. Li, H.; Gong, W. Study of Allosteric Transitions of Human P-Glycoprotein by Using the Two-State Anisotropic Network Model. *Front. Med.* **2022**, *9*, 26. [CrossRef]
205. Vahedi, S.; Chufan, E.E.; Ambudkar, S.V. Global alteration of the drug-binding pocket of human P-glycoprotein (ABCB1) by substitution of fifteen conserved residues reveals a negative correlation between substrate size and transport efficiency. *Biochem. Pharmacol.* **2017**, *143*, 53–64. [CrossRef]
206. Loo, T.W.; Clarke, D.M. A salt bridge in intracellular loop 2 is essential for folding of human P-glycoprotein. *Biochemistry* **2013**, *52*, 3194–3196. [CrossRef]
207. Loo, T.W.; Clarke, D.M. Location of the rhodamine-binding site in the human multidrug resistance P-glycoprotein. *J. Biol. Chem.* **2002**, *277*, 44332–44338. [CrossRef]
208. Loo, T.W.; Bartlett, M.C.; Clarke, D.M. Human P-glycoprotein contains a greasy ball-and-socket joint at the second transmission interface. *J. Biol. Chem.* **2013**, *288*, 20326–20333. [CrossRef]
209. Thangapandian, S.; Kapoor, K.; Tajkhorshid, E. Probing cholesterol binding and translocation in P-glycoprotein. *Biochim. Biophys. Acta (BBA)—Biomembr.* **2020**, *1862*, 183090. [CrossRef]
210. Clouser, A.F.; Alam, Y.H.; Atkins, W.M. Cholesterol Asymmetrically Modulates the Conformational Ensemble of the Nucleotide-Binding Domains of P-Glycoprotein in Lipid Nanodiscs. *Biochemistry* **2021**, *60*, 85–94. [CrossRef]
211. Yang, X.; Ye, W.; Qi, Y.; Ying, Y.; Xia, Z. Overcoming Multidrug Resistance in Bacteria Through Antibiotics Delivery in Surface-Engineered Nano-Cargos: Recent Developments for Future Nano-Antibiotics. *Front. Bioeng. Biotechnol.* **2021**, *9*, 696514. [CrossRef]

212. Chegini, Z.; Khoshbayan, A.; Vesal, S.; Moradabadi, A.; Hashemi, A.; Shariati, A. Bacteriophage therapy for inhibition of multi drug-resistant uropathogenic bacteria: A narrative review. *Ann. Clin. Microbiol. Antimicrob.* **2021**, *20*, 30. [CrossRef]
213. Nikaido, H. Multidrug Resistance in Bacteria. *Annu. Rev. Biochem.* **2009**, *78*, 119–146. [CrossRef]
214. Sodani, K.; Patel, A.; Kathawala, R.; Chen, Z.S. Multidrug resistance associated proteins in multidrug resistance. *Chin. J. Cancer* **2011**, *31*, 58–72. [CrossRef]
215. Kim, J.; Cater, R.; Choy, B.; Mancia, F. Structural Insights into Transporter-Mediated Drug Resistance in Infectious Diseases. *J. Mol. Biol.* **2021**, *433*, 167005. [CrossRef] [PubMed]
216. Velamakanni, S.; Yao, Y.; Gutmann, D.A.P.; van Veen, H.W. Multidrug Transport by the ABC Transporter Sav1866 from Staphylococcus aureus. *Biochemistry* **2008**, *47*, 9300–9308. [CrossRef] [PubMed]
217. Dashtbani-Roozbehani, A.; Brown, M.H. Efflux Pump Mediated Antimicrobial Resistance by Staphylococci in Health-Related Environments: Challenges and the Quest for Inhibition. *Antibiotics* **2021**, *10*, 1502. [CrossRef] [PubMed]
218. Zolnerciks, J.K.; Wooding, C.; Linton, K.J. Evidence for a Sav1866-like architecture for the human multidrug transporter P-glycoprotein. *FASEB J.* **2007**, *21*, 3937–3948. [CrossRef]
219. Göddeke, H.; Schäfer, L.V. Capturing Substrate Translocation in an ABC Exporter at the Atomic Level. *J. Am. Chem. Soc.* **2020**, *142*, 12791–12801. [CrossRef]
220. Stockner, T.; Gradisch, R.; Schmitt, L. The role of the degenerate nucleotide binding site in type I ABC exporters. *FEBS Lett.* **2020**, *594*, 3815–3838. [CrossRef]
221. Dastvan, R.; Rasouli, A.; Dehghani-Ghahnaviyeh, S.; Gies, S.; Tajkhorshid, E. Proton-driven alternating access in a spinster lipid transporter. *Nat. Commun.* **2022**, *13*, 5161. [CrossRef]
222. Mishra, S.; Verhalen, B.; Stein, R.A.; Wen, P.C.; Tajkhorshid, E.; Mchaourab, H.S. Conformational dynamics of the nucleotide binding domains and the power stroke of a heterodimeric ABC transporter. *eLife* **2014**, *3*, e02740. [CrossRef]
223. Göddeke, H.; Timachi, M.H.; Hutter, C.A.J.; Galazzo, L.; Seeger, M.A.; Karttunen, M.; Bordignon, E.; Schäfer, L.V. Atomistic Mechanism of Large-Scale Conformational Transition in a Heterodimeric ABC Exporter. *J. Am. Chem. Soc.* **2018**, *140*, 4543–4551. [CrossRef]
224. Oliveira, A.S.; Baptista, A.M.; Soares, C.M. Conformational changes induced by ATP-hydrolysis in an ABC transporter: A molecular dynamics study of the Sav1866 exporter. *Proteins Struct. Funct. Bioinform.* **2011**, *79*, 1977–1990. [CrossRef]
225. Nöll, A.; Thomas, C.; Herbring, V.; Zollmann, T.; Barth, K.; Mehdipour, A.R.; Tomasiak, T.M.; Brüchert, S.; Joseph, B.; Abele, R.; et al. Crystal structure and mechanistic basis of a functional homolog of the antigen transporter TAP. *Proc. Natl. Acad. Sci. USA* **2017**, *114*, E438–E447. [CrossRef]
226. Becker, J.P.; Van Bambeke, F.; Tulkens, P.M.; Prevost, M. Dynamics and Structural Changes Induced by ATP Binding in SAV1866, a Bacterial ABC Exporter. *J. Phys. Chem. B* **2010**, *114*, 15948–15957. [CrossRef]
227. Pollock, N.L.; Lloyd, J.; Montinaro, C.; Rai, M.; Dafforn, T.R. Conformational trapping of an ABC transporter in polymer lipid nanoparticles. *Biochem. J.* **2022**, *479*, 145–159. [CrossRef]
228. Aittoniemi, J.; de Wet, H.; Ashcroft, F.M.; Sansom, M.S. Asymmetric switching in a homodimeric ABC transporter: A simulation study. *PLoS Comput. Biol.* **2010**, *6*, e1000762. [CrossRef]
229. Zaitseva, J.; Oswald, C.; Jumpertz, T.; Jenewein, S.; Wiedenmann, A.; Holland, I.B.; Schmitt, L. A structural analysis of asymmetry required for catalytic activity of an ABC-ATPase domain dimer. *EMBO J.* **2006**, *25*, 3432–3443. [CrossRef]
230. Roberts, A.G. *The Structure and Mechanism of Drug Transporters BT—Enzyme Kinetics in Drug Metabolism: Fundamentals and Applications*; Springer: New York, NY, USA, 2021; pp. 193–234. [CrossRef]
231. Zeng, Y.; Charkowski, A.O. The role of ATP-binding cassette transporters in bacterial phytopathogenesis. *Phytopathology* **2021**, *111*, 600–610. [CrossRef]
232. Doerrler, W.T.; Gibbons, H.S.; Raetz, C.R. MsbA-dependent translocation of lipids across the inner membrane of *Escherichia coli*. *J. Biol. Chem.* **2004**, *279*, 45102–45109. [CrossRef]
233. Zhou, Z.; White, K.A.; Polissi, A.; Georgopoulos, C.; Raetz, C.R. Function of Escherichia coli MsbA, an essential ABC family transporter, in lipid A and phospholipid biosynthesis. *J. Biol. Chem.* **1998**, *273*, 12466–12475. [CrossRef]
234. Zgurskaya, H.I.; Lopez, C.A.; Gnanakaran, S. Permeability barrier of Gram-negative cell envelopes and approaches to bypass it. *ACS Infect. Dis.* **2015**, *1*, 512–522. [CrossRef]
235. Bertani, B.; Ruiz, N. Function and biogenesis of lipopolysaccharides. *EcoSal Plus* **2018**, *8*, 1–19. [CrossRef]
236. Lyu, J.; Liu, C.; Zhang, T.; Schrecke, S.; Elam, N.P.; Packianathan, C.; Hochberg, G.K.; Russell, D.; Zhao, M.; Laganowsky, A. Structural basis for lipid and copper regulation of the ABC transporter MsbA. *Nat. Commun.* **2022**, *13*, 7291. [CrossRef] [PubMed]
237. Kaur, H.; Lakatos, A.; Spadaccini, R.; Vogel, R.; Hoffmann, C.; Becker-Baldus, J.; Ouari, O.; Tordo, P.; Mchaourab, H.; Glaubitz, C. The ABC exporter MsbA probed by solid state NMR–challenges and opportunities. *Biol. Chem.* **2015**, *396*, 1135–1149. [CrossRef] [PubMed]
238. Furuta, T. Structural dynamics of ABC transporters: Molecular simulation studies. *Biochem. Soc. Trans.* **2021**, *49*, 405–414. [CrossRef] [PubMed]
239. Orelle, C.; Mathieu, K.; Jault, J.M. Multidrug ABC transporters in bacteria. *Res. Microbiol.* **2019**, *170*, 381–391. [CrossRef]
240. Thélot, F.; Orlando, B.J.; Li, Y.; Liao, M. High-resolution views of lipopolysaccharide translocation driven by ABC transporters MsbA and LptB2FGC. *Curr. Opin. Struct. Biol.* **2020**, *63*, 26–33. [CrossRef] [PubMed]

241. Moradi, M.; Tajkhorshid, E. Mechanistic picture for conformational transition of a membrane transporter at atomic resolution. *Proc. Natl. Acad. Sci. USA* **2013**, *110*, 18916–18921. [CrossRef]
242. Moradi, M.; Tajkhorshid, E. Computational recipe for efficient description of large-scale conformational changes in biomolecular systems. *J. Chem. Theory Comput.* **2014**, *10*, 2866–2880. [CrossRef]
243. Mittal, A.; Böhm, S.; Grütter, M.G.; Bordignon, E.; Seeger, M.A. Asymmetry in the homodimeric ABC transporter MsbA recognized by a DARPin. *J. Biol. Chem.* **2012**, *287*, 20395–20406. [CrossRef]
244. Siarheyeva, A.; Sharom, F.J. The ABC transporter MsbA interacts with lipid A and amphipathic drugs at different sites. *Biochem. J.* **2009**, *419*, 317–328. [CrossRef]
245. Spadaccini, R.; Kaur, H.; Becker-Baldus, J.; Glaubitz, C. The effect of drug binding on specific sites in transmembrane helices 4 and 6 of the ABC exporter MsbA studied by DNP-enhanced solid-state NMR. *Biochim. Biophys. Acta (BBA)—Biomembr.* **2018**, *1860*, 833–840. [CrossRef]
246. Woebking, B.; Velamakanni, S.; Federici, L.; Seeger, M.A.; Murakami, S.; van Veen, H.W. Functional role of transmembrane helix 6 in drug binding and transport by the ABC transporter MsbA. *Biochemistry* **2008**, *47*, 10904–10914. [CrossRef]
247. Furuta, T.; Yamaguchi, T.; Kato, H.; Sakurai, M. Analysis of the structural and functional roles of coupling helices in the ATP-binding cassette transporter MsbA through enzyme assays and molecular dynamics simulations. *Biochemistry* **2014**, *53*, 4261–4272. [CrossRef]
248. Doshi, R.; Ali, A.; Shi, W.; Freeman, E.V.; Fagg, L.A.; van Veen, H.W. Molecular disruption of the power stroke in the ATP-binding cassette transport protein MsbA. *J. Biol. Chem.* **2013**, *288*, 6801–6813. [CrossRef]
249. Westfahl, K.M.; Merten, J.A.; Buchaklian, A.H.; Klug, C.S. Functionally important ATP binding and hydrolysis sites in Escherichia coli MsbA. *Biochemistry* **2008**, *47*, 13878–13886. [CrossRef]
250. Randak, C.O.; Dong, Q.; Ver Heul, A.R.; Elcock, A.H.; Welsh, M.J. ATP and AMP mutually influence their interaction with the ATP-binding cassette (ABC) adenylate kinase cystic fibrosis transmembrane conductance regulator (CFTR) at separate binding sites. *J. Biol. Chem.* **2013**, *288*, 27692–27701. [CrossRef]
251. Williams, R.S.; Tainer, J.A. Learning our ABCs: Rad50 directs MRN repair functions via adenylate kinase activity from the conserved ATP binding cassette. *Mol. Cell* **2007**, *25*, 789–791. [CrossRef]
252. Yaginuma, H.; Kawai, S.; Tabata, K.V.; Tomiyama, K.; Kakizuka, A.; Komatsuzaki, T.; Noji, H.; Imamura, H. Diversity in ATP concentrations in a single bacterial cell population revealed by quantitative single-cell imaging. *Sci. Rep.* **2014**, *4*, 6522. [CrossRef]
253. Kaur, H.; Lakatos-Karoly, A.; Vogel, R.; Nöll, A.; Tampé, R.; Glaubitz, C. Coupled ATPase-adenylate kinase activity in ABC transporters. *Nat. Commun.* **2016**, *7*, 13864. [CrossRef]
254. Kaur, H.; Abreu, B.; Akhmetzyanov, D.; Lakatos-Karoly, A.; Soares, C.M.; Prisner, T.; Glaubitz, C. Unexplored nucleotide binding modes for the ABC exporter MsbA. *J. Am. Chem. Soc.* **2018**, *140*, 14112–14125. [CrossRef]

membranes

MDPI

Review

Ins and Outs of Rocker Switch Mechanism in Major Facilitator Superfamily of Transporters

Stephanie Sauve, Joseph Williamson, Adithya Polasa and Mahmoud Moradi *

Department of Chemistry and Biochemistry, University of Arkansas, Fayetteville, AR 72701, USA; ssauve@uark.edu (S.S.); jgw004@uark.edu (J.W.)
* Correspondence: moradi@uark.edu

Abstract: The major facilitator superfamily (MFS) of transporters consists of three classes of membrane transporters: symporters, uniporters, and antiporters. Despite such diverse functions, MFS transporters are believed to undergo similar conformational changes within their distinct transport cycles, known as the rocker-switch mechanism. While the similarities between conformational changes are noteworthy, the differences are also important since they could potentially explain the distinct functions of symporters, uniporters, and antiporters of the MFS superfamily. We reviewed a variety of experimental and computational structural data on a select number of antiporters, symporters, and uniporters from the MFS family to compare the similarities and differences of the conformational dynamics of three different classes of transporters.

Keywords: major facilitator superfamily (MFS); membrane protein; rocker-switch mechanism; transporters; secondary membrane transporter proteins; alternating-access mechanism

1. Introduction

Cells need specific concentrations of molecules and ions to be present on the inside and the outside of the cellular membrane for survival [1–3]. Movement in and out of the cell through the cellular membrane lipid bilayer can either be accomplished through passive diffusion or active transport. In passive diffusion, solutes move across their concentration gradient from a high concentration to a low concentration without requiring any additional energy to move to the other side [4–6]. On the other hand, in active transport, solutes move against their concentration gradients from a low concentration to a high concentration [7–9]. To do this, additional energy is required; usually, adenosine triphosphate (ATP) provides the energy needed to move the solutes across the opposing gradient [1,10,11]. However, the cellular membrane is semipermeable, meaning it allows for certain solutes to cross through the membrane to the inside or outside of the cell, but it does not allow for all solutes to cross on their own [12–14]. Therefore, in order to maintain adequate levels of solute concentrations on both sides of the membrane, transporters embedded in the membrane that help facilitate movement of molecules and ions across the membrane are essential [15–22]. In order to move the substrates across the membrane, the protein goes through a series of conformational changes. This review discusses the mechanism and corresponding conformational changes that are needed to facilitate this mechanism in the major facilitator superfamily (MFS).

MFS transporters are thought to be one of the oldest protein transporter families on the Earth and have existed for 3 billion years [23,24]. They are the largest group of secondary active membrane transporters in the cell and are found in almost all living organisms [20,24–29]. Therefore, they use an electrochemical gradient as a driving force for transport [22,30,31]. It is important to note that according to the Transporter Classification Database (TCDB, http://www.tcdb.org/ (accessed on 23 March 2023)) there is research that suggests the superfamily includes proteins whose main function is not to catalyze secondary transport. An example of such proteins are the major intrinsic proteins (MIPs) aquaporins and glycerol. Although these proteins contribute to the diversity found within this superfamily, they will not be discussed in this review. As a result of the diversity found within the superfamily due to it being one of the oldest protein transporter families, MFS transporters move a wide variety of substrates across the cellular membrane [23–25,27]. Some examples of the many substrates transported by the MFS include sugars (e.g., hexoses and pentoses), nitrite/nitrate, amino acids (e.g., histidine and lysine), oligopolypeptides (e.g., dipeptides and tripeptides), and organic cations [23,28,32–35].

Since the superfamily is so large with such wide diversity, it is easier to classify it into smaller groups. The superfamily can be divided into transporter types that are antiporters, symporters, and uniporters [28]. Antiporters move one or more substrates into the cell while simultaneously moving one or more substrates out of the cell or vice versa either with or against their concentration gradient. Symporters move two or more substrates in the same direction either with or against their gradients. Uniporters move one substrate at a time [36]. Although these transporters vary on how many substrates they carry and the direction in which they move them in, they undergo similar mechanisms within their distinct transport cycles, for example the rocker-switch mechanism. Although the similarities between these transporter types are important, their differences are also relevant since they could explain the distant functions of the superfamily. Therefore, we reviewed a variety of experimental and computational structural data on a select number of antiporters, symporters, and uniporters from the MFS family to compare the similarities and differences of the conformational dynamics of the three different classes of transporters.

In the following review, we discuss details of the MFS rocker-switch mechanism for each class of aforementioned transporters in depth. We focus on how the structure of the transporter type contributes to the mechanism of movement being carried out. Additionally, we highlight the importance of the formation and breakage of salt bridges in the movement of substrate from one side of the cellular membrane to the other. Each example discussed has a generalized mechanism for that specific transporter type along with a specific example of how salt bridges pertain to a transporter of interest of the transporter type. Such information has not been compiled to one review thus far.

1.1. Structure of MFS Transporters

The Transporter Classification Database (TCDB, http://www.tcdb.org (accessed on 23 March 2023)) identifies up to 106 different families within the major facilitator superfamily based on each transporters' function and phylogeny. Though there have been hundreds of transporter proteins identified for the MFS, little progress has actually been made in terms of structural progress in the last two decades [37]. Despite the wide variety of mechanistic roles each one can perform, a similarity between all MFS transporters is their structural folding paradigm [19,36]. The canonical MFS-fold consists of a total organization of 12 transmembrane (TM) segments that further organize into 2 bundles (each with 6 TMs) where Bundle 1 at the N-terminal domain (NTD) contains TM 1–6, and Bundle 2 at the C-terminal domain contains TM 7–12 [29,38–41] (Figure 1). The organization at these domains is taken 1 step further as each bundle of 6 can be distinguished as 2 groups of 3 TM segments (1–3 and 4–6 or 7–9 and 10–12). Nevertheless, these domains join together to form a "Λ"—(inward facing) or "V"—(outward facing) shaped transporter showing NTD and CTD symmetry [22,42]. Although most MFS transporters contain 12 TM segments, transporters can have more or less than 12 TM segments, meaning that the number of TM segments in each bundle must be adjusted accordingly for these transporters [38,43].

The domain bundles are mainly connected by a long and flexible intracellular loop, though particular TMs of each bundle also interconnect with each other. The interconnections between TMs have been found to vary across different types of transporters [23,43,44]. Consistently, the substrate binding pocket is located at the interface between the bundles [29]. The structural details described above were identified using crystal structures of many MFS transporters, such as lactose symporter (LacY) [45], the glycerol-3-phosphate-

phosphate antiporter (GlpT) of *Escherichia coli* [20], the oxalate transporter (OxlT) [46] of *Oxalobacter formigenes*, and the melibiose transporter (MelB) [47] of *Salmonella typhimurium*. It was also confirmed in these studies that structural differences between domains were largely due to the way substrates bind to the substrate binding pocket located between the two domains [48].

The domain bundles contain a structural similarity of perpendicular symmetry with respect to the plane of the membrane, also referred to as pseudo-two-fold symmetry [29,49]. This relationship has been identified as a 180° rotation parallel to the plane of the membrane [40,50] (Figure 1). Another common similarity amongst MFS transporters are the inverted fold repeats found within all transporter folds [51]. These repeats can be explained in terms of the MFS bundles discussed throughout this paper, wherein one is present in the upright position, while the other is inverted by 180° within the particular membrane. The term for this is "dual topology" which does not affect the membrane functionality as a homodimer [52].

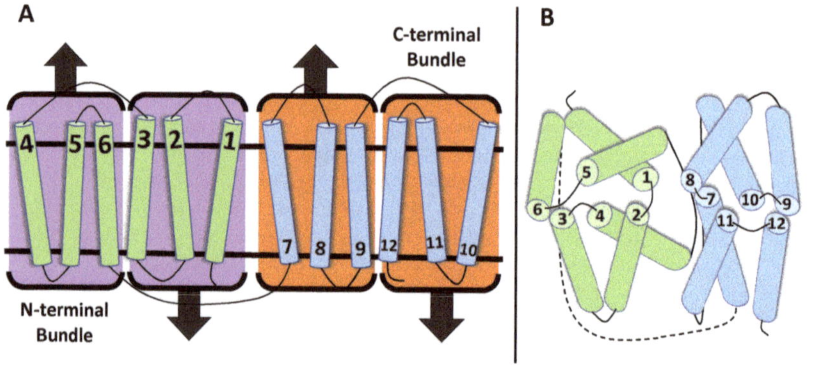

Figure 1. Schematic representation of the canonical MFS-fold transporter topology. (**A**) The canonical MFS topology is comprised of 12 TMs with a N and C orientation, forming two structurally similar six-helix bundles. The N-terminal bundle (TMs 1–6; green) and the C-terminal bundle (TMs 7–12; blue) are connected by a cytosolic loop, which can sometimes contain structural elements (black). Each of the bundles is made up from 3 TM structural-inverted repeats. (**B**) The first TM in each of the 3 TM repeats (TM1, TM4, TM7, and TM10) form the central cavity helices and often undergo local changes to bind and release the substrate during alternating access.

Typically, MFS transporters have been generalized to express three conformational states as they work to transport relevant substrates in and out of a membrane. The three states are commonly referred to as the inward open/facing (IF) state where the binding site is in a space between the N- and C-domain on the cytoplasmic side; the occluded (OCC) state where the transporter is not open to either side of the membrane as a result of the helices that surround the substrate being compact, thus the substrate is unable to enter or leave the binding site; and the outward open/facing (OF) state where the binding space is on the extracellular side [21,29,36,53]. As one could imagine, the helices responsible for these movements would need to be either highly flexible or able to withstand breaking [54]. Research has identified particular helices responsible for motioning the substrate through the membrane gradient, referred to as the central cavity helices. Due to well-conserved glycine and/or proline residues with a tendency to bend in response to substrate binding, in partially and fully occluded (discussed later) states, these substrate gating helices have proven difficult to distinguish in the OF and IF states provided through crystallization [55].

1.2. The Rocker-Switch Mechanism

In 1990, Peter Mitchell opposed the previously standing moving carrier model that said transporters physically move with their substrates to the opposite side of the membrane with what he called the moving barrier model. In this model, he stated that the substrate binds to a specific point on the transporter, but the permeability barrier between that binding site and the two sides of the membrane could change so that the site was only accessible from one side of the membrane at a time [56]. Since then, mechanisms for alternating access models have been investigated. These methods state that transport proteins alternate between being open to the extracellular or intracellular side of the membrane so that the substrate cavity in the middle of the protein is only open to one side of the membrane at a time [21,22,30,57–60].

There are currently three models demonstrating the alternating access mechanism to describe how transporters move their substrates across the membrane [22] (Figure 2). They are called the rocking bundle model, the elevator model, and the rocker-switch model. The rocking-bundle mechanism suggests that the conformational change from inward open and outward open arise from tilting the helical bundles of one domain, while the second domain stays fixed [61,62]. On the other hand, the elevator model, which consists of one moving domain that slides through the membrane and one fixed domain, requires pushing an entire domain through the membrane [16,62,63]. The rocker-switch model is the third current model of the alternating access mechanism and stands in contrast to the rocking-bundle model and the elevator model.

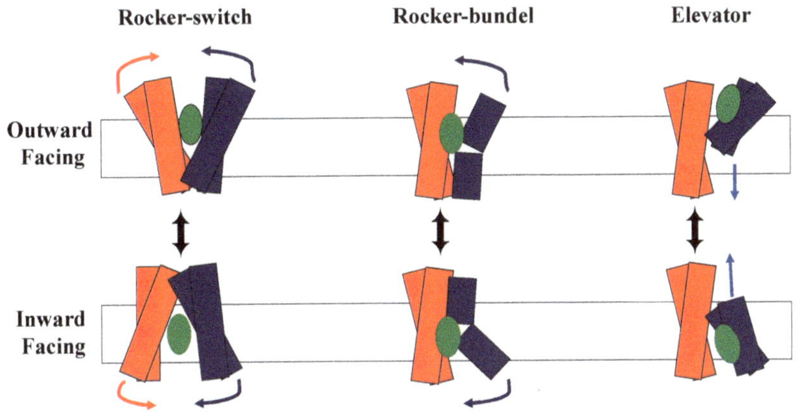

Figure 2. Schematic representation of the three types of current alternating access models with the substrate shown in green. The NTD is shown in blue and the CTD is shown in red.

The rocker-switch model explains the switching and thus alternating access between two conformational states, IF and OF open, seen in membrane transporters. The alternate access of the substrate binding site results from rocking the near symmetric N-domain and C-domain over a rotation axis that has the binding site at the interface of the domains [36,53]. However, there is research that suggests that although there is symmetry in the two domains, there can be non-symmetrical intra-domain movements that provide greater dynamic complexity [29]. As discussed previously, these domains are made up of helical bundles with helices 1–6 in the N-domain and helices 7–12 in the C-domain with the substrate binding site at the interface of the domains. The model states that the binding site is never open to both sides of the membrane at the same time [64]. Thus, once the rocking occurs, one side opens, while the other side closes, changing which side of the membrane the substrate binding site is open to allow substrates to move in and out of the cell (Figure 3). Therefore, since structure dictates function, a characteristic feature of rocker-switch proteins is the formation of two structurally similar domains that allow for the mechanism to be carried out [16,38]. Note that in the occluded conformation, the substrate binding site is not open to either the cytoplasm or periplasm [24,40,65] (Figure 4).

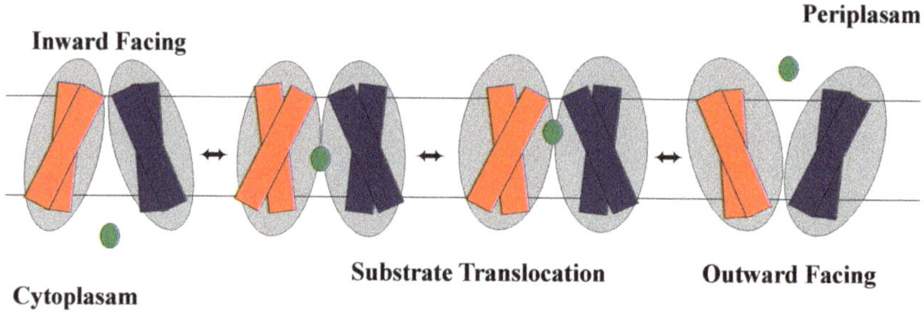

Figure 3. Schematic representation of the MFS facilitative diffusion of a substrate (green) across the cell membrane. The NTD is shown in blue and the CTD is shown in red.

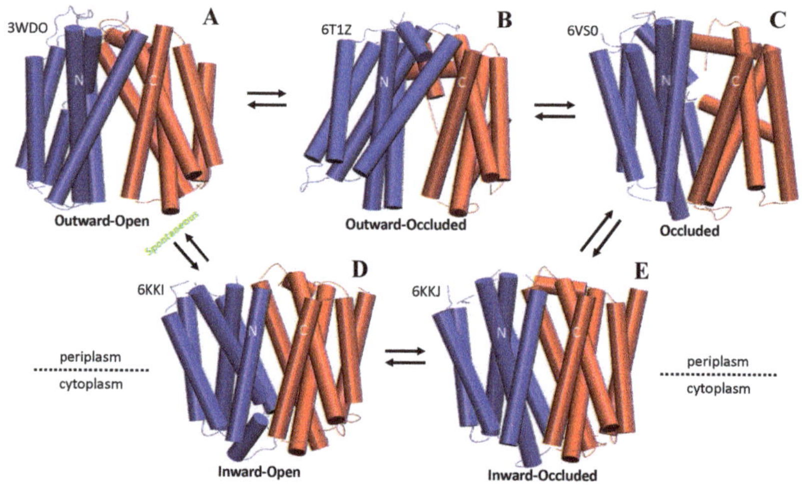

Figure 4. Crystallized depiction of the 5-state mechanism associated with many symporters and uniporters of the MFS that expand on the generalized cycle shown in Figure 3. Particular portions of some of the crystal structure of the proteins were excluded in order to capture the essence of the mechanism. (**A**) To display the outward-open state, the YajR (WT, PDB ID 3WDO) was used; residue IDs 387–455 were excluded. (**B**) The outward-occluded state, LmrP transporter from *L. lactis* (PDB ID 6T1Z), was used; residue IDs 501–603 were excluded. (**C**) The fully occluded state (with substrate), Protein B transporter of *E. coli* (PDB ID 6VS0), was used; residue IDs 501–680 were excluded. (**D**) The inward-occluded state was displayed using Drug:Proton Antiporter-1 (DHA1, PDB ID 6KKJ); residue IDs 401–403 were excluded. (**E**) The inward-open state was displayed using Drug:Proton Antiporter-1 (DHA1, PDB ID 6KKI).

1.3. Importance of Salt Bridges for the MFS

Salt bridges are non-covalent interactions that occur between oppositely charged residues in close contact to each other and within hydrogen bonding distance that are able to form electrostatic interactions [66–70]. These non-covalent interactions can be relatively weak when looked at individually, but they have an additive effect, meaning that they can have great influence on the overall system as the number of salt bridges found in the system increases. Additionally, the overall strength of their interaction depends on the environment in which the salt bridges are found because water and various ions can impact salt bridge formation and stability [69–73]. Salt bridges can provide stability to a protein structure, influence the interactions of proteins with other biomolecules, and can play many roles in protein function [66,68,69]. Therefore, the possible formation of salt bridges is an important interaction to take into consideration when viewing protein structure, and the making and breaking of these interactions should be acknowledged when trying to determine stability of the various conformational states of a protein.

It is known that salt bridges are formed within and between the domains of MFS transporters [20,24–26,28,42,45,55,74–82]. It is thought that interdomain salt bridges are what hold the NTD and CTD together in some MFS transporters, such as Glpt and LacY [20,45]. Therefore, they are vital for maintaining protein structure and for allowing for movement of the domains relative to each other, which is vital for the mechanism of action. Additionally, interbundle salt bridges are typically found proximal to the substrate binding site in MFS transporters [24,32,36,55,83,84]. This indicates that their placement in MFS transporters may have a similar purpose and could be connected to substrate binding. Overall, salt bridges form an energetic barrier that can be overcome by substrate binding since the binding of the substrate can either break or form a salt bridge [24]. Therefore, it is worthwhile to look at the role of salt bridge formation and breakage in MFS transporters. In the examples discussed in this paper, we consider the role of salt bridges in the rocker-switch mechanism for each transporter type.

2. Significant Structural and Conformational Differences in the Rocker-Switch Mechanism across the Major Facilitator Superfamily of Transporters

As discussed previously, certain structural patterns remain constant across all transporter types in the MFS, such as the helical bundle domains that help facilitate the movement of substrates from one side of the membrane to the other. Although there are many similarities between how MFS transporters use the rocker-switch mechanism, there are also noteworthy differences. Keep in mind that the different transporter types can move varying amounts of substrates across the membrane. Because of this, they use the rocker-switch mechanism uniquely to accomplish their goal. In the subsequent sections of this review, we discuss the rocker-switch mechanism as it pertains to antiporters, symporters, and uniporters, respectively, with examples of use for each category. We also highlight the roles of salt bridges in regard to the mechanism for each example described.

2.1. The Rocker-Switch Mechanism for Antiporters

Antiporters are proteins embedded in the membrane that move two or more substrates across the cellular membrane in opposite directions via a series of substrate binding and conformational changes. It is known that antiporters in the MFS can use the rocker-switch mechanism [24–26,28,74,75] (Figure 5). Although the rocker-switch mechanism is identical across the three different types of transporters in the MFS, the mechanism is not entirely the same due to the variation in the number of substrates the transporter can move and the direction that it moves them in. For example, unlike the symporters that use the rocker-switch mechanism, antiporters face inward when substrate concentration is low until a substrate binds from the cytosolic side of the membrane unless regulatory action is being performed on the membrane protein [75]. When a substrate does bind, it changes the local pH and makes it less stable for protons to bind [75]. This is important because many antiporters use the proton-motive force (PMF) to drive their action of transporting substrates across the membrane. There are many antiporters that use the rocker-switch mechanism, including nitrate/nitrite transporters [28], glycerol-3-phosphate transporters [25,64,74], and oxolate/formate transporters [25] to name a few.

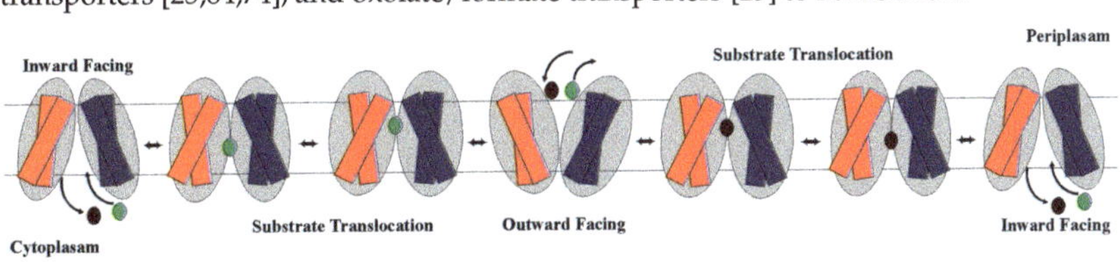

Figure 5. Schematic representation of the antiporter facilitating diffusion of two substrates (green and black) across the cell membrane with each substrate going in opposite directions. The NTD is shown in blue and the CTD is shown in red.

One example of an antiporter in the MFS is the glycerol-3-phosphate (GlpT) antiporter that uses the rocker-switch mechanism to move its substrates across the cellular membrane [74]. The glycerol-3-phosphate transporter (GlpT) moves one substrate, glycerol-3-phosphate, by using the inorganic phosphate concentration gradient on the opposite side of the membrane as its source of energy [26,64,85,86]. Binding of inorganic phosphate allows for the transition from the inward facing state to the outward facing state, allowing for movement to the opposite side [20,26,87]. To return to the inward facing state, glycerol-3-phosphate must bind [20,26,64,87,88]. Thus, a cycle of alternate binding occurs between the two substrates that changes which side of the membrane the binding site of the transporter protein is open to. Additionally, the GlpT antiporter is open to the cytosol side of the membrane with Helices 1 and 7 blocking substrate passage via aromatic residues that are contained on the helices that act as a blockade [25,26]. It has also been found that these helices are involved in the gating of the periplasmic side, while TM Helices H4, H5, H10, and H11 are more directly involved in gating of the cytoplasmic side [26]. Therefore, the gating ensures that substrates cannot freely flow through the transporter without the conformational shift occurring and correct substrate binding [29].

Mechanistically, it is thought that weak binding occurs by the substrate when binding residues are unprotonated [25,74]. Then, once protonation occurs there is greater interaction with the substrate pulling the helices closer and moving them, causing a conformational change [25,74]. Once the change is complete, the binding residues are deprotonated and therefore release the substrate on the opposite side of the membrane where it started [25,74]. This is consistent with the rocker-switch mechanism. The importance of protonation of specific residues has been seen in other MFS transporters as well, such as PiPT, a phosphate:H+ symporter, in which protonation of Asp 324 lowers the energy barrier for phosphate binding which provides a driving force for substrate translocation [29]. As a reminder, GlpT contains pseudo-twofold symmetry with the substrate binding site at the interface of the domains [20,26,87,89]. It is thought that because of the crystal structure of GlpT that the transitions between the inward facing and outward facing states is conducted by a $\sim 10^{\circ}$ rigid-body rotation of the two halves [20,25,26,87].

Additionally, it has been shown that antiporters can form salt bridges that stabilize the different facing conformations, and the making and breaking of the salt bridges on substrate binding and release can also cause movement of the helices and thus conformational changes [25,26,28,75,78,82]. For example, Law et al. found that when inorganic phosphate binds loosely to GlpT and H165 is unprotonated that there is a strong salt bridge formed between D274 and R45 and between E299 and K46, with an additional weaker salt bridge formed between D274 and K46 [25]. However, when the histidine becomes protonated and the substrate binds tighter, the D274–K46 salt bridge is broken, and the E299–K46 salt bridge is weakened, while a new salt bridge is formed between E299 and R269 [25]. (Figure 6). The formation of this competing salt bridge assists with domain rotation [20,25,87] (Figure 6).

The structure of antiporters in the MFS can be complex, which may help antiporters distinguish between the large range of substrates that they can carry. For example, the OxlT transporter which binds many different substrates has long helical regions composed of Helices 4, 5, 10, and 11 that extend deep into the cytoplasm with charged residues [25]. These residues aide in ensuring substrate specificity is present despite being able to transport more than one substrate. Additionally, although binding residues for substrates are typically found on both the C-domain and the N-domain of interacting helices, some antiporters in the MFS contain binding residues on only one domain. This can been seen in OxlT and the oxolate/formate transporters [90].

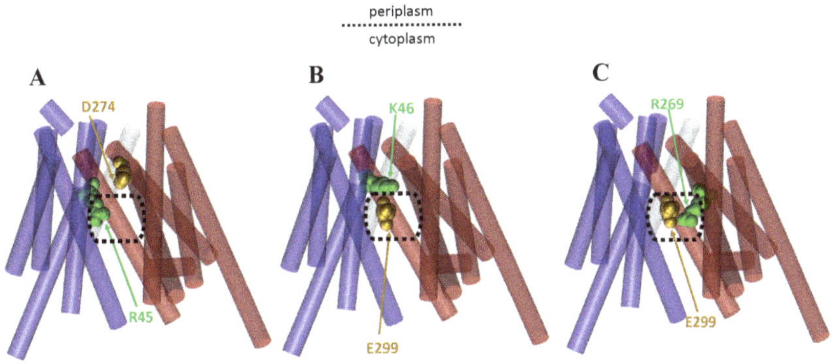

Figure 6. A cartoon representation of the GlpT transporter in its IF state based on its crystal structure (PDB ID: 1PW4), showing the salt bridges that form and break upon the binding of inorganic phosphate and the protonation of H165 in the Glpt transporter. The NTD is shown in blue and the CTD is shown in red. A dotted rectangle is provided to display the central cavities. The 'gold' and 'green' colors show the amino acid responsible for the negative and positive charge in the interaction, respectively: (**A**) the salt bridge interaction between D274 and R45 of the CTD and NTD, respectively; (**B**) the salt bridge interaction between E299 and K46 of the CTD and NTD, respectively; (**C**) the salt bridge interaction between E299 and R269 in both of the CTD.

2.2. The Rocker-Switch Mechanism for Symporters

The rocker-switch mechanism shown in Figure 7 is a well-known model for symporter conformational changes [91,92]. Symporters, which are membrane-integral proteins, are in charge of moving two or more molecules across the cell membrane at the same time (e.g., neurotransmitter transporters [91], sugar transporters [32,93], amino acid transporters [33,34], and nucleobase transporters [94,95]). According to many researchers, the symporter protein can exist in one of two major conformational states at any given time [32–34,64,91–97]. Two such states (Figure 2) are the outward-facing conformation and the inward-facing conformation [96,97], which are both maintained by various interactions between the protein and its substrates [96]. Proton-dependent oligopeptide transporter (POT) family proteins, also referred to as the PTR (peptide transport) family proteins, are additional members of the MFS symporters and are found in both Gram-positive and Gram-negative bacteria, yeast, plants, animals, and plants [42,64,98].

The POT family proteins mediate the absorption of a large variety of di- and tripeptides as well as peptide-like pharmacologically active compounds [64,76,99,100]. They use the inwardly directed proton (H+) electrochemical gradient to transport peptides across cell membranes in eukaryotes and prokaryotes [101,102]. Human peptide transporters 1 (PepT1), 2 (PepT2), and GkPOT are major POTs; PepT1 is mostly found in the small intestine, while PepT2 is found in the kidney, lungs, and central nervous system [77,102–104]. The human PepT protein contains a total of 14 TM helices, of which Helices H1–H12 adopt the general fold that was observed in other MFS transporters [20,41,45,105]. PepT1 and PepT2 encode the 14 common transmembrane (TM) helices, with both the N- and the C-terminal facing the cytoplasm [101–104,106].

The rocker-switch mechanism is associated with three steps in POT transporters (Figure 7). Step one is substrate binding-induced conformational change, step two involves protonation-induced conformational change, and step three involves release-induced conformational change [64,76,77,101–104,106–108]. A PepT crystal structure analysis revealed that the peptide-binding site is composed on the surfaces of both the N- and the C-terminal helix bundles, while the proton-binding site is situated close to the extracellular gate [76]. In the transport cycle's first step, the peptide substrate binds to the POTs transporter in its outward-facing conformation. Then, protonation (proton binding) induces conformational changes after binding, in which TM helices undergo similar conformational changes [64,76,77,107,108]. This binding induces a conformational change that closes the transporter, and the transporter goes into an inward-facing position and exposes the peptide to the cytoplasm [77,107,108]. Consistent with experimental findings, a simulation study indicated that conformational changes in GkPOT are coupled to the transport of substrates and protons [77,107,108]. Finally, protons and peptides are released on the intracellular side of the membrane in this inward-facing conformation. A molecular dynamics simulation study of PepT [42,76,77] predicted that there are several salt bridges in the different conformations of PepT that keep the protein stable in its inward and outward shapes. Specifically, K127–E24, R25–E21, and R32–D316 salt bridges were consistently observed in all transport cycle conformations, whereas K127–E21 was predicted to form infrequently in inward and occluded conformations [42] (Figure 8). Mutating any of the residues involved in these salt bridges has previously been shown to abolish or significantly reduce transport [42,55].

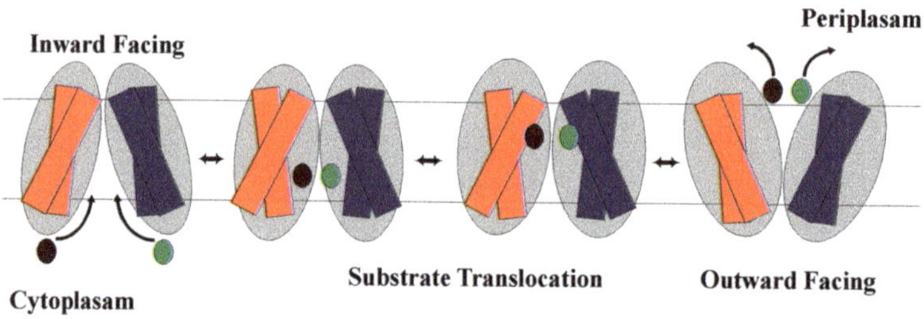

Figure 7. Schematic representation of the symporter facilitating diffusion of two substrates (green and black) across the cell membrane with each substrate going in the same direction. The NTD is shown in blue and the CTD is shown in red.

Overall, similar rocker-switch mechanism steps are observed in other symporters, including sugar transporters, serotonin transporter (SERT), dopamine transporter (DAT), glutamate transporter, excitatory amino acid transporter (EAAT3), amino acid transporters, lactose permease (LacY), maltose transporter (MalFGK2), nucleobase transporters, such as uracil transporter (UraA) and xanthine transporter (XanQ), phosphate transporter (PiPT), and purine transporter (PUP1) [29,34,93,109–117].

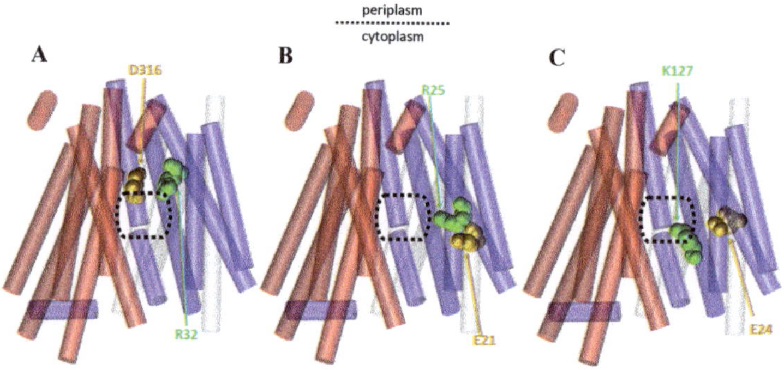

Figure 8. A cartoon representation of the PepT transporter in its IF state based on its crystal structure (PDB ID: 4UVM) showing the salt bridges that are consistently observed in all transport cycle conformations. The NTD is shown in blue and the CTD is shown in red. A dotted rectangle is provided to display the central cavities. The 'gold' and 'green' colors show the amino acid responsible for the negative and positive charge in the interaction, respectively: (**A**) the salt bridge interaction between D316 and R32 of the CTD and NTD, respectively; (**B**) the salt bridge interaction between E21 and R25, both of the NTD; (**C**) the salt bridge interaction between E24 and K127, both of the NTD.

2.3. The Rocker-Switch Mechanism for Uniporters

The uniporter transport system is likely the most identifiable transporter of the MFS transporters. Unlike symporters and antiporters, these systems function to move only a single substrate at a time either along or against the particular transporter gradient via facilitated diffusion [118] (Figure 9). Types of transporters that associate by facilitated diffusion are channel proteins and carrier proteins [119]. Glucose or GLUT transporters are a subset in the MFS TCBD Classification Database (TCBD ID 2.A.1, https://tcdb.org/search/result.php?tc=2.A.1 (accessed on 23 March 2023)) and actually among the first transporters to be functionally characterized with a substrate coordination map. GLUT1 is a particular 'passive' sugar porter (SP) uniporter found in a range of tissues essential for glucose transport into the brain [120]. The conformational transition from inside (without substrate) to outside the membrane was previously recorded as the slowest step of the cycle [121,122], likely due to the forming and breaking of the intracellular salt bridge network located at the ends of TM helices. Salt bridges within GLUT1 have not been recorded near the central cavity section for either outward or inward facing states [32,93,123–125].

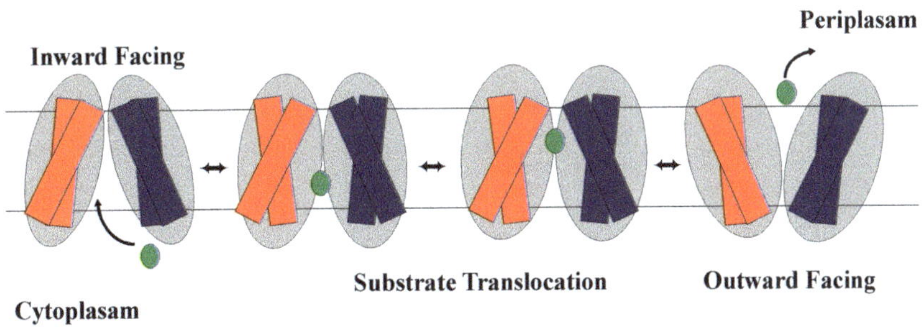

Figure 9. Schematic representation of the uniporter facilitating diffusion of one substrate across the cell membrane. The NTD is shown in blue and the CTD is shown in red.

A molecular dynamics study led by Dr. Etchebest studied aspects of the mechanics surrounding GLUT1 transfer and found two strongly connected salt bridge interactions that withstood nearly the entirety of the 1–1.5 microsecond MD simulations. The two main salt bridge interactions were reported as D240 with R232 (between intracellular Helices 2 and 3) and E247 with R212 (between intracellular Helices 3 and 1) (Figure 10). In turn, the intracellular portion of Helix 1 was found to have a strong connection to the intracellular portion of TM3 by residues E209 with R93 [97], as shown in Figure 10. These were not the only salt bridges identified through their simulations, but they establish a foundation for the underlying importance that salt bridges play in uniporter transportation.

A recent study on the human GLUT1 revealed that substrate binding causes the protein to "twist and tilt", resulting in the formation of a hydrophilic pathway for glucose transport [97]. These uniporters are believed to undergo five main conformational states beginning as outward facing, outward-occluded, occluded (with or without substrate), inward-occluded, and inward facing (Figure 4). They have also been found to spontaneously reset themselves from the inward facing to outward facing conformation in the absence of substrate binding. Interestingly, this indicates that the energy barriers separating the two states must be relatively low [124].

First, in the initial step of the transport cycle, glucose is seen to bind to the outward-facing conformation of GLUT1. This binding induces a conformational change in the transporter, closing it and exposing glucose to the cytoplasmic side of the cell. It is believed that the transmembrane helices rotate relative to one another during this conformational change [93]. In this same study, cryo-EM was used to crystallize the structure of GLUT1 in the presence of glucose. The structure demonstrated the rotation of the transmembrane helices, consistent with previous biochemical and biophysical studies [97,126–128].

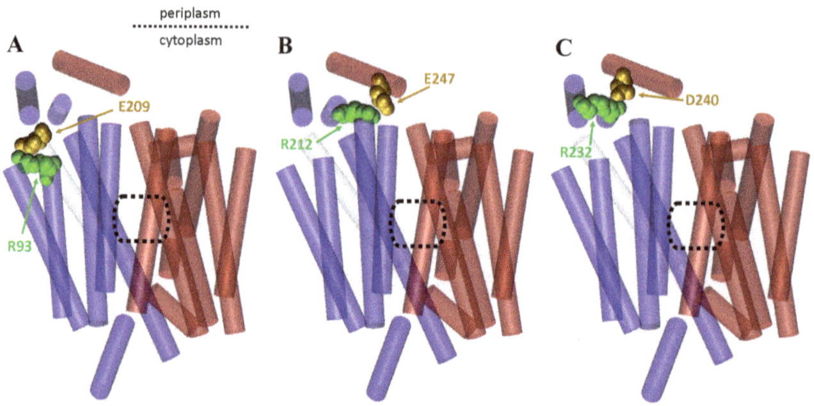

Figure 10. A cartoon representation of the GLUT1 transporter in its IF state based on its crystal structure (PDB ID: 4PYP), showing the salt bridges that are consistently observed in all transport cycle conformations. The NTD is shown in blue and the CTD is shown in red. A dotted rectangle is provided to display the central cavities. The 'gold' and 'green' colors show the amino acid responsible for the negative and positive charge in the interaction, respectively: (**A**) the salt bridge interaction between E209 and R93 of the CTD and NTD, respectively; (**B**) the salt bridge interaction between E247 and R212, both apart of the CTD; (**C**) the salt bridge interaction between D240 and R232, both apart of the CTD.

After binding to the outward-facing conformation of GLUT1, glucose molecules are transported across the membrane as the uniporter transitions to its first partial-occluded state (outward-occluded), (Figure 4). The helices typically responsible for these local TM movements are referred to as "gating helices" and are often apart of the central cavity helices. The responsibility of these helices is to 'occlude' the substrate from exiting during partial occlusion states [124]. In many cases, the responsible transmembranes interacting through the outward- and inward-facing state transitions are carried out by either TM4 or TM1 of the NTD with either TM10 or TM7 of the CTD (Figure 1) [24]. The fully occluded state has been hypothesized to be metastable for short-term occupation of the substrate during structural isomerization between the two partially occluded states [129].

Once the transporter has transitioned into the inward-facing state, the substrate will finally be released into the cytoplasm. This step involves an additional conformational change in which the transporter opens to expose the substrate-binding site on the extracellular side [123]. To allow the substrate release on the membrane's internal side, the aforementioned transmembrane helices must move, often by breaking or due to high flexibility. This was supported by a study in which the structure of GLUT1 in its inward-facing conformation was determined using X-ray crystallography which also observed that the helices moved considerably more than in the outward-facing conformation [32].

3. Conclusions

MFS transporters have long been studied to provide insight into alternating-access mechanisms. Through this review and the included figures, we have described the structure of MFS transporters, competing alternating-access mechanisms, the importance of salt bridges for MFS transporters, and the rocker-switch mechanism as it pertains to antiporters, symporters, and uniporters in major facilitator superfamily transporters. We have shown that the overall structure of MFS transporters usually consists of 12 TM helices that are divided evenly into the CTD and NTD with an interdomain loop connecting the helical bundles. It is the movement of these two bundles that vary in the three current alternating-access mechanisms that are available, called the rocking-bundle model, the elevator model, and the rocker-switch model. In addition to generally describing the rocker-switch mechanism, we have shown that antiporters, symporters, and uniporters vary on the use of the rocker-switch mechanism according to how many substrates each transporter can move and the direction they move them in. However, the general structure and foundation of the mechanism remain consistent. In particular, we have gone through the mechanisms for GlpT, POT transporters, and GLUT1 as examples for the antiporters, symporters, and uniporters, respectively. We expect that the information provided in this review will provide insight on the rocker-switch mechanism and on the major facilitator superfamily as a whole. However, there is still much to learn about the MFS, and as available structures improve, more information can be gained. In particular, more research needs to be conducted on individual helical movement and domain movement to obtain a better understanding of the mechanisms employed by MFS and which alternating-access mechanisms best characterize their dynamics.

Author Contributions: Conceptualization, S.S. and M.M.; methodology, S.S. and M.M.; software, S.S. and M.M.; validation, S.S., J.W., A.P. and M.M.; formal analysis, S.S., J.W. and A.P.; investigation, S.S., J.W., A.P. and M.M.; resources, S.S., J.W., A.P. and M.M.; data curation, S.S., J.W., A.P. and M.M.; writing—original draft preparation, S.S., J.W., A.P. and M.M.; writing—review and editing, S.S. and M.M.; visualization, S.S., J.W., A.P. and M.M.; supervision, M.M.; project administration, S.S. and M.M.; funding acquisition, M.M. All authors have read and agreed to the published version of the manuscript.

Funding: This work was supported by the National Institute of General Medical Sciences of the National Institutes of Health under award number R35GM147423. This work was also supported by the Arkansas Biosciences Institute.

Institutional Review Board Statement: Ethical review and approval were not applicable for this review.

Data Availability Statement: No new data is reported in this review article.

Conflicts of Interest: The authors declare no conflict of interest.

Abbreviations

The following abbreviations are used in this manuscript:

Major facilitator superfamily	MFS
Adenosine triphosphate	ATP
Inward facing	IF
Outward facing	OF
The Transporter Classification Database	TCDB
Major Intrinsic Protein	MIP
Transmembrane segments	TM
N-terminal domain	NTD
C-terminal domain	CTD
Lactose symporter	LacY
Glycerol-3-phosphate-antiporter	GlpT
Oxalate transporter	OxlT
Melibiose permease	MelB
Transmembrane domains	TMDs
Proton-dependent oligopeptide transporter	POT
Peptide transport	PTR
Peptide transporters 1	PepT1
Peptide transporters 2	PepT2
Serotonin transporter	SERT
Dopamine transporter	DAT
Excitatory Amino Acid Transporter	EAAT3
Lactose permease	LacY
Maltose transporter	MalFGK2
Uracil transporter	UraA

Xanthine transporter	XanQ
Purine transporter	PUP1
Proton-motive force	PMF
Sugar porter	SP
Glucose transporter 1	GLUT1

References

1. Stillwell, W. Membrane Transport. In *An Introduction to Biological Membranes*; Elsevier: Amsterdam, The Netherlands , 2016; pp. 423–451. [CrossRef]
2. Frans, M. Solute Transport: Basic Concepts. *Plant Physiol. Funct.* **2017**, *6*, 1–31. [CrossRef]
3. Xi, C.; Peter, P. The Balance of Fluid and Osmotic Pressures across Active Biological Membranes with Application to the Corneal Endothelium. *PLoS ONE* **2015**, *10*, e0145422. [CrossRef]
4. Timney, B.; Raveh, B.; Mironska, R.; Trivedi, J.; Kim, S.; Russel, D.; Wente, S.; Sali, A.; Rout, M. Simple rules for passive diffusion through the nuclear pore complex. *J. Cell Biol.* **2016**, *215*, 57–76. [CrossRef]
5. Camenisch, G.; Alsenz, J.; ven de Waterbeemd, H.; Folkers, G. Estimation of permeability by passive diffusion through Caco-2 cell monolayers using the drugs' lipophilicity and molecular weight. *Eur. J. Pharm. Sci.* **1998**, *6*, 313–319. [CrossRef]
6. Heidi, E.; Janet, E.; Locksley, M. Osmotic Transport across Cell Membranes in Nondilute Solutions: A New Nondilute Solute Transport Equation. *Biophys. J.* **2009**, *96*, 2559–2571. [CrossRef]
7. Forrest, L.; Krämer, R.; Ziegler, C. The structural basis of secondary active transport mechanisms. *Biochim. Et Biophys. Acta Bioenerg.* **2011**, *1807*, 167–188. [CrossRef] [PubMed]
8. Sweadner, K.; Goldin, S. Active Transport of Sodium and Potassium Ions—Mechanism, Function, and Regulation. *N. Engl. J. Med.* **1980**, *302*, 777–783. [CrossRef]
9. Kyte, J. Molecular considerations relevant to the mechanism of active transport. *Nature* **1981**, *292*, 201–204. [CrossRef] [PubMed]
10. Boyer, P. A Model For Conformational Coupling Of Membrane Potential and Proton Translocation To ATP Synthesis and To Active Transport. *Fed. Eur. Biochem. Soc. Lett.* **1975**, *58*, 1–6. [CrossRef] [PubMed]
11. Dunn, J.; Grider, M. *Physiology, Adenosine Triphosphate*; StatPearls: Tampa, FL, USA , 2023.
12. Kulkarni, T.; Slaughter, G. Application of Semipermeable Membranes in Glucose Biosensing. *Membranes* **2016**, *6*, 55. [CrossRef] [PubMed]
13. Park, H.; Kamcev, J.; Robeson, L.; Menachem, E.; Freeman, B. Maximizing the right stuff: The trade-off between membrane permeability and selectivity. *Science* **2017**, *356*, eaab0530. [CrossRef] [PubMed]
14. Starzak, M. Membranes, Synthetic (Chemistry). In *Encyclopedia of Physical Science and Technology*, 3rd ed.; Elsevier Science Ltd.: Cambridge, MA, USA, 2003; pp. 345–354. [CrossRef]
15. Lee, V. Membrane transporters. *Eur. J. Pharm. Sci.* **2000**, *11*, S41–S50. [CrossRef]
16. Drew, D.; Boudker, O. Shared Molecular Mechanisms of Membrane Transporters. *Annu. Rev. Biochem.* **2016**, *85*, 543–572. [CrossRef] [PubMed]
17. Stieger, B.; Steiger, J.; Locher, K. Membrane lipids and transporter function. *Biochim. Biophys. Acta (BBA)-Mol. Basis Dis.* **2021**, *1867*, 166079. [CrossRef] [PubMed]
18. He, Y.; He, G.; He, T. Specifically Targeted Transport of Plasma Membrane Transporters: From Potential Mechanisms for Regulating Cell Health or Disease to Applications. *Membranes* **2021**, *11*, 736. [CrossRef]
19. Denning, E.; Beckstein, O. Influence of lipids on protein-mediated transmembrane transport. *Chem. Phys. Lipids* **2013**, *169*, 57–71. [CrossRef]
20. Huang, Y.; Lemieux, M.; Song, J.; Auer, M.; Wang, D. Structure and Mechanism of the Glycerol-3-Phosphate Transporter from Escherichia coli. *Science* **2003**, *301*, 616–620. [CrossRef]
21. Chan, M.; Shukla, D. Markov state modeling of membrane transport proteins. *J. Struct. Biol.* **2021**, *213*, 107800. [CrossRef]
22. Beckstein, O.; Naughton, F. General principles of secondary active transporter function. *Biophys. Rev.* **2021**, *3*, 011307. [CrossRef]
23. Pao, S.; Paulsen, I.; Saier, M. Major Facilitator Superfamily. *Microbiol. Mol. Biol. Rev.* **1998**, *62*, 1–34. [CrossRef]
24. Drew, D.; North, R.; Nagarathinum, K.; Tanabe, M. Structures and General Transport Mechanisms by the Major Facilitator Superfamily (MFS). *Chem. Rev.* **2021**, *121*, 5289–5335. [CrossRef]
25. Law, C.; Maloney, P.; Wang, D.N. Ins and Outs of Major Facilitator Superfamily Antiporters. *Annu. Rev. Microbiol.* **2008**, *62*, 289–305. [CrossRef]
26. Moradi, M.; Babin, V.; Roland, C.; Sagui, C. The Adaptively Biased Molecular Dynamics method revisited: New capabilities and an application. *J. Phys. Conf. Ser.* **2015**, *640*, 12020. [CrossRef]
27. Madej, M.G. Function, Structure, and Evolution of the Major Facilitator Superfamily: The LacY Manifesto. *Adv. Biol.* **2014**, *2014*, 523591. [CrossRef]
28. Feng, J.; Selvam, B.; Shukla, D. How do antiporters exchange substrates across the cell membrane? An atomic-level description of the complete exchange cycle in NarK. *Structure* **2021**, *29*, 922–933. [CrossRef] [PubMed]
29. Penderson, B.; Kumar, H.; Waight, A.; Risenmay, A.; Roe-Zurz, Z.; Chau, B.; Schlessinger, A.; Bonomi, M.; Harries, W.; Sali, A.; et al. Crystal structure of a eukaryotic phosphate transporter. *Nature* **2013**, *496*, 533–536. [CrossRef] [PubMed]

30. Weyand, S.; Shimamura, T.; Beckstein, O.; Sansom, M.; Iwata, S.; Henderson, P.; Cameron, A. The alternating access mechanism of transport as observed in the sodium-hydantoin transporter Mhp1. *J. Synchrotron Radiat.* **2011**, *18*, 20–23. [CrossRef]
31. Shimamura, T.; Weyand, S.; Beckstein, O.; Rutherford, N.; Hadden, J.; Sharples, D.; Sansom, M.; Iwata, S.; Henderson, P.; Cameron, A. Molecular Basis of Alternating Access Membrane Transport by the Sodium-Hydantoin Transporter Mhp1. *Science* **2010**, *328*, 470–473. [CrossRef]
32. Nomura, N.; Verdon, G.; Kang, H.J.; Shimamura, T.; Nomura, Y.; Sonoda, Y.; Hussien, S.A.; Qureshi, A.A.; Coincon, M.; Sato, Y.; et al. Structure and mechanism of the mammalian fructose transporter GLUT5. *Nature* **2015**, *526*, 397–401. [CrossRef] [PubMed]
33. Kanai, Y.; Clémençon, B.; Simonin, A.; Leuenberger, M.; Lochner, M.; Weisstanner, M.; Hediger, M.A. The SLC1 high-affinity glutamate and neutral amino acid transporter family. *Mol. Asp. Med.* **2013**, *34*, 108–120. [CrossRef]
34. Singh, S.K.; Yamashita, A.; Gouaux, E. Antidepressant binding site in a bacterial homologue of neurotransmitter transporters. *Nature* **2007**, *448*, 952–956. [CrossRef]
35. Shechter, E. Secondary Active Transport. *Biochimie* **1986**, *86*, 357–365. [CrossRef]
36. Quistgaard, E.; Löw, C.; Guettou, F.; Norlund, P. Understanding transport by the major facilitator superfamily (MFS): Structures pave the way. *Nat. Rev. Mol. Cell Biol.* **2016**, *17*, 123–132. [CrossRef]
37. Yan, N. Structural Advances for the Major Facilitator Superfamily (MFS) Transporters. *Trends Biochem. Sci.* **2013**, *38*, 151–159. [CrossRef]
38. Lee, J.; Sands, Z.; Biggin, P. A Numbering System for MFS Transporter Proteins. *Front. Mol. Biosci.* **2016**, *3*, 21. [CrossRef] [PubMed]
39. Teruhisa, H.; Jürgen, H.; Peter, M.; Sriram, S. Structural Model for 12-Helix Transporters Belonging to the Major Facilitator Superfamily. *J. Bacteriol.* **2003**, *185*, 1712–1718. [CrossRef]
40. Madej M.; Kaback, H. Evolutionary Mix-and-Match with MFS Transporters II. *Proc. Natl. Acad. Sci. USA* **2013**, *110*, E4831–E4838. [CrossRef]
41. Mittal, S.; Dutta, S.; Shukla, D. Reconciling membrane protein simulations with experimental DEER spectroscopy data. *Phys. Chem. Chem. Phys.* **2023**, *25*, 6253–6262. [CrossRef] [PubMed]
42. Fowler, P.W.; Orwick-Rydmark, M.; Radestock, S.; Solcan, N.; Dijkman, P.M.; Lyons, J.A.; Kwok, J.; Caffrey, M.; Watts, A.; Forrest, L.R.; et al. Gating Topology of the Proton-Coupled Oligopeptide Symporters. *Structure* **2015**, *23*, 290–301. [CrossRef] [PubMed]
43. Reddy, V.; Maksim, A.; Castillo, R.; Sun, E.; Saier, M. The Major Facilitator Superfamily (MFS) Revisited. *Fed. Eur. Biochem. Soc. J.* **2012**, *279*, 2022–2035. [CrossRef]
44. Saier, M.; Beatty, J.; Goffeau, A.; Harley, K.; Heijne, W.; Jack, D.; Jähn, P.; Lew, K.; Liu, J.; Pao, S.; et al. The Major Facilitator Superfamily. *J. Mol. Microbiol. Biotechnol.* **1999**, *1*, 257–279. [PubMed]
45. Abramson, J.; Smirnova, I.; Vladimir, K.; Verner, G.; Kaback, R.; Iwata, S. Structure and Mechanism of the Lactose Permease of Escherichia coli. *Science* **2003**, *301*, 610–615. [CrossRef]
46. Heymann, J.; Sarker, R.; Teruhisa, H.; Shi, D.; Milne, J.; Maloney, P.; Subramaniam, S. Projection Structure and Molecular Architecture of OxlT, a Bacterial Membrane Transporter. *Eur. Mol. Biol. Organ. J.* **2001**, *20*, 4409–4413. [CrossRef]
47. Hacksell, I.; Rigaud, J.; Purhonen, P.; Pourcher, T.; Hebert, H.; Leblanc, G. Projection Structure at 8 Å Resolution of the Melibiose Permease, an Na-sugar Co-transporter from Escherichia coli. *Eur. Mol. Biol. Organ. J.* **2002**, *21*, 3569–3574. [CrossRef] [PubMed]
48. Mitchell, P. A General Theory of Membrane Transport from Studies of Bacteria. *Nature* **1957**, *180*, 134–136. [CrossRef] [PubMed]
49. Durana, A.; Meilera, J. Inverted Topologies in Membrane Proteins: A Mini-Review. *Comput. Struct. Biotechnol. J.* **2013**, *8*, e201308004. [CrossRef] [PubMed]
50. Radestock, S.; Forrest, L. The Alternating-Access Mechanism of MFS Transporters Arises from Inverted-Topology Repeats. *J. Mol. Biol.* **2011**, *407*, 698–715. [CrossRef]
51. Forrest, L. Structural Symmetry in Membrane Proteins. *Annu. Rev. Biophys.* **2015**, *44*, 311–337. [CrossRef]
52. Ubarretxena-Belandia, I.; Baldwin, J.; Schuldiner, S.; Tate, C. Three-Dimensional Structure of the Bacterial Multidrug Transporter EmrE Shows it is an Asymmetric Homodimer. *Eur. Mol. Biol. Organ. J.* **2003**, *22*, 6175–6181. [CrossRef]
53. Selvam, B.; Mittal, S.; Shukla, D. Free Energy Landscape of the Complete Transport Cycle in a Key Bacterial Transporter. *Am. Chem. Soc. Cent. Sci.* **2018**, *4*, 1146–1154. [CrossRef]
54. Mittal, S.; Shukla, D. Predicting Optimal DEER Label Positions to Study Protein Conformational Heterogeneity. *J. Phys. Chem.* **2017**, *121*, 9761–9770. [CrossRef]
55. Solcan, N.; Kwok, J.; Fowler, P.W.; Cameron, A.D.; Drew, D.; Iwata, S.; Newstead, S. Alternating access mechanism in the POT family of oligopeptide transporters. *Eur. Mol. Biol. Organ. J.* **2012**, *31*, 3411–3421. [CrossRef]
56. Mitchell, P. Osmochemistry of solute translocation. *Res. Microbiol.* **1990**, *141*, 286–289. [CrossRef]
57. Patlak, C. Contributions to the theory of active transport: II. The gate type non-carrier mechanism and generalizations concerning tracer flow, efficiency, and measurement of energy expenditure. *Bull. Math. Biophys.* **1957**, *19*, 209–235. [CrossRef]
58. Vidaver, G. Inhibition of parallel flux and augmentation of counter flux shown by transport models not involving a mobile carrier. *J. Theor. Biol.* **1966**, *10*, 301–306. [CrossRef] [PubMed]
59. Jardetzky, O. Simple Allosteric Model for Membrane Pumps. *Nature* **1966**, *211*, 969–970. [CrossRef]
60. Renae, R.; Robert, V. Elevating the alternating-access model. *Nat. Struct. Mol. Biol.* **2016**, *23*, 187–189. [CrossRef]
61. Forrest, L.; Rudnick, G. The Rocking Bundle: A Mechanism for Ion-Coupled Solute Flux by Symmetrical Transporters. *Physiology* **2009**, *24*, 377–386. [CrossRef] [PubMed]

62. Alisa, G.; Dirk, S. Elevator-type mechanisms of membrane transport. *Viochem. Soc. Trans.* **2020**, *48*, 1227–1241. [CrossRef]
63. Lee, C.; Kang, H.J.; von Ballmoos, C.; Newstead, S.; Uzdavinys, P.; Dotson, D.; Iwata, S.; Beckstein, O.; Cameron, A.; Drew, D. A two-domain elevator mechanism for sodium/proton antiport. *Nature* **2013**, *501*, 573–577. [CrossRef]
64. Ogden, D.; Immadisetty, K.; Moradi, M. Conformational Transition Pathways in Major Facilitator Superfamily Transporters. *bioRxiv* **2019**. [CrossRef]
65. Lukas, S.; Philip, F.; Mark, S.; Oliver, B. Flexible Gates Generate Occluded Intermediates in the Transport Cycle of LacY. *J. Mol. Biol.* **2014**, *426*, 735–751. [CrossRef]
66. Kumar, S.; Nussinov, R. Salt bridge stability in monomeric proteins. *J. Mol. Biol.* **1999**, *293*, 1241–1255. [CrossRef]
67. Kumar, S.; Nussinov, R. Close-Range Electrostatic Interactions in Proteins. *ChemBioChem* **2002**, *3*, 604–617. [CrossRef] [PubMed]
68. Bosshard, H.R.; Marti, D.; Jelesarov, I. Protein stabilization by salt bridges: Concepts, experimental approaches and clarification of some misunderstandings. *J. Mol. Recognit.* **2004**, *17*, 1–16. [CrossRef] [PubMed]
69. Donald, J.; Kulp, D.; DeGrado, W. Salt bridges: Geometrically specific, designable interactions. *Proteins* **2010**, *79*, 898–915. [CrossRef]
70. Pylaeva, S.; Brehm, M.; Sebastiani, D. Salt Bridge in Aqueous Solution: Strong Structural Motifs but Weak Enthalpic Effect. *Sci. Rep.* **2018**, *8*, 13626. [CrossRef]
71. Fox, J.; Kang, K.; Sherman, W.; Héroux, A.; Sastry, G.M.; Baghbanzadeh, M.; Lockett, M.; Whitesides, G. Interactions between Hofmeister Anions and the Binding Pocket of a Protein. *J. Am. Chem. Soc.* **2015**, *137*, 3859–3866. [CrossRef]
72. Smith, M.D.; Cruz, L. Effect of Ionic Aqueous Environments on the Structure and Dynamics of the Aβ21–30 Fragment: A Molecular-Dynamics Study. *J. Phys. Chem.* **2013**, *117*, 6614–6624. [CrossRef]
73. Petrauskas, V.; Maximowitsch, E.; Matulis, D. Thermodynamics of Ion Pair Formations Between Charged Poly(Amino Acid)s. *J. Phys. Chem.* **2015**, *199*, 12164–12171. [CrossRef]
74. Law, C.; Yang, Q.; Soudant, C.; Maloney, P.; Wang, D.N. Kinetic Evidence Is Consistent with the Rocker-Switch Mechanism of Membrane Transport by GlpT. *Biochemistry* **2007**, *46*, 12190–12197. [CrossRef]
75. Zhang, X.; Zhao, Y.; Heng, J.; Jiang, D. Energy coupling mechanisms of MFS transporters. *Protein Sci.* **2015**, *24*, 1560–1579. [CrossRef] [PubMed]
76. Newstead, S.; Drew, D.; Cameron, A.D.; Postis, V.L.G.; Xia, X.; Fowler, P.W.; Ingram, J.C.; Carpenter, E.P.; Sansom, M.S.P.; McPherson, M.J.; et al. Crystal structure of a prokaryotic homologue of the mammalian oligopeptide–proton symporters, PepT1 and PepT2. *Eur. Mol. Biol. Organ. J.* **2011**, *30*, 417–426. [CrossRef]
77. Immadisetty, K.; Polasa, A.; Shelton, R.; Moradi, M. Elucidating the molecular basis of spontaneous activation in an engineered mechanosensitive channel. *Comput. Struct. Biotechnol. J.* **2022**, *20*, 2539–2550. [CrossRef] [PubMed]
78. Abramson, J.; Iwata, S.; Kaback, H.R. Lactose permease as a paradigm for membrane transport proteins (Review). *Mol. Membr. Biol.* **2004**, *21*, 227–236. [CrossRef] [PubMed]
79. Guellec, J.; Elbahnsi, A.; Le Tertre, M.; Uguen, K.; Gourlaouen, I.; Férec, C.; Ka, C.; Callebaut, I.; Le Gac, G. Molecular model of the ferroportin intracellular gate and implications for the human iron transport cycle and hemochromatosis type 4A. *Fed. Am. Soc. Exp. Biol.* **2019**, *33*, 14625–14635. [CrossRef]
80. Tortosa, V.; Bonaccorsi di Patti, C.; Iacovelli, F.; Pasquadibisceglie, A.; Falconi, M.; Musci, G.; Polticelli, F. Dynamical Behavior of the Human Ferroportin Homologue from Bdellovibrio bacteriovorus: Insight into the Ligand Recognition Mechanism. *Int. J. Mol. Sci.* **2020**, *21*, 6785. [CrossRef]
81. Mohan, S.; Sheena, A.; Poulose, N.; Anilkuma, G. Molecular Dynamics Simulation Studies of GLUT4: Substrate-Free and Substrate-Induced Dynamics and ATP-Mediated Glucose Transport Inhibition. *PLoS ONE* **2010**, *5*, e14217. [CrossRef] [PubMed]
82. Dastvan, R.; Rasouli, A.; Dehghani-Ghahnaviyeh, S.; Gies, S.; Tajkhorshid, E. Proton-driven alternating access in a spinster lipid transporter. *Nat. Commun.* **2022**, *13*, 5161. [CrossRef]
83. Ethayathulla, A.; Yousef, M.; Amin, A.; Leblanc, G.; Kaback, R.; Guan, L. Structure-based mechanism for Na+/melibiose symport by MelB. *Nat. Commun.* **2014**, *5*, 3009. [CrossRef]
84. Fukuda, M.; Takeda, H.; Kato, H.; Doki, S.; Ito, K.; Maturana, A.; Ishitani, R.; Nureki, O. Structural basis for dynamic mechanism of nitrate/nitrite antiport by NarK. *Nat. Commun.* **2015**, *6*, 7097. [CrossRef] [PubMed]
85. Lawrence, M.; Lindahl, L.; Zengel, J. Effects on translation pausing of alterations in protein and RNA components of the ribosome exit tunnel. *J. Bacteriol.* **2008**, *190*, 5862–5869. [CrossRef]
86. Elvin, C.M.; Hardy, C.M.; Rosenberg, H. Pi exchange mediated by the GlpT-dependent sn-glycerol-3-phosphate transport system in *Escherichia coli*. *J. Bacteriol.* **1985**, *161*, 1054–1058. [CrossRef] [PubMed]
87. Lemieux, J.; Huang, Y.; Weng, D. Glycerol-3-phosphate transporter of Escherichia coli: Structure, function and regulation. *Res. Microbiol.* **2004**, *188*, 623–629. [CrossRef]
88. Law, C.; Almqvist, J.; Berstein, A.; Goetz, R.; Huang, Y.; Soudant, C.; Laaksonen, A.; Hovmöller, S.; Wang, D. Salt-bridge Dynamics Control Substrate-induced Conformational Change in the Membrane Transporter GlpT. *J. Mol. Biol.* **2008**, *378*, 828–839. [CrossRef]
89. Lemieux, J. Eukaryotic major facilitator superfamily transporter modeling based on the prokaryotic GlpT crystal structure (Review). *Mol. Membr. Biol.* **2007**, *24*, 333–341. [CrossRef] [PubMed]
90. Iyalomhe, O.; Khantwal, C.; Kang, D.C. The Structure and Function of OxlT, the Oxalate Transporter of Oxalobacter formigenes. *J. Membr. Biol.* **2015**, *248*, 641–650. [CrossRef]

91. Forrest, L.R.; Zhang, Y.W.; Jacobs, M.T.; Gesmonde, J.; Xie, L.; Honig, B.H.; Rudnick, G. Mechanism for alternating access in neurotransmitter transporters. *Proc. Natl. Acad. Sci. USA* **2008**, *105*, 10338–10343. [CrossRef]
92. Adhikary, S.; Deredge, D.J.; Nagarajan, A.; Forrest, L.R.; Wintrode, P.L.; Singh, S.K. Conformational dynamics of a neurotransmitter:sodium symporter in a lipid bilayer. *Proc. Natl. Acad. Sci. USA* **2017**, *114*, E1786–E1795. [CrossRef] [PubMed]
93. Deng, D.; Xu, C.; Sun, P.; Wu, J.; Yan, C.; Hu, M.; Yan, N. Crystal structure of the human glucose transporter GLUT1. *Nature* **2014**, *510*, 121–125. [CrossRef] [PubMed]
94. Weyand, S.; Shimamura, T.; Yajima, S.; Suzuki, S.; Mirza, O.; Krusong, K.; Carpenter, E.P.; Rutherford, N.G.; Hadden, J.M.; O'Reilly, J.; et al. Structure and Molecular Mechanism of a Nucleobase–Cation–Symport-1 Family Transporter. *Science* **2008**, *322*, 709–713. [CrossRef]
95. Lu, F.; Li, S.; Jiang, Y.; Jiang, J.; Fan, H.; Lu, G.; Deng, D.; Dang, S.; Zhang, X.; Wang, J.; et al. Structure and mechanism of the uracil transporter UraA. *Nature* **2011**, *472*, 243–246. [CrossRef] [PubMed]
96. Krishnamurthy, H.; Gouaux, E. X-ray Structures of LeuT in Substrate-Free Outward-Open and apo Inward-Open States. *Nature* **2012**, *481*, 469–474. [CrossRef]
97. Galochkina, T.; Ng Fuk Chong, M.; Challali, L.; Abbar, S.; Etchebest, C. New insights into GluT1 mechanics during glucose transfer. *Sci. Rep.* **2019**, *9*, 998. [CrossRef]
98. Chen, X.Z.; Zhu, T.; Smith, D.E.; Hediger, M.A. Stoichiometry and Kinetics of the High-affinity H+-coupled Peptide Transporter PepT2. *J. Biol. Chem.* **1999**, *274*, 2773–2779. [CrossRef]
99. Agu, R.; Cowley, E.; Shao, D.; MacDonald, C.; Kirkpatrick, D.; Renton, K.; Massoud, E. Proton-Coupled Oligopeptide Transporter (POT) Family Expression in Human Nasal Epithelium and Their Drug Transport Potential. *Mol. Pharm.* **2011**, *8*, 664–672. [CrossRef] [PubMed]
100. Doki, S.; Kato, H.E.; Solcan, N.; Iwaki, M.; Koyama, M.; Hattori, M.; Iwase, N.; Tsukazaki, T.; Sugita, Y.; Kandori, H.; et al. Structural basis for dynamic mechanism of proton-coupled symport by the peptide transporter POT. *Proc. Natl. Acad. Sci. USA* **2013**, *110*, 11343–11348. [CrossRef]
101. Ganapathy, V.; Leibach, F.H. Role of pH gradient and membrane potential in dipeptide transport in intestinal and renal brush-border membrane vesicles from the rabbit. Studies with L-carnosine and glycyl-L-proline. *J. Biol. Chem.* **1983**, *258*, 14189–14192. [CrossRef] [PubMed]
102. Daniel, H.; Spanier, B.; Kottra, G.; Weitz, D. From Bacteria to Man: Archaic Proton-Dependent Peptide Transporters at Work. *Physiology* **2006**, *21*, 93–102. [CrossRef]
103. Fei, Y.J.; Kanai, Y.; Nussberger, S.; Ganapathy, V.; Leibach, F.H.; Romero, M.F.; Singh, S.K.; Boron, W.F.; Hediger, M.A. Expression cloning of a mammalian proton-coupled oligopeptide transporter. *Nature* **1994**, *368*, 563–566. [CrossRef] [PubMed]
104. Covitz, K.M.Y.; Amidon, G.L.; Sadée, W. Membrane Topology of the Human Dipeptide Transporter, hPEPT1, Determined by Epitope Insertions. *Biochemistry* **1998**, *37*, 15214–15221. [CrossRef] [PubMed]
105. Yin, Y.; He, X.; Szewczyk, P.; Nguyen, T.; Chang, G. Structure of the Multidrug Transporter EmrD from *Escherichia coli*. *Science* **2006**, *312*, 741–744. [CrossRef] [PubMed]
106. Daniel, H.; Kottra, G. The proton oligopeptide cotransporter family SLC15 in physiology and pharmacology. *Pflüg. Arch. Eur. J. Physiol.* **2004**, *447*, 610–618. [CrossRef]
107. Weitz, D.; Harder, D.; Casagrande, F.; Fotiadis, D.; Obrdlik, P.; Kelety, B.; Daniel, H. Functional and Structural Characterization of a Prokaryotic Peptide Transporter with Features Similar to Mammalian PEPT1. *J. Biol. Chem.* **2007**, *282*, 2832–2839. [CrossRef] [PubMed]
108. Killer, M.; Wald, J.; Pieprzyk, J.; Marlovits, T.C.; Löw, C. Structural snapshots of human PepT1 and PepT2 reveal mechanistic insights into substrate and drug transport across epithelial membranes. *Sci. Adv.* **2021**, *7*, eabk3259. [CrossRef]
109. Coleman, J.A.; Green, E.M.; Gouaux, E. X-ray structures and mechanism of the human serotonin transporter. *Nature* **2016**, *532*, 334–339. [CrossRef] [PubMed]
110. Wang, H.; Elferich, J.; Gouaux, E. Structures of LeuT in bicelles define conformation and substrate binding in a membrane-like context. *Nat. Struct. Mol. Biol.* **2012**, *19*, 212–219. [CrossRef] [PubMed]
111. Grewer, C.; Watzke, N.; Wiessner, M.; Rauen, T. Glutamate translocation of the neuronal glutamate transporter EAAC1 occurs within milliseconds. *Proc. Natl. Acad. Sci. USA* **2000**, *97*, 9706–9711. [CrossRef] [PubMed]
112. Kaback, H.R.; Sahin-Tóth, M.; Weinglass, A.B. The kamikaze approach to membrane transport. *Nat. Rev. Mol. Cell Biol.* **2001**, *2*, 610–620. [CrossRef]
113. Oldham, M.L.; Khare, D.; Quiocho, F.A.; Davidson, A.L.; Chen, J. Crystal structure of a catalytic intermediate of the maltose transporter. *Nature* **2007**, *450*, 515–521. [CrossRef]
114. Wadiche, J.I.; Amara, S.G.; Kavanaugh, M.P. Ion fluxes associated with excitatory amino acid transport. *Neuron* **1995**, *15*, 721–728. [CrossRef]
115. Singh, S.K.; Piscitelli, C.L.; Yamashita, A.; Gouaux, E. A Competitive Inhibitor Traps LeuT in an Open-to-Out Conformation. *Science* **2008**, *322*, 1655–1661. [CrossRef]
116. Yan, R.; Zhao, X.; Lei, J.; Zhou, Q. Structure of the human LAT1–4F2hc heteromeric amino acid transporter complex. *Nature* **2019**, *568*, 127–130. [CrossRef] [PubMed]
117. Johnson, Z.L.; Cheong, C.G.; Lee, S.Y. Crystal structure of a concentrative nucleoside transporter from Vibrio cholerae at 2.4 Å. *Nature* **2012**, *483*, 489–493. [CrossRef] [PubMed]

118. Xuejun, Z.; Lei, H. Uniporter substrate binding and transport: Reformulating mechanistic questions. *Biophys. Rep.* **2016**, *2*, 45–54. [CrossRef]
119. Alberts, B.; Johnson, A.; Lewis, J.; Raff, M.; Roberts, K.; Walter, P. *Molecular Biology of the Cell*; Garland Science: New York, NY, USA, 2002; Volume 4.
120. Simpson, I.; Vannucci, S.; Maher, F. Glucose Transporters in Mammalian Brain. *Biochem. Soc. Trans.* **1994**, *22*, 671–675. [CrossRef] [PubMed]
121. Vollers, S.; Carruthers, A. Sequence Determinants of GLUT1-Mediated Accelerated-Exchange Transport. *J. Biol. Chem.* **2012**, *287*, 42533–42544. [CrossRef] [PubMed]
122. Longo, N.; Griffin, L.; Elsas, L. Influx and Efflux of 3-O-methyl-D-glucose by Cultured Human Fibroblasts. *Am. J. Physiol.* **1988**, *254*, 628–633. [CrossRef]
123. Sun, L.; Zeng, X.; Yan, C.; Sun, X.; Gong, X.; Rao, Y.; Yan, N. Crystal structure of a bacterial homologue of glucose transporters GLUT1–4. *Nature* **2012**, *490*, 361–366. [CrossRef]
124. Qureshi, A.; Suades, A.; Matsuoka, R.; Brock, J.; McComas, S.; Nji, E.; Orellana, L.; Claesson, M.; Delemotte, L.; Drew, D. The Molecular Basis for Sugar Import in Malaria Parasites. *Nature* **2020**, *578*, 321–325. [CrossRef]
125. Deng, D.; Sun, P.; Yan, C.; Ke, M.; Jiang, X.; Xiong, L.; Ren, W.; Hirata, K.; Yamamoto, M.; Fan, S.; et al. Molecular Basis of Ligand Recognition and Transport by Glucose Transporters. *Nature* **2015**, *526*, 391–396. [CrossRef]
126. Kapoor, K.; Finer-Moore, J.S.; Pedersen, B.P.; Caboni, L.; Waight, A.; Hillig, R.C.; Bringmann, P.; Heisler, I.; Müller, T.; Siebeneicher, H.; et al. Mechanism of inhibition of human glucose transporter GLUT1 is conserved between cytochalasin B and phenylalanine amides. *Proc. Natl. Acad. Sci. USA* **2016**, *113*, 4711–4716. [CrossRef] [PubMed]
127. Salas-Burgos, A.; Iserovich, P.; Zuniga, F.; Vera, J.C.; Fischbarg, J. Predicting the Three-Dimensional Structure of the Human Facilitative Glucose Transporter Glut1 by a Novel Evolutionary Homology Strategy: Insights on the Molecular Mechanism of Substrate Migration, and Binding Sites for Glucose and Inhibitory Molecules. *Biophys. J.* **2004**, *87*, 2990–2999. [CrossRef] [PubMed]
128. Deng, D.; Yan, N. GLUT, SGLT, and SWEET: Structural and mechanistic investigations of the glucose transporters. *Protein Sci.* **2016**, *25*, 546–558. [CrossRef] [PubMed]
129. Yaffe, D.; Forrest, L.; Schuldiner, S. The Ins and Outs of Vesicular Monoamine Transporters. *J. Gen. Physiol.* **2018**, *150*, 671–682. [CrossRef] [PubMed]

membranes

MDPI

Article

Properties and Crystal Structure of the *Cereibacter sphaeroides* Photosynthetic Reaction Center with Double Amino Acid Substitution I(L177)H + F(M197)H

Tatiana Yu. Fufina [1], Georgii K. Selikhanov [1,2], Azat G. Gabdulkhakov [2] and Lyudmila G. Vasilieva [1,*]

1 Federal Research Center Pushchino Scientific Center for Biological Research PSCBR, Institute of Basic Biological Problems, Russian Academy of Sciences, Institutskaya Street 2, 142290 Pushchino, Russia
2 Institute of Protein Research, Russian Academy of Sciences, Institutskaya Street 4, 142290 Pushchino, Russia
* Correspondence: vsyulya@mail.ru; Tel.: +7-(4967)-732-680

Abstract: The photosynthetic reaction center of the purple bacterium *Cereibacter sphaeroides* with two site-directed mutations Ile-L177–His and M197 Phe–His is of double interest. The substitution I(L177)H results in strong binding of a bacteriochlorophyll molecule with L-subunit. The second mutation F(M197)H introduces a new H-bond between the C2-acetyl carbonyl group of the bacteriochlorophyll P_B and His-M197, which is known to enhance the stability of the complex. Due to this H-bond, π -electron system of P finds itself connected to an extensive H-bonding network on the periplasmic surface of the complex. The crystal structure of the double mutant reaction center obtained with 2.6 Å resolution allows clarifying consequences of the Ile L177 – His substitution. The value of the P/P^+ midpoint potential in the double mutant RC was found to be ~20 mV less than the sum of potentials measured in the two RCs with single mutations I(L177)H and F(M197)H. The protein environment of the BChls P_A and B_B were found to be similar to that in the RC with single substitution I(L177)H, whereas an altered pattern of the H-bonding networks was found in the vicinity of bacteriochlorophyll P_B. The data obtained are consistent with our previous assumption on a correlation between the bulk of the H-bonding network connected with the π-electron system of the primary electron donor P and the value of its oxidation potential.

Keywords: purple bacteria; membrane pigment-protein complex; photosynthetic reaction center; site-directed mutagenesis; crystal structure

1. Introduction

Photosynthetic reaction centers are integral membrane complexes responsible for the extremely efficient conversion of light energy into the chemical energy of separated charges. The crystal structure of the reaction center (RC) of the purple nonsulfur bacterium *Cereibacter* (*C.*) *sphaeroides* shows that this complex consists of three subunits and ten cofactors (Figure 1A) [1]. The L and M protein subunits and cofactors are related to each other by an approximate two-fold symmetry axis [2]. Cofactors are represented by four bacteriochlorophylls (BChl), two bacteriopheophytins (BPeo), two quinones, one carotenoid molecule, and a non-heme iron atom and are arranged in two membrane-spanning branches, A and B. Only one, A branch, is active in photosynthetic electron transfer. Two BChls, P_A and P_B, are combined into a dimer P that serves as a primary electron donor. An important property of P is the midpoint potential (E_m) P/P^+ which is related to the energy level of P* after light excitation. In the wild-type RCs of purple bacteria, cofactors interact with integral L- and M-subunits through relatively weak non-covalent contacts. Figure 1B shows four BChl molecules in the *C. sphaeroides* RC and some amino acid residues from their environment. Histidine residues, L173 and M202, coordinate central Mg^{2+} atoms of BChls P_A and P_B, respectively, and His-M182 is the ligand of BChl B_B. In native *C. sphaeroides* RC, the C2-acetyl carbonyl group of the BChl P_A forms an

H-bond with His-L168. The C2-acetyl carbonyl group of BChl P_B is free of H-bonds with its environment since the residue closest to this group and symmetry related to His-L168 is Phe-M197 [3]. Crystal structures of photosynthetic RCs also contain a considerable amount of bound water molecules with a few of them located in the hydrophobic core of the complexes [4] (not shown, discussed below).

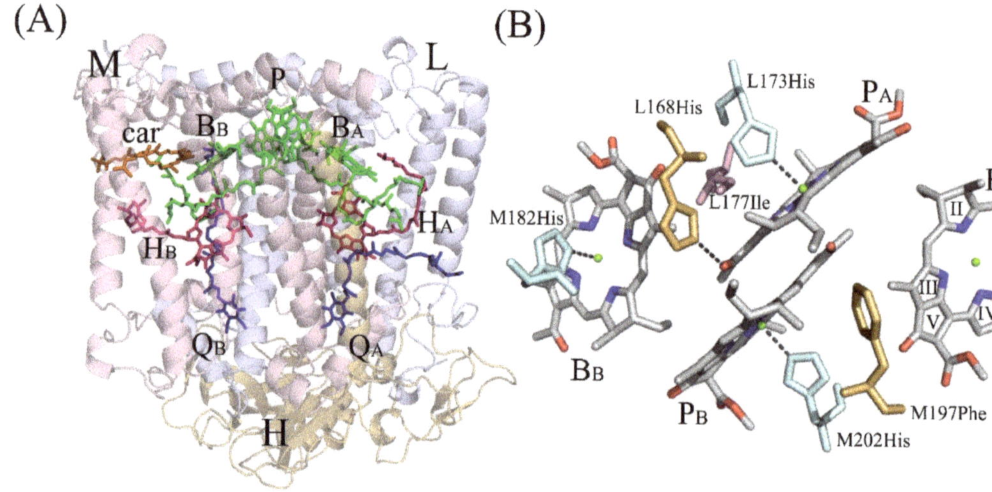

Figure 1. (**A**)–The overall structure of the reaction center of *C. sphaeroides* (PDB ID: 3v3y) and (**B**)-protein environment of bacteriochlorophylls. L, M, and H–correspondent RC subunits; P-BChl dimer of P_A and P_B; B_A and B_B-monomeric BChls; H_A and H_B - monomeric BPheos; Q_A and Q_B-ubiquinones; car-carotenoid. Shown in pale-blue: His-M182, His-L173, and His-M202- ligands of BChls B_B, P_A, and P_B, correspondently. Symmetry related residues His-L168 and Phe-M197 are shown in pale-yellow, Ile-L177 is shown in light-pink. Pyrrole rings numbers are shown on the structure of BChl B_A.

Previously, it was shown that substitution of Ile-L177 for His in the RC of *C. sphaeroides* near BChl B_B and P_A resulted in unexpectedly strong binding between one of these chromophores and L-subunit, which was stable during SDS-PAGE under a variety of denaturing conditions [5]. The appearance of a new band at 638 nm in the absorption spectrum of the I(L177)H mutant RC indicated that the central Mg^{2+} atom of one out of four BChls became hexa-coordinated, while the other BChls remained penta-coordinated [6]. Treatment of the isolated RC with sodium borohydride revealed that the pigment strongly bound to the protein is BChl P_A [7]. Later, the crystal structure of RC I(L177)H was obtained with a relatively low resolution of 2.9 Å [6]. The difference electron density map showed low sigma peaks adjacent to the central Mg^{2+} atom of BChl B_B, indicating the potential presence of a water molecule. Computer modeling proposed the involvement of two water molecules in the β-coordination of the BChl B_B, however, a higher resolution structure is needed for a conclusion on this issue. The question concerning the nature of the strong attachment of BChl P_A to the RC protein also remains open.

We suggested that decreased thermal stability of the I(L177)H mutant RC structure, shown in previous work [7], could be one of the possible reasons for the poor quality of the RC crystals and the resulting insufficient resolution of the crystal structure [6]. It is known that H-bond formation between the protein environment and the C2-acetyl carbonyl groups of the BChl dimer contributes to the strengthening of the local structure of the *C. sphaeroides* RC, increasing its resistance to elevated temperatures and pressure [8,9]. Based on the recently resolved crystal structure of the RC F(M197)H, it was found that the increased stability of this complex is associated with the incorporation of the C2-acetyl carbonyl group of BChl P_B into the H-bonding network surrounding this cofactor. It was also suggested that the size and pattern of the H-bonding networks near P can affect the oxidation potential of the BChl dimer [10]. However, more experimental data are needed to support this assumption. While the H-bonding networks associated with proton transfer and secondary quinone reduction are well characterized, the H-bonding clusters on the periplasmic side of the purple bacteria RCs are less studied and their role requires further investigation.

We have combined I(L177)H and F(M197)H mutations in *C. sphaeroides* RC to increase the stability of its structure and thereby increase the probability of obtaining the crystal structure of this complex with higher resolution. This work introduces the new double mutant RC and describes its spectral properties and oxidation potential of the primary electron donor P. X-ray crystal structure of the RC I(L177)H + F(M197)H, obtained with a resolution of 2.6 Å, clarifies some details of unusually strong pigment-protein interaction resulting from the Ile→His substitution at L177 position and shows that a new water molecule appears from β–side of BChl B_B, which is responsible for hexa-coordination of the central Mg^{2+} atom of this cofactor. In addition, the crystal structure provides data in support of the previously stated assumption about the correlation between the bulk of the H-bonding network in the vicinity of the BChl dimer and the value of the oxidation potential of the primary electron donor.

2. Experimental

2.1. Construction of the Mutant Strain

The I(L177)H and F(M197)H mutations previously introduced [11,12] were combined together in *puf*-operon by simple cloning using available unique restriction sites [11]. Polyhistidine tag (6His) was introduced from the 3′- end of the *puf*-M gene, similar to what was previously described [13]. The modified *puf*-operon with the mutations was cloned into the broad-host-range vector pRK415, containing DNA fragment that included the *puf*LMX genes. The resulting plasmid was transferred into *C. sphaeroides* strain DD13 [14] through conjugative crossing to give a recombinant strain with the RC-only phenotype. The control pseudo wild-type strain was similarly made using the DD13 strain, complemented by a wild-type copy of the *puf*LMX genes.

2.2. Bacterial Growth and Reaction Center Purification

The growth of recombinant bacterial strains under dark semi-aerobic conditions has been described elsewhere [11]. The cells were harvested and then broken by ultrasonication, and membranes were sedimented by ultracentrifugation. Reaction centers were solubilized using lauryldimethylamine oxide (LDAO) (Sigma-Aldrich, Missouri, USA) and then purified by passage through Ni-NTA agarose column (Serva, Heidelberg, Germany). After the purification of the RCs, the detergent was replaced as previously described [15]. The purified RCs were dissolved in 20 mM Tris-HCl buffer (pH 8.0) (Helicon, Moscow, Russia) containing 0.2% sodium cholate (Sigma-Aldrich, Missouri, USA) or 0.05 % Triton X-100 (Helicon, Moscow, Russia).

2.3. Thermal Stability, Adsorption Spectroscopy, and Redox Potential Measurements

Thermal stability of the RCs was studied in Tris-Triton buffer with 250 mM NaCl by monitoring thermo-dependent changes of the Q_Y band amplitude at 48°C as previously described [15]. Absorption spectra were recorded using a Shimadzu UV-1800 spectrophotometer (Shimadzu, Kyoto, Japan) at room temperature. Sodium ascorbate (1 mM) was added to samples to maintain the primary electron donor in a reduced state. The P/P^+ midpoint potential was determined by chemical titrations using potassium ferricyanide and sodium ascorbate as previously described [6].

2.4. Reaction Center Crystallization and Data Analysis

Trigonal crystals of the reaction center with the double mutation I(L177)H + F(M197)H, space group $P3_121$, were grown using the hanging drop vapor diffusion method as previously described in detail [6]. X-ray diffraction data were collected at the P11 beamline

PETRA III electron storage ring (Hamburg, Germany) using cryo-cooled crystals. The crystal used for data collection diffracted to a higher resolution limit of 2.6 Å. The diffraction data were processed and scaled using XDS [16]. Molecular replacement was performed with Phaser [17], using the coordinates of the I(L177)H mutant RC structure from *C. sphaeroides* (PDB ID: 3V3Z) [6] as the search model. Data collection and refinement statistics are given in Table 1. Rigid body refinement was carried out before refinement using restrained maximum likelihood refinement in REFMAC 5.0 [18]. Manual rebuilding of the model was carried out in Coot [19]. The final cycle with an occupancy refinement was performed in Phenix [20]. The coordinates and structure factors of double mutant RC have been deposited in the Protein Data Bank (PDB ID: 8C3F). Figures were prepared using the program PyMol [21].

Table 1. Data collection and processing.

Data Collection	
Diffraction source	PETRA III, beamline P11
Wavelength (Å)	1.03312
Temperature (K)	100
Detector	Pilatus 6M
Rotation range per image (°)	0.1
Total rotation range (°)	180
Space group	$P3_121$
a, b, c (Å)	139.3 139.3 184.6
Resolution range (Å)	50.00–2.60 (2.67–2.60)
Total No. of reflections	637 789 (43 719)
No. of unique reflections	64 208 (4 564)
Completeness (%)	99.7 (97.2)
Redundancy	9.9 (9.6)
$\langle I/\sigma(I) \rangle$	19.21 (1.04)
$R_{r.i.m.}$ ‡	10.7 (216.8)
$CC_{1/2}$	99.9 (49.0)
Structure solution and refinement	
Resolution range (Å)	50.00–2.60 (2.63–2.60)
Completeness (%)	99.9 (100.0)
No. of reflections, working set	64 207 (3 995)
No. of reflections, test set	5 901 (200)
R_{cryst}	19.43 (34.92)
R_{free}	21.98 (38.88)
R.m.s. deviations	
Bonds (Å)	0.008
Angles (°)	1.15
Average B factors (Å^2)	64.95
Protein	64.13

Table 1. *Cont.*

Data Collection	
Ligand	72.15
Water	59.05
Calculated DPI (Å)	0.17
Maximal estimated error (Å)	0.31
Ramachandran plot	
Most favoured (%)	96.57
Allowed (%)	3.43

3. Results

3.1. Absorption Spectra

The mutant RC with the double substitution I(L177)H + F(M197)H was stably assembled in photosynthetic membranes and the yield of the RC after purification was similar to that of the wild-type RC from the same amount of the cell mass. The absorption spectrum of the wild-type RC measured at room temperature shows the Q_Y P band at 865 nm, the Q_Y B band at 804 nm, and the Q_Y H band at 759 nm [22]. In the short wavelength region, the BChls and BPheos of the wild-type RC give rise to asymmetric absorbance bands with main peaks at 599 and 533 nm, respectively. Previously, it was shown that the absorption spectrum of RC F(M197)H was similar to the spectrum of the wild-type RC [23]. As described by [11], the room temperature absorption spectrum of RC I(L177)H showed a noticeable 18 nm blue shift and a diminished amplitude of the Q_Y P band near 847 nm, a diminished amplitude of the Q_Y B band, as well as decreased absorption in the Q_X region of BChl absorption near 600 nm. The absorption spectrum of RC I(L177)H + F(M197)H showed similarity to that of RC I(L177)H, except that the blue shift of the Q_Y P band was less pronounced and the maximum of this band was observed at 854 nm (Figure 2).

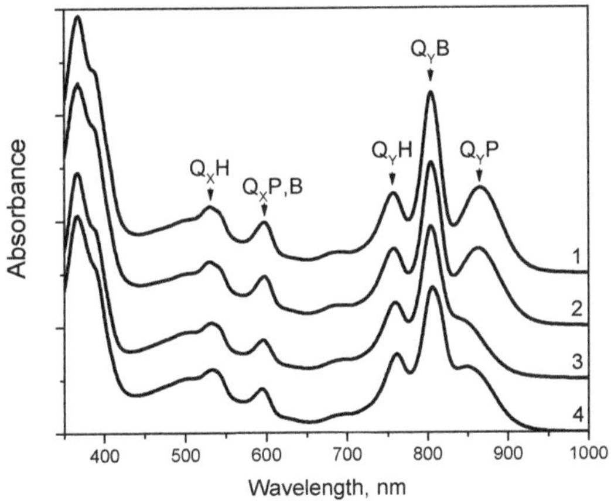

Figure 2. Electronic ground-state absorption spectra of *C. sphaeroides* reaction center from wild-type (1) and mutant strains F(M197)H (2), I(L177)H (3), and I(L177)H+F(M197)H (4). Spectra were measured at room temperature and normalized at Q_YH absorption band.

3.2. Thermal Stability

Thermal stability of the wild-type RC as well as mutant RCs I(L177)H, F(M197)H, and I(L177)H+F(M197)H was determined by recording thermo-dependent changes of the Q_Y B band amplitude at 48°C in Tris-Triton buffer. It was shown that after 60 min of incubation in the spectrum of the wild-type RC, the amplitude of the Q_Y B band at 804 nm decreased by 50%, whereas the spectra of the RCs I(L177)H, F(M197)H, and I(L177)H+F(M197)H demonstrated 55%, 40%, and 45% decreases, respectively (Figure 3). These data show that, as expected, the addition of the F(M197)H mutation increases the thermal stability of RC I(L177)H, however, it remains lower than that of the single mutant RC F(M197)H.

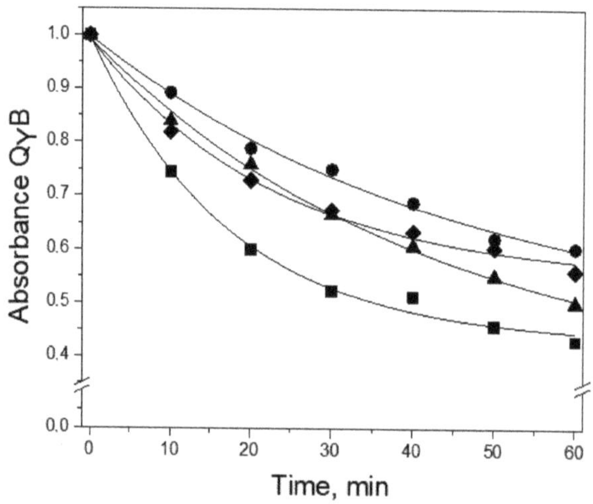

Figure 3. Changes in Q_YB absorption band amplitude in the spectra of the wild-type (▲), I(L177)H (■), F(M197)H (•), and I(L177)H+F(M197)H (♦) RCs during their incubation in Tris-Triton buffer at 48 °C.

3.3. *P/P⁺ Midpoint Potential*

The value of the P/P^+ midpoint potential (E_m P/P^+) of the double mutant RC was determined by redox titrations as 555 ± 10 mV (Figure S1, Supplementary). Previously, it was shown that the single mutation F(M197)H raises the value of E_m P/P^+ by 125 mV [24] due to the formation of an H-bond between the C2-acetyl carbonyl group of BChl P_B and the imidazole group of His-M197, whereas the substitution I(L177)H results in a decrease of the E_m P/P^+ value by 50 mV [6]. Assuming that in the double mutant RC each of the mutations causes the same structural changes as in each RC with single substitutions, one would expect the resulting E_m P/P^+ value to be ~575 mV. The crystal structure of RC I(L177)H + F(M197)H resolved in this study allows us to suggest a possible reason for the discrepancy between expected and experimentally observed mid-point potential P/P^+ (discussed below).

3.4. *Crystal Structure*

The crystal structure of the double mutant reaction center was obtained with a resolution of 2.6 Å and an estimated maximal coordinate error of 0.31 Å (Table 1). After refinement, the structural model of RC I(L177)H + F(M197)H was compared with the structures of the wild-type RC (PDB ID: 3V3Y), RC F(M197)H (PDB ID: 7OD5), and I(L177)H (PDB ID: 3V3Z). In the structure of RC I(L177)H + F(M197)H, the protein environment of BChls B_B and P_A is similar to that in the previously described structure of RC I(L177)H [3]. A new electron density appears from the β-side of B_B, which suggests the incorporation of a new water molecule (Figure 4A). The structure shows that the water is located at a distance of 2.7 Å from the central Mg^{2+} atom of B_B, which is typical for the BChl axial ligands. This water is also 3.1 Å close to the imidazole group of His-L177, which is within the typical range for H-bond formation (Figure 4).

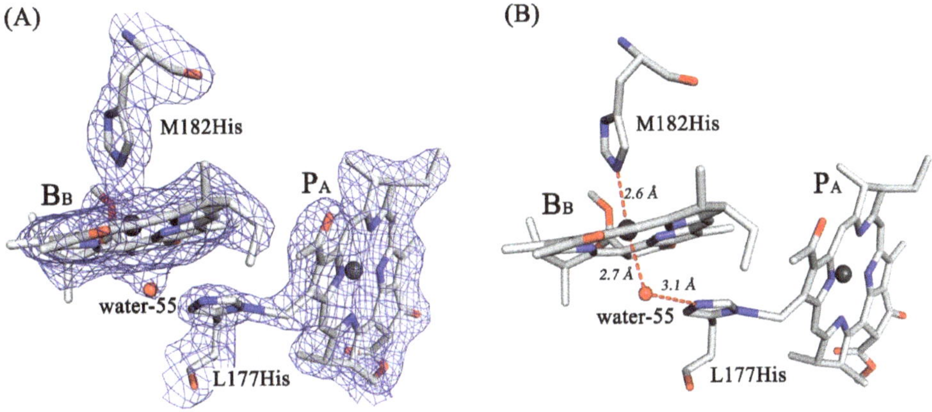

Figure 4. Electron density maps for RC I(L177)H + F(M197)H attributable to the P_A and B_B BChls as well as the His-L177 residue (**A**) and the structural model fitted to the density (**B**) (PDB ID: 8C3F).

The imidazole group of His-L177 points towards the methyl carbon of the pyrrole ring I of P_A with the center-to-center distance between N and C atoms of 2.06 Å. According to the electron density map, imidazole forms a covalent bond with the methyl carbon of P_A with a high degree of probability. Based on the structural data, we modeled pigment-protein interactions that occur as a result of Ile→His substitution at the L177 site (Figure 4B).

Comparison of the structures of the double mutant RC and RC F(M197)H in the proximity of BChl P_B shows structural conservation in the main body of the protein as well as in the cofactor system. In both complexes, the imidazole group of His-M197 is located within the H-bond distance from the C2-acetyl carbonyl group of BChl P_B. It was noted that the C2-acetyl group shows 29.7° out of plane rotation (Figure 5), which is more significant compared with that observed in RC F(M197)H (20.7°) [10]. The data are consistent with spectral properties of RC I(L177)H + F(M197)H (as discussed below).

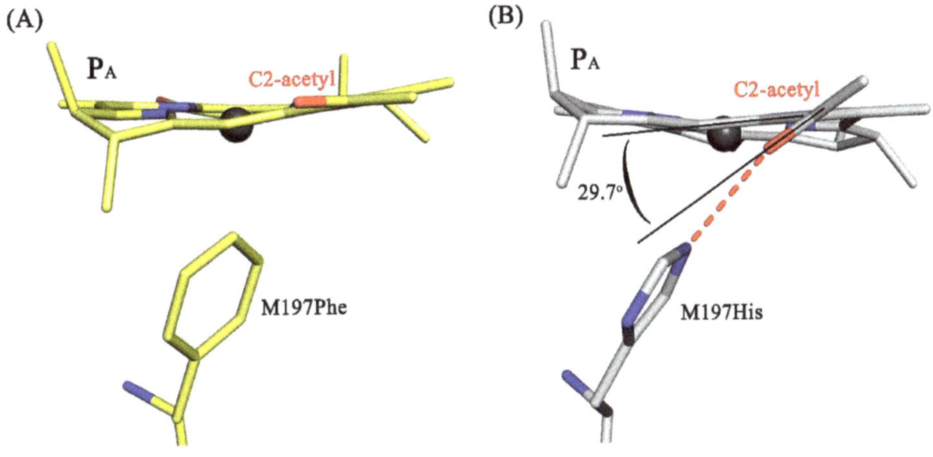

Figure 5. Position of the C2-acetyl group of BChl P_B in the structures of the wild-type RC (PDB ID: 3V3Y) (**A**) and RC I(L177)H + F(M197)H (PDB ID: 8C3F) (**B**). The view is along the BChl plane.

Analysis of the RC I(L177)H + F(M197)H structure in the vicinity of P_B (Figure 6) shows that the main pattern of the previously described H-bonding network, which includes the C2-acetyl carbonyl group of BChl P_B, the imidazole group of His-M197, side groups of Asn-M195, Ser-L158, Asp-L155, and Tyr-M198 and two water molecules (assigned as waters-C and - E in Figure 6 and [10]), is similar to that observed in the RC with the single substitution F(M197)H. The sigma level of the electron density map peak for the water-C is low (1.7) and not sufficient to assert its presence with confidence. However, this water is observed in the structures of the wild-type RC and RC F(M197)H [10] as part of their H-bonding networks. As shown in Figure 6, there is a high degree of probability that water-C is also present in the structure of the double mutant RC. The distances between the related side groups in the network are all within the range of 2.5– 3.03 Å consistent with H-bond formation (Figure 6). The only exception is the distance between water-C and the side group of Asp-L155, which exceeds 3.7 Å. Taking into account the coordinate error, it suggests that no H-bond forms between these groups in RC I(L177)H + F(M197)H. The absence of this H-bond is consistent with some shortening of the distance between water-C and the imidazole group of His-M197 (2.7 Å), compared with the structure of RC F(M197)H (3.1 Å) (Figure 6). In addition, Figure 6B shows a change in the mutual arrangement of the Tyr-M198 and Asp-L155 side groups, indicating that Asp-L155 is shifted away from water-C in the double mutant RC structure.

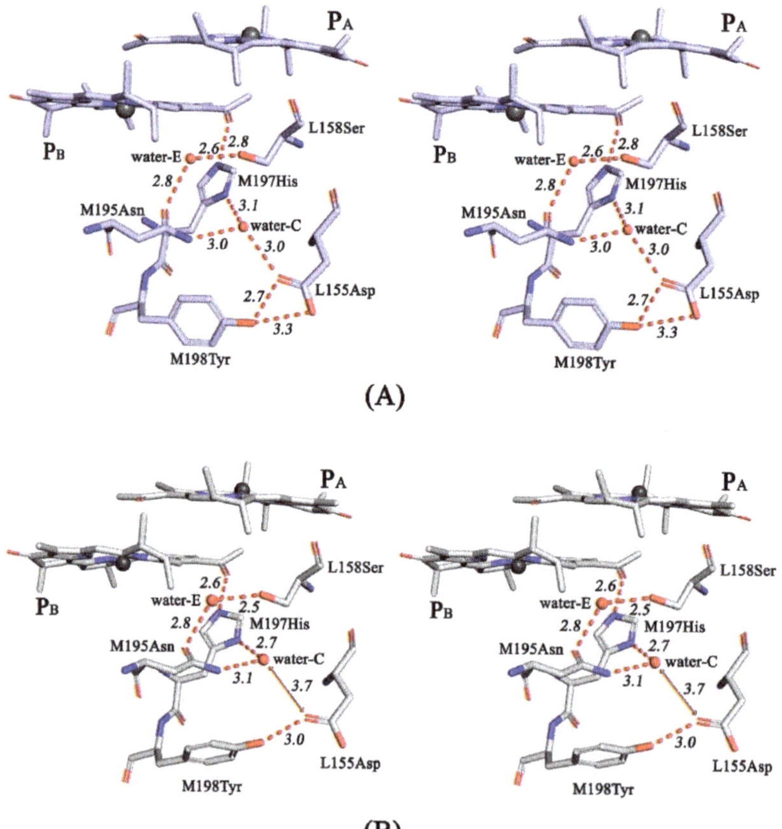

Figure 6. Stereoviews of the hydrogen-bonding networks in the proximity of the M197 residue in RC F(M197)H (**A**) and RC I(L177)H + F(M197)H (**B**). All distances are shown in Å.

It has been shown that in the wild-type RC of *C. sphaeroides*, an H-bonding network also presents on the periplasmic side of BChl P_A [10]. The network includes the C2-acetyl carbonyl group of BChl P_A, the imidazole group of His-L168, side groups of Asn-L166, Asp-

M184, and Tyr-L169 and a water molecule assigned as water-D. In the crystal structure of RC I(L177)H + F(M197)H, the positions of the side groups of the aforementioned residues did not change significantly, compared with those in the wild-type and F(M197)H RC structures (Figure S2 in Supplementary section). Although the sigma level of the electron density map peak for the water-D is low (1.5), its presence is indicated by the distance of this water from the side groups of Asn-L166 and Asp-M184, which is within the range of H-bond distances, similar to those in the wild-type and F(M197)H RCs. Thus, the data obtained show that the protein environment of BChls P_A and B_B is similar to that in the RC with single substitution I(L177)H.

4. Discussion

We have characterized spectral properties and mid-point potential P/P^+ of the double mutant RC I(L177)H + F(M197)H as well as presented the crystal structure of the complex with 2.6 Å resolution. It is shown that the spectral properties of the double mutant RC are similar to those of RC I(L177)H, except that the blue shift of the Q_Y P band is less pronounced, namely 11 nm compared with 18 nm for RC I(L177)H. It was noticed that while a single mutation F(M197)H did not lead to evident changes in the spectral properties of RC, the addition of this mutation to RC I(L177)H led to a significant red shift of the Q_Y P band. This spectral shift of the Q_Y P band is likely a result of a much more substantial out-of-plane rotation of the C2-acetyl carbonyl group of BChl P_B (29.7°) compared with that observed in the structure of RC F(M197)H (20.7°) under the same conditions. As discussed previously in detail [10], the energy of the P Q_Y transition in the mutant RC F(M197)H appears to be affected both by the rotation of the acetyl group of P_B, which causes a red shift of the P Q_Y band, and by a small increase in the distance between the two BChls of the dimer that leads to a blue shift of the band. In RC F(M197)H, these opposing effects compensate for each other, and as a result, the absorption spectrum remained practically the same as in the wild-type RC [23]. The resolution of the double mutant RC crystal structure does not allow us to estimate possible changes in the distance between the BChl planes in the dimer P but we see much more pronounced out-of-plane rotation of the acetyl group of BChl P_B and suggest that it causes an additional 7 nm red-shift of the P Q_Y band compared to the absorption spectrum of RC I(L177)H.

The original goal of this work was to combine RC I(L177)H and F(M197)H mutations to increase the stability of the mutant RC to improve the resolution of its crystal structure. The data obtained demonstrate that RC I(L177)H + F(M197)H is slightly more thermostable than the wild-type RC but not as stable as RC F(M197)H. Nevertheless, after the introduction of an additional H-bond of P with its protein environment, we succeeded in improving the resolution of the crystal structure up to 2.6 Å. For the first time, an analysis of the structural consequences of the I(L177)H mutation in the *C. sphaeroides* RC was performed based on its crystal structure obtained with 2.9 Å resolution (PDB ID: 3V3Z) [6]. The model of pigment-protein interactions that arose as a result of the mutation was somewhat refined in this work. It is shown that only one new water molecule, rather than two, appeared on the β-side of B_B and is suggested to be the 6th –ligand of the Mg^{2+}atom of this BChl. This water appears to donate H-bond to the imidazole group of His-L177, which is also in contact with the methyl group of the pyrrole ring-1 of BChl P_A (Figure 4). The distance between the N and C atoms of these groups in the structure is 2.06 Å, which exceeds known distance characteristic for C-N covalent bond (~1.5 Å) [25]. The resolution of the double mutant RC crystal structure is not yet sufficient to unequivocally confirm the covalent nature of BChl – protein binding and to reveal the origin of this bond and needs further improvement. However, it should be noted that the potential reactivity of this (B)Chl methyl group is supported by the fact that in the Chl *f* structure in the pyrrole ring-I, instead of a methyl group, a vinyl group is present [26].

The mutant RC I(L177)H + F(M197)H presented in this work turned out to be of interest as an object that has a modified H-bonding network near the BChl dimer P. Recently, the presence of H-bonding clusters on the periplasmic surface of the *C. sphaeroides* RC was reported [10]. Comparative analysis of such networks in the structures of natural and mutant RCs of different purple bacteria allowed us to assume that the polar protein environment can specifically affect the oxidative potential of the primary electron donor. In particular, it was suggested that when the π-electron system of P is connected with these networks, there is a correlation between the value of E_m P/P^+ and the patterns, sizes, and, apparently, the total strength of H-bonds in the clusters. The crystal structure of RC I(L177)H + F(M197)H provides data in support of this assumption.

The main pattern of the H-bonding network in the protein environment of BChl P_B in the double mutant RC structure (Figure 4) was found to be similar to that observed in the F(M197)H mutant RC [10], with one exception. The 3.7 Å remoteness of the water-C from the side group of Asp-L155 exceeds the H-bond distance (Figure 6B). If the H-bond between Asp-L155 and water-C is not formed, one H-bond falls out of the H-bonding network. As a result, the bulk of the network connected with BChl P_B through imidazole His-M197 became not as extended as in the F(M197)H RC. We assume that this can explain why the value of E_m P/P^+ in double mutant RC is ~20 mV lower than the sum of changes in the mid-point potentials caused by each of the two mutations. One should also not exclude the possibility that a more significant out-of-plane rotation of the acetyl group of P_B reduces its conjugation into the π-electron system of the macrocycle, affecting both the total strength of H-bonds in the network and the value of E_m P/P^+. The loss of one H-bond in the network may also serve as an explanation for the fact that the thermal stability of RC I(L177)H + F(M197)H turned out to be less than that of RC F(M197)H [10]. A possible reason for the non-formation of the H-bond between Asp-L155 and water-C in RC I(L177)H + F(M197)H may be associated with the effect of the above-mentioned strong binding of BChl P_A with His L177, which is expected to shift P towards BChl B_B. This assumption is consistent with more significant out-of-plane rotation of the C2-acetyl carbonyl group of P_B, compared with that of RC F(M197)H.

H-bond networks on the periplasmic surface of purple bacterial RCs have long been noticed. The periplasmic surface of the *C. spheroids* RC is an area of short-term binding of soluble cytochrome c2, which donates electrons for oxidized P [27,28]. Acidic residues Asp-L155 and Asp-M184 (Figure 6 and Figure S2) were shown to be among a few others on the RC surface that interact with basic residues on cytochrome c_2. In the recent work of Allen et al., it was proposed that the wide-spread network on the periplasmic side of RC participate in a hydrogen-bonded linear pathway that transfers protons between the environment of P^+ and the membrane surface [29]. This pathway includes the amino acid residues Asp-L155, Ser-L158, Tyr-M198, Asn M195, and a few water molecules in the vicinity of P_B (Figure 6 and [10]). The possible role of the symmetry related cluster of H-bonds near BChl P_A remains to be clarified.

To summarize, the data obtained in this study clarify the details of the pigment-protein interactions that arose as a result of mutation I(L177)H. In addition, the crystal structure of RC I(L177)H + F(M197)H illustrates that protein surroundings can affect the oxidation potential of the primary electron donor by altering the parameters of the H-bonding networks connected with the π-electron system of P.

Supplementary Materials: The following supporting information can be downloaded at: https://www.mdpi.com/article/10.3390/membranes13020157/s1, Figure S1: Redox titrations of the wild-type and the I(L177)H+F(M197)H RCs; Figure S2: The hydrogen-bonding networks in the proximity of the His L168 in the wild-type RC and the RC I(L177)H + F(M197)H.

Author Contributions: L.G.V.—concept of the study and supervision of the work; T.Y.F., G.K.S. and A.G.G.—performing experiments; L.G.V., T.Y.F., G.K.S. and A.G.G.—discussion of the results of the study, writing and editing of the article. All authors have read and agreed to the published version of the manuscript.

Funding: This research received no external funding.

Institutional Review Board Statement: The authors declare no conflict of interest in financial or any other sphere. This article does not contain any studies with human participants or animals performed by any of the authors.

Informed Consent Statement: Not applicable.

Data Availability Statement: MDPI Research Data Policies.

Acknowledgments: We acknowledge PETRA III for provision of synchrotron radiation facilities and beamline P11 for collecting data for this experiment, and A. Zabelin for fruitful discussion. This work was financially supported by the State Budget Project no. 122041100204-3.

Conflicts of Interest: The authors declare no conflict of interest.

References

1. Allen, J.P.; Feher, G.; Yeates, T.O.; Komiya, H.; Rees, D.C. Structure of the reaction center from *Rhodobacter sphaeroides R-26*: The cofactors. *Proc. Natl. Acad. Sci. USA* **1987**, *84*, 5730–5734. [CrossRef] [PubMed]
2. Feher, G.; Allen, J.P.; Okamura, M.Y.; Rees, D.C. Structure and function of bacterial photosynthetic reaction centres. *Nature* **1989**, *339*, 111–116. [CrossRef]
3. Yeates, T.O.; Komiya, H.; Chirino, A.; Rees, D.C.; Allen, J.P.; Feher, G. Structure of the reaction center from *Rhodobacter sphaeroides R-26* and *2.4.1*: Protein-cofactor (bacteriochlorophyll, bacteriopheophytin, and carotenoid) interactions. *Proc. Natl. Acad. Sci. USA* **1988**, *85*, 7993–7997. [CrossRef] [PubMed]
4. Jones, M.R. Structural plasticity of reaction centers from purple bacteria. In *Advances in Photosynthesis and Respiration*; Hunter, C.N., Daldal, F., Thurnauer, M.C., Beatty, J.T., Eds.; Springer: Dordrecht, The Netherlands, 2009; Volume 28, pp. 295–321.
5. Fufina, T.Y.; Vasilieva, L.G.; Khatypov, R.A.; Shkuropatov, A.Y.; Shuvalov, V.A. Substitution of isoleucine L177 by histidine in *Rhodobacter sphaeroides* reaction center results in the covalent binding of P_A bacteriochlorophyll to the L subunit. *FEBS Lett.* **2007**, *581*, 5769–5773. [CrossRef]
6. Vasilieva, L.G.; Fufina, T.Y.; Gabdulkhakov, A.G.; Leonova, M.M.; Khatypov, R.A.; Shuvalov, V.A. The site-directed mutation I(L177)H in *Rhodobacter sphaeroides* reaction center affects coordination of P_A and B_B bacteriochlorophylls. *Biochim. Biophys. Acta* **2012**, *1817*, 1407–1417. [CrossRef] [PubMed]
7. Fufina, T.Y.; Vasilieva, L.G.; Shuvalov, V.A. Examination of stability of mutant photosynthetic reaction center of *Rhodobacter sphaeroides* I(L177)H and determination of location of bacteriochlorophyll covalently bound to the protein. *Biochemistry* **2010**, *75*, 208–213. [CrossRef]
8. Holden-Dye, K.; Crouch, L.I.; Williams, C.M.; Bone, R.A.; Cheng, J.; Böhles, F.; Heathcote, P.; Jones, M.R. Opposing structural changes in two symmetrical polypeptides bring about opposing changes to the thermal stability of a complex integral membrane protein. *Arch. Biochem. Biophys.* **2011**, *505*, 160–170. [CrossRef]
9. Kangur, L.; Jones, M.R.; Freiberg, A. Hydrogen bonds in the vicinity of the special pair of the bacterial reaction center probed by hydrostatic high-pressure absorption spectroscopy. *Biophys. Chemistr.* **2017**, *231*, 27–33. [CrossRef]
10. Selikhanov, G.K.; Fufina, T.Y.; Günther, S.; Meents, A.; Gabdulkhakov, A.G.; Vasilieva, L.G. X-Ray structure of *Rhodobacter sphaeroides* reaction center with M197 Phe-His substitution clarifies properties of the mutant complex. *IUCrJ* **2022**, *9*, 261–271. [CrossRef]
11. Khatypov, R.A.; Vasilieva, L.G.; Bolgarina, T.I.; Shuvalov, V.A. Substitution of isoleucine L177 by histidine affects the pigment composition and properties of the reaction center of the purple bacterium *Rhodobacter sphaeroides*. *Biochemistry* **2005**, *70*, 1256–1261. [CrossRef]
12. Khmelnitskiy, A.Y.; Khatypov, R.A.; Khristin, A.M.; Leonova, M.M.; Vasilieva, L.G.; Shuvalov, V.A. Charge separation in *Rhodobacter sphaeroides* mutant reaction centers with increased midpoint potential of the primary electron donor. *Biochemistry* **2013**, *78*, 60–67. [CrossRef] [PubMed]
13. Goldsmith, J.O.; Boxer, S.G. Rapid isolation of bacterial photosynthetic reaction centers with an engineered poly-histidine tag. *Biochim. Biophys. Acta* **1996**, *1276*, 171–175. [CrossRef]
14. Jones, M.R.; Visschers, R.W.; van Grondelle, R.; Hunter, C.N. Construction and characterization of a mutant strain of *Rhodobacter sphaeroides* with the reaction center as the sole pigment-protein complex. *Biochemistry* **1992**, *31*, 4458–4465. [CrossRef] [PubMed]
15. Fufina, T.Y.; Vasilieva, L.G. Effects of detergents and osmolytes on thermal stability of native and mutant *Rhodobacter sphaeroides* reaction centers. *Biochemistry* **2021**, *86*, 517–524. [CrossRef]
16. Kabsch, W. *XDS* . *Acta Crystallogr. Sect. D Biol. Crystallogr.* **2010**, *66*, 125–132. [CrossRef]
17. McCoy, A.J.; Grosse-Kunstleve, R.W.; Adams, P.D.; Winn, M.D.; Storoni, L.C.; Read, R.J. Phaser crystallographic software. *J. Appl. Crystallogr.* **2007**, *40*, 658–674. [CrossRef]
18. Murshudov, G.N.; Skubák, P.; Lebedev, A.A.; Pannu, N.S.; Steiner, R.A.; Nicholls, R.A.; Winn, M.D.; Long, F.; Vagin, A.A. REFMAC5 for the refinement of macromolecular crystal structures. *Acta Crystallogr. Sect. D Biol. Crystallogr.* **2011**, *67*, 355–367. [CrossRef]
19. Emsley, P.; Lohkamp, B.; Scott, W.G.; Cowtan, K. Features and development of Coot. *Acta Crystallogr. Sect. D Biol. Crystallogr.* **2010**, *66*, 486–501. [CrossRef]

20. Afonine, P.V.; Grosse-Kunstleve, R.W.; Echols, N.; Headd, J.J.; Moriarty, N.W.; Mustyakimov, M.; Terwilliger, T.C.; Urzhumtsev, A.; Zwart, P.H.; Adams, P.D. Towards automated crystallographic structure refinement with *phenix.refine*. *Acta Crystallogr. Sect. D Biol. Crystallogr.* **2012**, *68*, 352–367. [CrossRef]
21. DeLano, W.L. The PyMOL Molecular Graphics System v2.5. Available online: https://pymol.org/2/ (accessed on 18 December 2021).
22. Hoff, A.J.; Deisenhofer, J. Photophysics of photosynthesis. Structure and spectroscopy of reaction centers of purple bacteria. *Phys. Rep.* **1997**, *287*, 242–247. [CrossRef]
23. Ridge, J.P.; Fyfe, P.K.; McAuley, K.E.; Van Brederode, M.; Robert, B.; Van Grondelle, R.; Isaacs, N.W.; Cogdell, R.J.; Jones, M.R. An examination of how structural changes can affect the rate of electron transfer in a mutated bacterial photoreaction centre. *Biochem. J.* **2000**, *351*, 567–578. [CrossRef] [PubMed]
24. Lin, X.; Murchison, H.A.; Nagarajan, V.; Parson, W.W.; Allen, J.P.; Williams, J.C. Specific alteration of the oxidation potential of the electron donor in reaction centers from *Rhodobacter sphaeroides*. *Proc. Natl. Acad. Sci. USA* **1994**, *91*, 10265–10269. [CrossRef] [PubMed]
25. Finkelshtein, A.V.; Ptitsyn, O.B. *Protein Physics*; Knizhnyi Dom Universitet Publisher: Moscow, Russia, 2005. (In Russian)
26. Chen, M.; Schliep, M.; Willows, R.D.; Cai, Z.-L.; Neilan, B.A.; Scheer, H. A Red-Shifted Chlorophyll. *Science* **2010**, *329*, 1318–1319. [CrossRef] [PubMed]
27. Axelrod, H.L.; Okamura, M.Y. The structure and function of the cytochrome c2: Reaction center electron transfer complex from *Rhodobacter Sphaeroides*. *Photosynth. Res.* **2005**, *85*, 101–114. [CrossRef] [PubMed]
28. Vasilev, C.; Mayneord, G.E.; Brindley, A.A.; Johnson, M.P.; Hunter, C.N. Dissecting the cytochrome c2-reaction centre interaction in bacterial photosynthesis using single molecule force spectroscopy. *Biochem. J.* **2019**, *476*, 2173–2190. [CrossRef] [PubMed]
29. Allen, J.P.; Chamberlain, K.D.; Williams, J.C. Identification of amino acid residues in a proton release pathway near the bacteriochlorophyll dimer in reaction centers from *Rhodobacter Sphaeroides*. *Photosynth. Res.* **2023**, *155*, 23–34. [CrossRef]

membranes

MDPI

Review

The Re-Localization of Proteins to or Away from Membranes as an Effective Strategy for Regulating Stress Tolerance in Plants

Yee-Shan Ku *, Sau-Shan Cheng, Ming-Yan Cheung, Cheuk-Hin Law and Hon-Ming Lam *

School of Life Sciences and Centre for Soybean Research of the State Key Laboratory of Agrobiotechnology, The Chinese University of Hong Kong, Hong Kong SAR, China
* Correspondence: ysamyku@cuhk.edu.hk (Y.-S.K.); honming@cuhk.edu.hk (H.-M.L.)

Abstract: The membranes of plant cells are dynamic structures composed of phospholipids and proteins. Proteins harboring phospholipid-binding domains or lipid ligands can localize to membranes. Stress perception can alter the subcellular localization of these proteins dynamically, causing them to either associate with or detach from membranes. The mechanisms behind the re-localization involve changes in the lipidation state of the proteins and interactions with membrane-associated biomolecules. The functional significance of such re-localization includes the regulation of molecular transport, cell integrity, protein folding, signaling, and gene expression. In this review, proteins that re-localize to or away from membranes upon abiotic and biotic stresses will be discussed in terms of the mechanisms involved and the functional significance of their re-localization. Knowledge of the re-localization mechanisms will facilitate research on increasing plant stress adaptability, while the study on re-localization of proteins upon stresses will further our understanding of stress adaptation strategies in plants.

Keywords: protein re-localization; protein lipidation; co-translational modification; post-translational modification; plant stress adaptability; transcriptional regulation; myristoylation; palmitoylation; prenylation

1. Introduction

In the natural environment, plants are constantly facing various abiotic and biotic stresses. The cell wall is the front line for stress perception. For example, mechanical stresses caused by injury or insect bites lead to the deformation of the cell wall, which then alters the contact between the cell wall and the cell membrane [1]. Hyperosmotic stress can cause plasmolysis, which is reversible in living plant cells. In addition to the cell membrane, within the cell, organelles are also surrounded by membranes. The dynamic properties of membrane proteins allow the sensing and relay of signals upon stresses.

Microscopic technologies and protein fractionation techniques enable the tracing and detection of protein re-localization upon stresses. Alterations in subcellular localization could be achieved by mechanisms including the cleavage of the signal peptide, interactions with membrane-localized proteins, and the lipidation of proteins. Some proteins possess the potential to bind to membranes due to features such as the pleckstrin homology (PH) domain and the protein kinase C conserved region 2 (C2) domain.

Common lipidation modifications of proteins include myristoylation, prenylation, palmitoylation, oleation, and glycosylphosphatidylinositol (GPI) anchoring. Among these mechanisms, myristoylation is the major protein lipidation mechanism in eukaryotes [2]. It is an irreversible co-translational or post-translational modification of proteins [2]. Myristoylation refers to the attachment of a myristoyl group to the N-terminal glycine of the protein through the formation of an amide bond mediated by N-myristoyltransferase (NMT) [3,4]. This modification is especially common in plants [5,6]. Prenylation refers to the attachment of a lipophilic farnesyl or geranylgeranyl group to a cysteine residue near the C-terminus of the protein [7]. Three types of prenyltransferases have been identified. Farnesyltransferase (FT) and geranylgeranyltransferase (GT) mediate the attachment of the farnesyl group and the geranylgeranyl group, respectively [8]. In addition, the attachment of two geranylgeranyl groups to the C-terminus of Rab proteins is mediated by Rab geranylgeranyltransferase (RGT) [9]. This modification allows the attachment of hydrophilic proteins to the hydrophobic membrane [7]. Palmitoylation refers to the attachment of fatty acids, usually palmitic acid, to the cysteine, serine, or threonine residues of a protein. The major type of palmitoylation is S-palmitoylation, which is the attachment of fatty acids to the cysteine residue [10]. Palmitoylation regulates the association of proteins with the plasma membrane, which is important for intracellular signaling [11]. For example, receptor-like kinases (RLKs) and heterotrimeric G proteins are major types of palmitoylated proteins [10]. The modification is catalyzed by protein S-acyltransferases (PATs), in which the DHHC-CRD domain is critical for the action of PATs [10,12,13]. Oleate is one of the most common monounsaturated fatty acids in plant cells, composed of 18 carbons with a cis double bond in the c-9 position [14]. The biosynthesis of oleate begins with acetyl coenzyme A (acetyl-CoA), which is converted to a saturated C18 product. After that, an acyl-carrier protein is conjugated to the C18 product by fatty acid desaturase to form the double bond [15]. Glycosylphosphatidylinositol (GPI) is a phosphoglyceride linked to the C-terminus of a protein. The structure of GPI mainly consists of three parts, including three mannoses, a glucosamine, and a phosphatidylinositol. GPI serves as a bridge between the protein and the lipid molecule [16–19]. The modification allows the attachment of proteins to lipid molecules, which then enables the protein to be anchored to membranes even if the protein does not contain any transmembrane domains. GPI anchoring is a post-translational modification that happens in the endoplasmic reticulum (ER) [16,18,20]. Various enzymes are involved in anchor synthesis and protein substrate modification [21].

Some proteins harbor functional domains for phospholipid binding, which is regulated by several factors, including the cellular level of calcium, the abundance of specific phospholipid species, and the local curvature of the membrane [22]. The two major groups of phospholipid-binding domains in proteins are the PH and protein C2 domains [22]. PH domain-containing proteins share low sequence homology. However, structurally, the proteins exhibit similar small modular structures including two perpendicular anti-parallel β-sheets followed by a C-terminal amphipathic helix. The PH domain binds to the phosphatidylinositols of biological membranes so that it helps to recruit proteins to different cellular compartments with specific phosphatidylinositols in their membrane composition under different cellular conditions [23]. The C2 domain comprises about 116 amino acid residues and is so named because of its location between the two copies of the C1 domain in protein kinase C and the catalytic domain. It is shown to have an affinity for a wide range of lipid components of cell membranes, including phosphatidylserine and phosphatidylcholine.

The functional significance of the protein re-localization includes the regulation of the signal relay, gene expression, water transport, and the promotion of cell integrity (Table 1). The understanding of protein re-localization upon stresses will facilitate the delineation of stress-coping strategies in plants at the cellular level.

2. Protein Re-Localization upon Water-Related Stress

2.1. Enhanced Localization of Annexin 1 at the Cell Membrane upon Plasmolysis

Plasmolysis is an immediate cellular response under hypertonic conditions. In Arabidopsis, upon NaCl- or mannitol-induced plasmolysis of root epidermal cells, annexin 1-green fluorescent protein (ANN1-GFP) was found to be enriched at the plasma membrane and remained at the plasma membrane even after de-plasmolysis [24]. Annexins have been reported as positive regulators of abiotic and biotic stresses in plants [25]. It was suggested that the drop in cellular pH upon osmotic stress altered the hydrophobicity of annexins and their relocation, and the subsequent formation of oligomeric ion channels in the membrane [24]. The enhanced curvature of the plasma membrane upon plasmolysis was also hypothesized to be related to the clustering of ANN1-GFP and its interaction with the membrane [24]. It was suggested that the ANN1-GFP accumulation at the plasma membrane upon osmotic stress could promote the association with Hechtian strands and the reticulum, which could then encourage the attachment of protoplasts to the cell wall [26,27]. However, it is not clear whether the drop in cellular pH due to other stresses can also result in the plasma membrane localization of annexins.

2.2. De-S-Palmitoylation of MfNACsa to Activate Its Transcriptional Regulatory Function

NAC [NAM (no apical meristem), ATAF (Arabidopsis transcription activator factor) 1/2, and CUC (cup-shaped cotyledon) 2] transcription factors are known as major regulators of drought and salinity responses in plants [28]. *MfNACsa* was identified from a drought- and cold-tolerant diploid variety of *Medicago falcata*. The overexpression of *MfNACsa* in *Medicago truncatula* promoted drought tolerance [29]. In the unstressed condition, *MfNACsa* is *S*-palmitoylated and localized at the plasma membrane [29]. Under drought stress, *MfNACsa* is induced, with *MfNACsa* being de-*S*-palmitoylated and localized in the nucleus, where it activates the expression of *MtGly1* [29]. MtGly1 in turn promotes drought tolerance by maintaining the glutathione pool in a reduced state [29].

2.3. Redistribution of Aquaporins under Water-Related Stresses

Aquaporins are transmembrane proteins that facilitate the intercellular or intracellular movement of water and small neutral solutes [30,31]. Aquaporin isoforms include plasma membrane intrinsic proteins (PIPs), tonoplast intrinsic proteins (TIPs), nodulin 26-like intrinsic proteins (NIPs), small basic intrinsic proteins (SIPs), and uncharacterized intrinsic proteins (XIPs) [30]. Besides rice and Arabidopsis, which are the model plants for monocots and dicots, respectively, aquaporins are widely studied in poplar, which is a tree model for studying hydraulics [32]. Re-localizations of rice, Arabidopsis, and poplar aquaporins upon stress have been reported. In rice, the redistribution of OsPIP1;1, OsPIP2;4, and OsPIP2;5 away from the plasma membrane was observed in exodermal and mesodermal cells under salt stress or PEG-induced osmotic stress [33]. In Arabidopsis, upon salt stress, TIP1;1 was found to re-locate from the tonoplast to intracellular spherical structures [34]. However, the re-localization of poplar aquaporins to or away from membranes upon water-related stresses is relatively unclear. Instead, upon hypotonic stress, PtoPIP1;1 from poplar was found to exhibit polar-like localization at the plasma membrane compared to the relatively even distribution at the plasma membrane under normal conditions [35]. In addition, the ice plant has been employed as a model for studying abiotic stress in plants [36]. Upon osmotic stress, McTIP1;2 from *Mesembryanthemum crystallinum* was found to have promoted localization in the tonoplast [37]. The re-localization of aquaporins facilitates the regulation of water transport to cope with water-related stresses. However, the mechanism of the aforementioned redistribution is largely unknown. Nevertheless, in poplar, under excessive Zn, AQUA1 was found to be re-localized in new-forming pro-vacuoles while localizing on different membranes destined to form aggregates related to autophagic multivesicular bodies [38]. The post-translational phosphorylation of AQUA1 was suggested as the possible mechanism behind the re-localization, as the re-localization was disturbed by phosphatases and kinase inhibitors [38]. Although the re-localization of AQUA1 was not reported to be associated with water-related stresses, such a mechanism may set a reference for studying the re-localization of aquaporins under other stresses.

3. Protein Re-Localization upon Salt Stress

3.1. Reduction of Aquaporins in the Plasma Membrane upon Salt Stress

Salt treatment has been demonstrated to induce plasma membrane internalization [39,40]. This suggests that the trafficking of membrane proteins could be induced by salt stress. Using GFP as the reporter, the exocytosis and endocytosis of AtPIP1;2 and AtPIP2;1 were shown to be induced by salt stress, though the hypothesis of salt-induced massive internalization of PIPs was not supported [40]. However, in other studies in *Arabidopsis thaliana*, sorbitol-induced osmotic stress, NaCl-induced salt stress, and salicylic acid (SA) treatment were shown to reduce PIP accumulation in the plasma membrane [41,42]. Using GFP as the reporter, the movement of GFP-PIP2;1 during salt treatment was studied. Under salt stress, the Brownian diffusion, directed diffusion, and mixed trajectory of PIP2;1 were decreased compared to the unstressed condition, while restricted diffusion was increased [41]. Under normal conditions, the internalization of GFP-PIP2;1 was found to be dependent on endocytic pathways, including the clathrin pathway and, to a lesser extent, the membrane raft-associated pathway [41]. A higher density of GFP-PIP2;1 on the plasma membrane was detected upon the interruption of the endocytic pathway by TyrA23 (tyrphostin A23, a clathrin-mediated endocytic pathway inhibitor [43]), but not by other endocytic pathway inhibitors such as MβCD (methyl-β-cyclodextrin: a sterol inhibitor), Fen (fenpropimorph: a sterol synthesis inhibitor), and PPMP (DL-threo-1-phenyl-2-palmitoylamino-3-morpholino-1-propanol: a sphingolipid biosynthesis inhibitor) [41]. Upon salt treatment, the correlation between the GFP-PIP2;1 density and the endocytic pathways was enhanced [41]. When the endocytic pathway was interrupted by TyrA23 together with salt treatment, compared to salt treatment alone without TyrA23, the density of GFP-PIP2 on the plasma membrane was significantly increased [41]. A similar phenomenon was observed when the membrane raft-associated endocytic pathway was interrupted by MβCD, Fen, or PPMP [41]. The internalization of PIP proteins is likely related to the prevention of water loss from cells [44]. In addition, upon salt stress, using GFP as the reporter, the re-localization of TIP1;1 from the tonoplast into intracellular spherical structures was observed in Arabidopsis [34]. Interestingly, in the same study, the subcellular localization of TIP2;1 was found to be unchanged but remained associated with the tonoplast upon salt treatment [34]. It therefore appears that the re-localizations of proteins, even those with similar functions, are regulated by different mechanisms.

3.2. Recruitment of SOS2 to the Plasma Membrane upon Salt Stress

The SOS (salt overly sensitive) signaling cascade has been widely known to regulate salt tolerance [45]. SOS1, SOS2, and SOS3, a plasma membrane Na^+/H^+ exchanger [46], a protein kinase [47,48], and a calcium-binding protein [48], respectively, are the major components of the SOS signaling cascade. SOS3 could be detected in both soluble and membrane protein fractions [49]. Although N-myristoylation is essential for the function of SOS3, it does not dictate the membrane-association property of SOS3 [49]. SOS3 activates the protein kinase activity of SOS2 in a Ca^{2+}-dependent manner and recruits SOS2 to the plasma membrane [48,50]. *SOS3* is only expressed in the root but not the shoot, while *SOS2* is expressed in both roots and shoots [51]. Such expression patterns imply that SOS2 may be regulated by other proteins in tissues in which *SOS3* is not expressed. Using yeast as the model, it was found that SCABP8 (SOS3-LIKE CALCIUM BINDING PROTEIN8) can also recruit SOS2 to the plasma membrane and activate SOS1 [51]. Later, it was found that, upon salt stress, VPS23A (vacuolar protein sorting 23A) positively regulates the plasma membrane localization of SOS2 [52]. Without stress, SOS2 is distributed throughout the whole cell [52]. Upon salt stress, SOS2 has enhanced localization on the plasma membrane [52]. However, when *VPS23A* was mutated, SOS2 did not show enhanced plasma membrane localization upon salt stress [52]. The mutation of *VPS23A* also led to the increased salt sensitivity of Arabidopsis plants [52]. However, in the *vps23A* mutant, when SOS2 was artificially designed to localize to the plasma membrane by overexpressing *SOS2* with the myristoylation sequence, the transgenic plants had better tolerance to salt compared to the *vps23A* mutant background [52]. The results suggest the importance of the plasma membrane localization of SOS2 in conferring salt tolerance [52]. The phenomenon is consistent with the notion that SOS2 activates SOS1, a membrane-bound Na^+/H^+ exchanger. Although the SOS2 protein does not have a canonical membrane localization signal, the regulated localization of SOS2 upon stress enables the functional plasticity of SOS2, which is a protein kinase that can activate other proteins [52].

3.3. Stabilization of MdCBL1 at the Plasma Membrane by Palmitoylation upon Salt Stress

In apples, upon salt stress, the expression of *MdPAT16* (*Malus domestica palmitoyltransferase 16*) is induced and is a positive regulator of salt tolerance and sugar accumulation [53]. MdPAT16 palmitoylates MdCBL1 (*Malus domestica* calcineurin B-like 1) to mediate the plasma membrane localization of MdCBL1, which is a positive regulator of sugar accumulation [53]. MdCBL1^{C3S}, which had a mutated palmitoylation site, was found to be mislocalized to the cytoplasm and nucleus [53]. It was therefore concluded that MdPAT16 stabilizes the plasma membrane localization of MdCBL1 by palmitoylation upon salt stress [53].

4. Protein Re-Localization upon Heat/Cold Stress

4.1. Promotion of DnaJ Lipidation by Heat Shock

Heat stress induces endoplasmic reticulum (ER) stress, resulting from the aggregation of misfolded proteins [54], which are bound and stabilized by heat shock proteins (HSPs). The heat stress factor (HSF) is localized in the nucleus and regulates gene expression by recognizing heat stress promoter elements [55,56]. Protein modifications, especially those on HSPs, are found to be important for heat stress tolerance [57]. DnaJ, which is also known as HSP40, is localized to the membrane by farnesylation and geranylgeranylation [58,59]. Heat shock promotes the prenylation of DnaJ and thus the enrichment of DnaJ proteins at the membrane [58,59]. The enrichment of prenylated DnaJ was suggested to be positively correlated to heat tolerance [58,59].

In Arabidopsis, AtJ3 is identified as a cytosolic HSP40 family member that contains a CaaX box for farnesylation and mediates the protein farnesylation-dependent response to heat stress [60]. AtJ3 farnesylation facilitates its association with the membrane [61]. Arabidopsis *j3* mutants with abolished CaaX boxes failed to produce farnesylated AtJ3. Compared to the wild type, the *j3* mutant exhibited intolerance to prolonged exposure at 37 °C for 4 days but improved tolerance to sudden heat shock at 44 °C for 30 min. Such responses to heat stress are similar to those of the Arabidopsis *hit5/era1* (*heat-intolerant 5/enhanced response to aba*) mutant that has a mutated β-subunit of the protein farnesyltransferase (PFT), which mediates protein farnesylation [60,62]. Moreover, the mutant form of protein farnesyltransferase encoded by *hit5/era1* resulted in an abscisic acid (ABA)-independent but temperature-dependent phenotype [63]. Previous findings have suggested that J-proteins improve thermotolerance in Arabidopsis by functioning as a component of HSP70/HSP40-based chaperones under heat stress [62,64]. Compared to the wild type, the *j3* mutant had an increased accumulation of insoluble proteins in cells when subjected to heat stress [62]. It has been reported that AtJ3 and HSP70-4 co-operate to mediate heat stress tolerance in Arabidopsis [62]. A subcellular localization study showed that EYFP^{N}-HSP70-4 and EYFP^{C}-J3 interacted to reconstitute EYFP, which was condensed in the membrane-less heat stress granules under heat stress while showing diffused localization in the cell under physiological temperature [62]. Although the interaction between AtJ3 and HSP70 was farnesylation-independent, it was proposed that the farnesylated site of AtJ3 is required for directing HSP70 to the hydrophobic residues on the heat-denatured proteins, and assisting in protein refolding for heat stress alleviation [62].

Other than assisting in protein refolding in conjunction with HSP70-4, AtJ3 is also a component of the membrane-bound RISC (RNA-induced silencing complex) in Arabidopsis. AtJ3 farnesylation promotes membrane association by interacting with AGO1, an effector protein in microRNA (miRNA)- and small interfering RNA (siRNA)-mediated gene silencing in Arabidopsis [65]. It was reported that *era1* or the *AtJ2/AtJ3* mutant had increased levels of the miRNA-associated membrane-bound polysomes with impaired noncell autonomously acting siRNA gene silencing, suggesting the potential function of the post-translational modification of AtJ3 in regulating translation and leading to the differential stress tolerance among different Arabidopsis genotypes [65].

4.2. Protein Re-Localization by Transmembrane Domain Cleavage to Regulate Gene Expressions upon Heat Stress

Under heat stress, the membrane-tethered bZIP28 is re-localized to the nucleoplasm as a result of the heat-induced cleavage of the C-terminus by site 1 or site 2 proteases to remove its transmembrane domain [66]. In the nucleoplasm, the re-localized bZIP28 up-regulates heat stress-coping genes such as the one encoding the ER-localized chaperone, BiP2, in Arabidopsis [66,67]. A similar proteolytic activation of transcription factors under heat stress was also reported in rice. A rice NAC transcription factor, OsNTL3, carries a predicted C-terminal transmembrane domain [68]. Under heat stress and ER stress, the membrane-localized OsNTL3 is re-localized to the nucleus and binds to the promoter region of *OsbZIP74* for stress response activation [68].

4.3. Alteration of Protein Localization by mRNA cleavage upon Heat Stress

The inositol-requiring enzyme 1 (IRE1)-mediated bZIP mRNA splicing for the proteolytic activation of bZIP transcription factors is conserved in plants [69,70]. In Arabidopsis, the ER-localized inositol-requiring enzyme 1 (AtIRE1) is responsible for the unconventional mRNA splicing of *AtbZIP60* under heat stress, which leads to a shift in the open reading frame that promotes the nuclear localization of the protein instead of localization at the membrane [69]. Nuclear localization enables the transcriptional regulatory function of AtZIP60 to modulate gene expression [69]. In rice, the mRNA of *OsbZIP74* was also spliced by OsIRE1 when under heat and ER stress [70]. The cleavage of the C-terminal transmembrane domain from the transcript of *OsbZIP74* promotes nuclear localization and thus enables the transcriptional regulatory function of the protein [70].

4.4. Translocation of Proteins to the Nucleus to Regulate Gene Expressions upon Cold Stress

In Arabidopsis, a cytosolic redox protein, thioredoxin h2 (Trx-h2), was reported to be involved in redox-mediated regulation and structural switching upon exposure to cold stress [71]. Under normal conditions, Trx-h2 is anchored to the cytoplasmic endomembrane via the myristoyl group that is covalently attached to Gly2. However, under cold exposure, the de-myristoylation of Trx-h2 facilitates nuclear localization by exposing the nuclear localization signal (NLS) located at the C-terminus [71]. Trx-h2 reduces the disulfide bonds of the inactive CBF oligomers to release the active monomers, which then bind the promoter regions of cold-responsive (COR) genes to activate their expression [71,72]. Therefore, the translocation of Trx-h2 into the nucleus upon exposure to cold stress favors the expression of cold-related genes and promotes cold stress tolerance [71]. The myristoylated plasma membrane-localized Clade-E growth-regulating 2 (EGR2) phosphatase is also involved in regulating the expression of cold stress-coping genes in Arabidopsis [73]. EGR2 interacts with Open Stomata 1 (OST1), which is a positive regulator of *CBF* gene expressions and inhibits the activity of OST1 under normal conditions [73]. At 22 °C, EGR2 was found to be N-myristoylated by N-myristoyl-transferase 1 (NMT1) and detected in the plasma membrane but not in the soluble fraction [73]. When EGR2 was mutated to inhibit N-myristoylation, the mutated egr2 lost the membrane-binding specificity and was detected in the plasma membrane, cytosol, and nucleus at 22 °C [73]. At 4 °C, EGR2 was detected in all plasma membranes, the cytosol, and the nucleus [73]. It was therefore concluded that the plasma membrane localization of EGR2 was dependent on N-myristoylation, which was hampered at 4 °C. The N-myristoylation of EGR2 is important for its interaction with OST1. Upon binding to OST1, the N-myristoylated EGR2 dephosphorylates OST1 to repress its kinase activity, which is important for the activation of CBF pathways to achieve freezing tolerance. Under cold stress, the interaction between EGR2 and NMT1 is weakened, and the N-myristoylation of EGR2 by NMT1 is suppressed. Without being bound by N-myristoylated EGR2, the kinase activity of OST1 is not repressed. Consequently, the CBF pathway is activated to achieve freezing tolerance [73].

The Salt Tolerance Related Protein (STRP) could be found in the cytosol, nucleus, or plasma membrane [74]. However, under cold stress, the amount of STRP in the membrane fraction decreases while it increases in the cytosol and nucleus [74]. The association of the STRP with membranes is proposed to be regulated by lipid attachment or by anchoring to plasma membrane-localized proteins [74]. As the *strp* mutant is more susceptible to oxidative damage with increased lipid peroxidation and altered membrane integrity compared to the wild type, it is proposed that the increased localization to the nucleus might help modulate the expression of the specific cold-activated gene or represent a protection mechanism in response to reactive oxygen species (ROS) production upon cold stress [74].

4.5. Changes in Protein Subcellular Localization in Response to Heat/Cold Stress-Induced Ca^{2+} Signaling

In response to the sudden change in temperature, Ca^{2+}-permeable channels mediate signals that lead to an influx of Ca^{2+} into plant cells [75–77]. By regulating the calcium signaling, the tolerance of the plant to temperature fluctuations could be improved [78,79]. An important function of calcium-dependent protein kinase (CDPK or CDK) is to perceive changes in the cytosolic calcium concentration in response to external stimuli [80]. The expression level of the membrane-localized calcium-dependent protein kinase is often correlated to the responses to heat, cold, and wounding stress in the plant [81–87]. The activation of the downstream cold-regulated (COR) gene expressions for better cold stress adaptation requires the cytosolic Ca^{2+} signal [88]. The membrane-localized ZmCDPK7 is a heat-response kinase in maize that participates in ABA signaling and heat stress tolerance by phosphorylating sHSP17.4 [89]. ZmCDPK7 contains N-terminal myristoylation and palmitoylation sites with no transmembrane region for anchoring to the membrane. It is reported that there is a shift in the membrane-localized ZmCDPK7 to the cytoplasm under heat-stress conditions [90]. It is proposed that the ZmCDPK7 activates the sHSP17.4 via phosphorylation in the cytoplasm for maintaining protein stability, while the sHSP17.4 was previously reported to be heavily phosphorylated under heat and drought conditions in maize [89,91].

5. Protein Re-Localization upon Mechanical Stress for Protein Activation

Several calcium-dependent protein kinases are systemically induced upon wounding, suggesting the possible involvement of these proteins in wounding and herbivory responses. From a genome-wide analysis of calcium-dependent protein kinases in *Glycine max*, two genes encoding membrane-localized CDPKs, *GmCPK3* and *GmCPK31*, showed enhanced expression under wounding and herbivory stresses [92]. In Arabidopsis, a herbivory-induced phytohormone-independent pathway mediated by the CPK cascade was associated with better defense against wounding and herbivory [93]. AtCPK3 and AtCPK13 activate a member of the heat stress transcription factor family, HsfB2a, via phosphorylation, which then activates the expression of herbivory-inducible defense-related genes [93]. In tomatoes, the expression of *LeCDPK1* was rapidly and transiently enhanced both locally at the site of the injury and systemically in the distant non-wounded leaves [94]. Similar to tomatoes, both the kinase activity and the mRNA of *ZmCPK11* increased as a systemic response to wounding in maize [95]. Unlike other CPKs such as AtCPK1 and DcCPK1 (*Daucus carota* CPK1), ZmCPK11 lacks a myristoylation/palmitoylation site [95]. However, it was found that ZmCPK11 was activated via interactions with phospholipids [95]. After injury, the activity of the membrane-bound CDPK relative to the total CDPK activity was increased two-fold [94,96]. The translocation of ZmCPK11 from the cytosol to membranes and the increased activity of ZmCPK11 upon its binding to phospholipids were proposed to be the possible mechanisms behind the wounding response in maize [95].

In rice, a GTPase-activating protein 1 (OsGAP1) was reported to activate the GTPase activity of OsYchF1 [97]. Under mechanical stress and in the presence of Ca^{2+}, OsYchF1 was localized to the plasma membrane via its interaction with OsGAP1 [98]. It is proposed that the re-localization of OsGAP1 is associated with the functional activation of OsYchF1 to promote resistance to wounding and pathogen challenges [97,98].

6. Protein Re-Localization upon Biotic Stress

Rice *OsERG1* contains a single C2 domain, which is responsible for calcium-dependent phospholipid binding. In one study, the transcript level of *OsERG1* was induced by the elicitors of a fungal blast, *Magnaporthe grisea*. In addition to inducing transcription, the elicitor treatment also led to the subcellular re-localization of OsERG1 from the cytosol to the plasma membrane, although the mechanism of the re-localization was unclear. By either elevating the subcellular calcium level or applying a calcium-mobilizing agonist (A23187), the plasma membrane localization of OsERG1 was either induced or suppressed, respectively [99]. In wheat, TaERG3 also harbors a C2 domain. The transcript level of *TaERG3* was inducible by ABA treatment, high salinity, cold treatment, an increased level of calcium, and infection by *Puccinia striiformis* f. sp. *tritici*, which causes stripe rust. *TaERG3* is characterized as a positive regulator of ABA signaling and salt and cold stress. *TaERG3* also enhances the resistance towards *Puccinia striiformis* f. sp. *tritici*. *TaERG3* is localized in the plasmalemma and nucleus [100]. However, the effect of stress on the subcellular localization of TaERG3 remains unclear.

Upon biotic stress, chloroplasts are the major sites for the production of antimicrobial reactive oxygen species (ROS) and the biosynthesis of defense hormones, including SA and jasmonic acid (JA) [101]. Plant cells usually perceive biotic threats at the cell surface. Upon pathogen perception, the chloroplast thylakoid membrane-associated calcium-sensing receptor (CAS) is stimulated and activates pattern-triggered immunity (PTI) [102]. CPKs are major components of calcium signaling [103]. In Arabidopsis, CPK16 localizes along the plasma membrane through N-myristoylation under normal conditions. The N-myristoylation site of CPK16 overlaps with the chloroplast transit peptide (cTP). Upon flagellin 22 (flg22) treatment, CPK16 re-localizes from the plasma membrane to the chloroplast to promote chloroplast-dependent defenses [101]. At the same time, pathogen effectors also re-localize from the plasma membrane of the infected plant cell to the chloroplast to suppress defense response including hormone biosynthesis [101]. In a study on the tomato—*tomato yellow leaf curl virus* (TYLCV) interactions, the C4 protein from TYLCV was found to have the N-myristoylation site overlap with the chloroplast transit peptide [101]. Upon the activation of the defense response triggered by the replication-associated viral protein (Rep), the bacterial elicitor peptide flg22, or the plant peptide Pep1, the TYLCV C4 protein is re-localized to the chloroplast from the plasma membrane [101]. Once it has entered the chloroplast, the TYLCV C4 protein interacts with and suppresses the function of the thylakoid membrane-associated CAS, which is involved in PTI-induced transcriptional reprogramming, SA biosynthesis, callose deposition, and anti-bacterial and anti-fungal resistance [101]. The simultaneous re-localization of both the protective protein from the plant and the pathogenic protein from the virus to the chloroplast shows the continuous battle between the plant and the virus.

7. Protein Re-Localization upon Oxidative Stress

Oxidation is a secondary stress resulting from both abiotic and biotic stresses [104]. Hydrogen peroxide (H_2O_2) is one of the most common reactive oxygen species (ROS) in plants [105]. The transmembrane domains, TM2 and TM3, of PIP2 dictate the plasma membrane localization of PIP2. Although PIP1s lack plasma membrane localization signaling and are unable to localize to the plasma membrane on their own, their interactions with PIP2 allow them to be localized to the plasma membrane [106]. In *A. thaliana*, H_2O_2 treatment on roots led to the reduced accumulation of AtPIP1 and AtPIP2 in the plasma membrane [42,107], but they were re-localized to intracellular structures tentatively identified as vesicles and small vacuoles instead [42]. In the root, catalase treatment could counteract the effect of SA or salt on the internalization of PIP2;1 from the plasma membrane, suggesting the involvement of H_2O_2 in PIP re-localization upon stress [42].

8. Protein Re-Localization in Response to Stress Hormones

8.1. Regulation of NMT1 Expression by ABA

ABA regulates numerous plant growth, development, and stress response mechanisms, including seed dormancy, leaf senescence, responses against pathogen infection, and osmotic, drought, oxidative, and salt stresses [108–111]. When the plant is under stress, the ABA level increases and is perceived by pyrabactin resistance 1 (PYR1)/PYR1-like (PYL)/regulatory components of ABA receptors (RCAR) proteins, leading to the degradation of the negative regulators of ABA, ABI1, and PP2CA, by ubiquitination with PUB12/13 and RGLG1/5 E3 ligase, respectively [112]. In normal conditions, RGLG1 is N-myristoylated and bound to the plasma membrane. Upon ABA or salt treatment, the expression of *NMT1*, which is responsible for the N-myristoylation of RGLG1, is down-regulated. As a result, without being myristoylated, RGLG1 is shuttled from the plasma membrane to the nucleus, where it mediates the ubiquitination of the nuclear-localized PP2CA [113].

8.2. Mediation of Brassinosteroid Signaling by Myristoylated BSK1

Brassinosteroid (BR) plays roles in diverse cellular biological processes, including cell division, the cell cycle, morphogenesis, reproduction, and various stress responses [114]. BR is thought to act as a switch between normal metabolic activities and stress responses [114]. This role of BR as a master switch relies on brassinosteroid signaling kinase 1 (BSK1), which associates with the receptor kinases, BR-insensitive 1 (BRI1) or flagellin-sensing 2 (FLS2), and thus functions in both BR-regulated plant growth and flg22-triggered immunity. Upon BR perception, BRI1 phosphorylates BSK1 in the membrane raft. The phosphorylated BSK1 in turn activates BRI1 suppressor 1 (BSU1) and dephosphorylates BR-insensitive 2 (BIN2) to trigger downstream signaling. When flg22 molecules from invading pathogens are recognized by FLS2, BSK1 is relocated to the non-membrane raft with mitogen-activated protein kinase kinase kinase (MAPKKK) to trigger defense responses [115]. Therefore, the lining of BSK1 along the cytosolic side of the plasma membrane is crucial for interacting with BRI1 or FLS2 and triggering its master key function. This plasma membrane association is achieved by N-myristoylation. The $BSK1^{G2A}$ mutant proteins, with the loss of the N-myristoylation site, are mainly distributed in the cytoplasm and endoplasmic reticulum [116]. Moreover, the $BSK1^{G2A}$ mutant protein is unstable and subject to degradation by autophagy [116].

9. Protein Re-Localization in Response to Other Signaling Events

9.1. Regulation of SnRK Signaling by N-Myristoylation

Sucrose nonfermenting 1-related kinase 1 (SnRK1) is a key signaling molecule regulating cellular metabolism under both stress and growth-promoting conditions [117]. It is a heterotrimeric complex comprising an α-catalytic, and β- and γ-regulatory subunits [117]. SnRK1 functions are induced by stress conditions (including nutrient starvation and biotic and abiotic stresses) and switch the cellular metabolism from anabolism to catabolic pathways such as autophagy or stress responses [117]. SnRK1 functions by directly phosphorylating cytosolic metabolic enzymes or by phosphorylating the transcription factors in the nucleus that regulate nuclear-encoded plastid and mitochondrial genes, which in turn alter cellular metabolic states. Transcription factor genes targeted by SnRK1 include *bZIP63*, *FUS3*, *IDD8*, *EIN3*, *WRI1*, and *MYC2*, and they are involved in diverse metabolic and hormonal signaling pathways, such as sugar signaling, amino acid starvation, seed maturation and germination, oil synthesis, flowering time, ethylene, and ABA and JA responses [117]. When phosphorylated by SnRK1, these transcriptional factors either lose their ability to regulate transcription or bind DNA or are even degraded. To phosphorylate the transcription factors, SnRK1 has to be localized in the nucleus. The cellular localization of SnRK1 relies on its N-myristoylation status [118]. In Arabidopsis, the SnRK1 β1 subunit is the target site for N-myristoylation by NMT1 [118]. Myristoylated SnRK1 is sequestered to the plasma membrane and is thus unable to phosphorylate its target transcription factors [118]. During plant development, the different levels of *NMT1* expression in different tissues exert tissue-specific regulation on the activity of SnRK1 to coordinate the proper overall development of the plant [118].

9.2. Light/Sugar Sensing

Calmodulin (CaM53) from *Petunia hybrida* Mitchell is found to be geranylated by geranylgeranyl transferases (GGTase-I) and is thus plasma membrane-bound. However, under dark or low sugar level conditions, CaM53 is not geranylated and localizes in the nucleus. This nuclear localization of CaM53 leads to stunted growth, decreased stem internode length, leaf curling, chlorosis, and necrosis [119]. However, the detailed role of CaM53 in signaling remains unclear [119].

10. Discussion and Conclusions

Various proteins re-localize to or away from membranes upon stresses. They include aquaporin water channels for regulating water transport, structural proteins for regulating cell integrity, chaperones for assisting in protein folding, signaling proteins for regulating Ca^{2+} and hormone signaling, and transcription-related proteins for regulating gene expression. An example of the transcriptional regulation resulting from protein re-localization is shown in Figure 1. The specific protein re-localizations and their functional significance are summarized in Table 1. Although myristoylation is irreversible, most of the other protein re-localization mechanisms, such as interacting with other biomolecules and lipidation, are reversible. Such reversibility enables the timely and plastic functional regulation of stress responses.

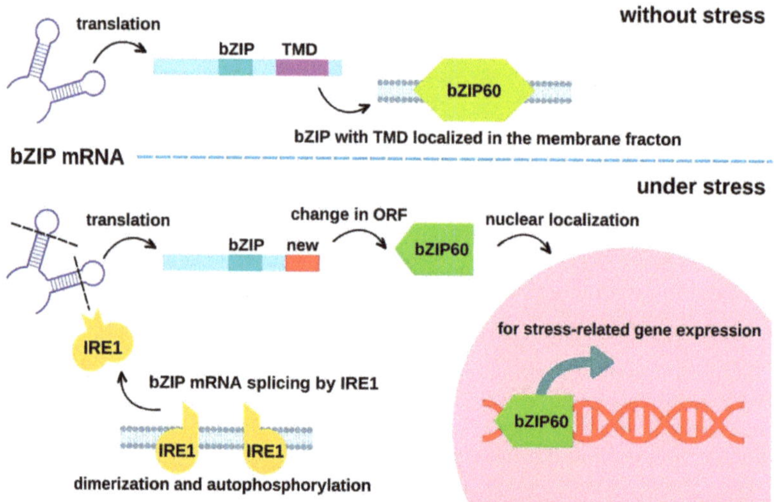

Figure 1. A diagram illustrating an example of the transcriptional regulations by re-localized proteins upon stress. TMD, transmembrane domain; ORF, open reading frame.

Upon re-localization, several proteins function by modulating up-stream cellular events. For example, in *A. thaliana*, the de-myristoylation of the SnRK1 β1 subunit facilitates the transcriptional regulation of genes including *bZIP63*, *FUS3*, *IDD8*, *EIN3*, *WRI1*, and *MYC2*. Among these targets, *FUS3* is annotated as a kinase for signal transduction, while the other genes are annotated as transcription factors. The re-localization of SnRK1 probably initiates a vast number of cellular responses. Other examples of the modulation of up-stream cellular events include the regulation of BR signaling by the re-localization of BSK1 upon biotic stress and the regulation of ABA signaling by the re-localization of RGLG1 upon ABA or salt treatment. Since BR and ABA are stress hormones that regulate various cellular events, the re-localization of BSK1 and RGLG1 probably results in the regulation of a vast number of cellular events. These examples show that the effects on cellular events from the re-localization of certain proteins can be greatly amplified in the downstream pathways. The modification of sub-cellular localization, therefore, appears to be an effective strategy for regulating the stress adaptability of plants. However, the mechanisms underlying the re-localization of a number of proteins upon stress have remained unknown. On the other hand, a number of proteins, such as ANJ, EGR2, HvFP1, and HIPP26, are lipidated upon stress (Table 1). However, the mechanisms for the re-localization of the proteins upon stress have remained unclear. Investigations of the re-localization mechanisms will facilitate the research on plant stress adaptability, while the study on the re-localization of proteins upon stress will further our understanding of stress adaptation strategies in plants.

Studies on the re-localization of proteins to or away from membranes are largely biased toward model crops such as Arabidopsis and rice. Moreover, many of the studies on the re-localization of proteins upon stresses are on aquaporins, which are involved in the transport of various molecules including water and neutral solutes. It is therefore reasonable that aquaporins are associated with the adaptation to various stresses including water-related stress and salt stress. Although poplar has been employed as the model for studies on aquaporins, the specific studies on the re-localization of poplar aquaporins upon stresses are less popular than those in Arabidopsis and rice. As mentioned above, although the functions of the proteins may appear to be similar, the re-localization upon stress could be different. It is therefore worthwhile to expand the studies on different proteins and different plant species. Scattered studies on the re-localization of proteins to or away from membranes were carried out in other plants, such as ice plants (*M. crystallinum*), apples (*M. domestica*), saltbushes (*A. nummularia*), maize (*Z. mays*), barley (*H. vulgare*), wheat (*T. aestivum*), and garden petunias (*P. hybrida*) (Table 1). Although these species are less common for studies on stresses compared to Arabidopsis, rice, and poplar, they are advantageous for studying specific mechanisms. For example, apples have been employed to study sugar accumulation in fruit while garden petunias have been employed to study flower color modification.

In conclusion, a vast number of proteins re-localize in the cell upon abiotic and biotic stresses. Mechanisms for such re-localization include the cleavage of the signaling peptide, interactions with membrane-localized proteins, and the lipidation of proteins. The functional significance of such re-localization includes regulating molecular transport, cell integrity, protein folding, signaling, and gene transcription. By regulating signaling and transcription, the effects brought about by protein re-localization can be greatly amplified. Thus, protein re-localization appears to be an effective strategy for promoting plant stress adaptability. The different re-localization mechanisms of protein homologs and the largely unexplored protein re-localization in non-model plants upon stress leave considerable room for future research.

Table 1. Summary of proteins that re-localize in plant cells upon stress.

Type of Stress	Species	Protein	Description of the Relocation	Mechanism of the Relocation	Functional Significance of the Relocation	References
Water-related stress	*Arabidopsis thaliana*	ANN1	Accumulation of ANN1 upon osmotic stress (plasmolysis)	Unknown	Association with Hechtian strands and reticulum at the plasma membrane for protection against osmotic stress	[24]
	Medicago falcata	MfNACsa	Relocated from the plasma membrane to the nucleus upon drought stress	De-S-palmitoylation of MfNACsa	MfNACsa activates the expression of *MtGly1* in the model plant *Medicago truncatula* for maintaining the glutathione pool in the reduced state to achieve drought tolerance	[29]
	Oryza sativa	OsPIP1;1, OsPIP2;4, and OsPIP2;5	Relocated away from the plasma membrane	Endocytosis of OsPIP2;5 is enhanced by salt stress	Regulation of water transport	[33]
	Populus tomentosa	PtoPIP1;1	Exhibited polar-like localization at the plasma membrane compared to the relatively even distribution at the plasma membrane under normal conditions	Not mentioned	Regulation of water transport	[35]
	Mesembryanthemum crystallinum	McTIP1;2	Exhibited promoted localization in the tonoplast	Not mentioned	Promotion of osmotic adjustment in the cell	[37]
Salt stress	*Arabidopsis thaliana*	SOS2	Enhanced plasma membrane localization upon salt stress	The plasma membrane localization is enhanced by VPS23A	For the activation of SOS1, a membrane-bound Na^+/H^+ exchanger	[52]
	Malus domestica	MdCBL1	Stabilized plasma membrane localization upon salt stress	The expression of *MdPAT16* is induced upon salt stress. MdPAT16 mediates the plasma membrane localization of MsCBL1 by palmitoylation	Regulation of sugar accumulation	[53]
	Arabidopsis thaliana	AtPIP2;1	Reduced plasma membrane localization upon osmotic stress and salt stress	Enhanced internalization of AtPIP2;1 upon water-deficit stress through endocytic pathways	Regulation of the water permeability of plasma membrane to protect cells from water-deficit stress	[41]
	Arabidopsis thaliana	TIP1;1	From the tonoplast to intracellular spherical structures	Not mentioned	Regulation of water transport inside the cell	[34]

Table 1. *Cont.*

Type of Stress	Species	Protein	Description of the Relocation	Mechanism of the Relocation	Functional Significance of the Relocation	References
Heat stress	*Arabidopsis thaliana*	AtJ3	From cytosol to membrane-less heat stress granules	Not mentioned	For the formation of HSP70/HSP40-based chaperones. Mutants failed to undergo AtJ3 farnesylation, leading to heat stress intolerance. AtJ3 farnesylation is responsible for directing HSP70 to the misfolded protein	[52,54,56]
	Arabidopsis thaliana	AGO1; AtJ2/AtJ3	From cytosol to membrane	Proposed model of farnesylation promotes the AtJ3-membrane interaction and AGO1-membrane interaction via J3, which further alters the loading of AGO1-miRNA to the polysome	Farnesyl transferase-deficient and farnesylation-deficient *j2/j3* mutants had increased levels of the miRNA-associated membrane-bound polysomes	[65]
	Arabidopsis thaliana	SKD1	From cytosol to messenger ribonucleoproteins	Not mentioned	Possibly involved in the selection of proteins to be localized to mRNP under stress conditions	[120]
	Arabidopsis thaliana	bZIP28	From membrane to endoplasmic reticulum and cytosol	Heat-induced cleavage at the membrane-tethering C-terminus	The re-localized bZIP28 up-regulates heat stress-coping genes such as the ER-localized chaperone BiP2 in Arabidopsis for coping with heat stress	[66,67]
	Arabidopsis thaliana	AtbZIP60	From membrane to nucleus	The ER-localized inositol-requiring enzyme 1 (AtIRE1) mediates the unconventional mRNA splicing of *AtbZIP60* by open reading frame shift	Membrane-localized active AtbZIP6 promotes the expression of stress-related genes	[69]
	Atriplex nummularia	ANJ1	Not mentioned	Not mentioned	Heat shock enhances the amount of the prenylated DnaJ protein in the membrane fraction; potentially functions in heat tolerance	[58]
	Oryza sativa	OsNTL3	From membrane to nucleus	Not mentioned	OsNTL3 binds to the promoter region of *OsbZIP74* for stress response activation	[68]
	Oryza sativa	OsbZIP74	From membrane to nucleus	The ER-localized inositol-requiring enzyme 1 (OsIRE1) cleaves off the C-terminal transmembrane domain of OsbZIP74	Membrane-localized active OsbZIP74 promotes the expression of stress-related genes	[70]
	Zea mays	ZmCDPK7	From membrane to cytosol	Not mentioned	ZmCDPK7 activates sHSP17.4 via phosphorylation in the cytoplasm for maintaining protein stability	[90,91]

Table 1. *Cont.*

Type of Stress	Species	Protein	Description of the Relocation	Mechanism of the Relocation	Functional Significance of the Relocation	References
Cold stress	*Arabidopsis thaliana*	Trx-h2	From membrane to nucleus	De-myristoylation of Trx-h2	Trx-h2 reduces the disulfide-bonded inactive CBF oligomers to form the active monomers that bind the promoter regions of cold-responsive (COR) genes	[71,72]
	Arabidopsis thaliana	EGR2	Not mentioned	Not mentioned	Low temperature attenuates the formation of the NMT1-EGR2 protein complex, leading to the suppression of the myristoylation of EGR2, and releasing its inhibition on OST1 for the proper activation of the CBF pathway and freezing tolerance	[73]
	Arabidopsis thaliana	STRP	Decrease in the membrane fraction of STRP	Not mentioned	Potentially affects the expressions of cold-activated genes, protects the chromatin structure, and stabilizes the membrane structure	[74]
	Hordeum vulgare	HvFP1	Not mentioned	Not mentioned	Farnesylation of HvFP1 is important for its precise nuclear localization	[121]
	Arabidopsis thaliana	HIPP26	Not mentioned	Not mentioned	Isoprenylation of HIPP26 is important for its precise nuclear localization	[122]
Mechanical stress	*Oryza sativa*	OsYchF1	From cytosol to membrane	Interact with the membrane-anchored interacting partner, OsGAP1	Proposed re-localization of the negative regulator of stress to alleviate stress susceptibility	[97,98]
Biotic stress	*Arabidopsis thaliana*	CPK16	From plasma membrane to chloroplast	Removal of N-myristoylation of RGLG1	Allows CPK16 to work in chloroplast and enhances the resistance towards *Pseudomonas syringae* pv. *tomato* DC3000 and *tomato yellow leaf curl virus*	[101]
	Oryza sativa	OsERG1	From cytosol to plasma membrane	Elevation of cellular calcium level	*OsERG1* is induced by elicitor from the fungal blast *Magnaporthe grisea* and is believed to play roles in fungal disease defense responses. The OsERG1 protein is relocated from cytosol to plasma membrane upon fungal elicitor treatment and calcium signals	[99]
	Arabidopsis thaliana	BSK1	From plasma membrane to cytoplasm and endoplasmic reticulum	Loss of N-myristoylation	BSK1 is plasma membrane-bound and interacts with BRI1 and FLS2 for triggering BR signaling or defense response. When flagellin is perceived, BSK1 would relocate to the non-membrane raft for functioning. If BSK1 fails to be modified by N-myristoylation, it would no longer associate with the plasma membrane and would be degraded through the autophagic pathway	[115,116]
Biotic stress and hormone	*Triticum aestivum*	TaERG3	From nucleus to plasma membrane	Increase in cellular calcium level	TaERG3 plays roles in ABA signaling and acts as a positive regulator of responses to high salt and low temperature. It also enhances resistance towards *Puccinia striiformis* f. sp. *tritici* (the pathogen causing stripe rust). It is predominantly localized in plasmalemma and nucleus	[100]
Stress hormone	*Arabidopsis thaliana*	RGLG1	From plasma membrane to nucleus	Reduction of N-myristoylation of RGLG1	Allows the binding of RGLG1 to PP2CA in the nucleus which is a negative regulator of ABA signaling	[113]

Table 1. *Cont.*

Type of Stress	Species	Protein	Description of the Relocation	Mechanism of the Relocation	Functional Significance of the Relocation	References
Nutritional starvation, biotic and abiotic stresses	*Arabidopsis thaliana*	SnRK1	From plasma membrane to nucleus	Removal of N-myristoylation of SnRK1 β1 subunit	Allows the binding of SnRK1 to its transcription factor targets including bZIP63, FUS3, IDD8, EIN3, WRI1, MYC2 in the nucleus. Upon phosphorylation by SnRK1, these transcription factors would either have reduced activities or be degraded. In turn, sugar and amino acid metabolism, oil synthesis, seed maturation and germination, flowering, jasmonic acid, ethylene, and abscisic acid signaling controlled by these transcription factors would be affected	[118]
Sugar sensing and signal trans-duction	*Petunia hybrida*	CaM53	From plasma membrane to nucleus	Loss of geranylation by geranylgeranyl transferases (GGTase-I)	In darkness or at low sugar levels, CaM53 is not geranylated and is localized in the nucleus. With light and sugar accumulation, CaM53 is geranylated by GGTase-I and becomes plasma membrane-bound	[119]

Author Contributions: Y.-S.K. put together the first complete draft. Y.-S.K. and H.-M.L. prepared the final version. H.-M.L. acquired the funding. All authors contributed to the search of the literature and to the writing. All authors have read and agreed to the published version of the manuscript.

Funding: This work was supported by grants from the Hong Kong Research Grants Council: General Research Fund (14164617) and Area of Excellence Scheme (AoE/M-403/16).

Institutional Review Board Statement: Not applicable.

Informed Consent Statement: Not applicable.

Data Availability Statement: Not applicable.

Acknowledgments: J.Y. Chu copy-edited this manuscript. Any opinions, findings, conclusions, or recommendations expressed in this publication do not reflect the views of the Government of the Hong Kong Special Administrative Region or the Innovation and Technology Commission.

Conflicts of Interest: The authors declare no conflict of interest.

References

1. Novaković, L.; Guo, T.; Bacic, A.; Sampathkumar, A.; Johnson, K.L. Hitting the wall—Sensing and signaling pathways involved in plant cell wall remodeling in response to abiotic stress. *Plants* **2018**, *7*, 89. [CrossRef] [PubMed]
2. Giglione, C.; Fieulaine, S.; Meinnel, T. N-terminal protein modifications: Bringing back into play the ribosome. *Biochimie* **2015**, *114*, 134–146. [CrossRef] [PubMed]
3. Gordon, J.I.; Duronio, R.J.; Rudnick, D.A.; Adams, S.P.; Gokel, G.W. Protein N-myristoylation. *J. Biol. Chem.* **1991**, *266*, 8647–8650. [CrossRef]
4. Bhatnagar, R.S.; Ashrafi, K.; Fütterer, K.; Waksman, G.; Gordon, J.I. Biology and enzymology of protein N-myristoylation. *Enzymes* **2001**, *21*, 241–290.
5. Boisson, B.; Giglione, C.; Meinnel, T. Unexpected protein families including cell defense components feature in the N-myristoylome of a higher eukaryote. *J. Biol. Chem.* **2003**, *278*, 43418–43429. [CrossRef]
6. Marmagne, A.; Ferro, M.; Meinnel, T.; Bruley, C.; Kuhn, L.; Garin, J.; Barbier-Brygoo, H.; Ephritikhine, G. A high content in lipid-modified peripheral proteins and integral receptor kinases features in the *Arabidopsis* plasma membrane proteome. *Mol. Cell. Proteom.* **2007**, *6*, 1980–1996. [CrossRef]
7. Zhang, F.L.; Casey, P.J. Protein prenylation: Molecular mechanisms and functional consequences. *Annu. Rev. Biochem.* **1996**, *65*, 241–469. [CrossRef]
8. Casey, P.J.; Seabra, M.C. Protein prenyltransferases. *J. Biol. Chem.* **1996**, *271*, 5289–5292. [CrossRef]
9. Kinsella, B.T.; Maltese, W.A. Rab GTP-binding proteins implicated in vesicular transport are isoprenylated *in vitro* at cysteines within a novel carboxyl-terminal motif. *J. Biol. Chem.* **1991**, *266*, 8540–8544. [CrossRef]
10. Hemsley, P.A.; Weimar, T.; Lilley, K.; Dupree, P.; Grierson, C. Palmitoylation in plants. *Plant Signal. Behav.* **2013**, *8*, 8–10. [CrossRef]

11. Akimzhanov, A.M.; Boehning, D.; Snyder, S.H. Rapid and transient palmitoylation of the tyrosine kinase Lck mediates Fas signaling. *Proc. Natl. Acad. Sci. USA* **2015**, *112*, 11876–11880. [CrossRef]
12. Roth, A.F.; Feng, Y.; Chen, L.; Davis, N.G. The yeast DHHC cysteine-rich domain protein Akr1p is a palmitoyl transferase. *J. Cell Biol.* **2002**, *159*, 23–28. [CrossRef]
13. Hemsley, P.A.; Weimar, T.; Lilley, K.S.; Dupree, P.; Grierson, C.S. A proteomic approach identifies many novel palmitoylated proteins in Arabidopsis. *New Phytol.* **2013**, *197*, 805–814. [CrossRef]
14. Harwood, J. Fatty acid metabolism. *Annu. Rev. Plant Physiol. Plant Mol. Biol.* **1988**, *39*, 101–138. [CrossRef]
15. Shanklin, J.; Cahoon, E.B. Desaturation and related modifications of fatty acids. *Annu. Rev. Plant Biol.* **1998**, *49*, 611–641. [CrossRef]
16. Vidugiriene, J.; Menon, A.K. Biosynthesis of glycosylphosphatidylinositol anchors. *Methods Enzymol.* **1995**, *250*, 513–535.
17. Oxley, D.; Bacic, A. Structure of the glycosylphosphatidylinositol anchor of an arabinogalactan protein from *Pyrus communis* suspension-cultured cells. *Proc. Natl. Acad. Sci. USA* **1999**, *96*, 14246–14251. [CrossRef]
18. Kinoshita, T.; Fujita, M. Biosynthesis of GPI-anchored proteins: Special emphasis on GPI lipid remodeling. *J. Lipid Res.* **2016**, *57*, 6–24. [CrossRef]
19. Zhou, K. Glycosylphosphatidylinositol-anchored proteins in *Arabidopsis* and one of their common roles in signaling transduction. *Front. Plant Sci.* **2019**, *10*, 1022. [CrossRef]
20. Takeda, J.; Kinoshita, T. GPI-anchor biosynthesis. *Trends Biochem. Sci.* **1995**, *20*, 367–371. [CrossRef]
21. Eisenhaber, B.; Maurer-Stroh, S.; Novatchkova, M.; Schneider, G.; Eisenhaber, F. Enzymes and auxiliary factors for GPI lipid anchor biosynthesis and post-translational transfer to proteins. *Bioessays* **2003**, *25*, 367–385. [CrossRef] [PubMed]
22. Lemmon, M.A. Membrane recognition by phospholipid-binding domains. *Nat. Rev. Mol. Cell Biol.* **2008**, *9*, 99–111. [CrossRef] [PubMed]
23. Cozier, G.E.; Carlton, J.; Bouyoucef, D.; Cullen, P.J. Membrane targeting by pleckstrin homology domains. *Curr. Top. Microbiol. Immunol.* **2003**, *282*, 49–88.
24. Tichá, M.; Richter, H.; Ovečka, M.; Maghelli, N.; Hrbáčková, M.; Dvořák, P.; Šamaj, J.; Šamajová, O. Advanced microscopy reveals complex developmental and subcellular localization patterns of ANNEXIN 1 in *Arabidopsis. Front. Plant Sci.* **2020**, *11*, 1153. [CrossRef] [PubMed]
25. Ben Saad, R.; Ben Romdhane, W.; Ben Hsouna, A.; Mihoubi, W.; Harbaoui, M.; Brini, F. Insights into plant annexins function in abiotic and biotic stress tolerance. *Plant Signal. Behav.* **2020**, *15*, e1699264. [CrossRef] [PubMed]
26. Pont-Lezica, R.F.; McNally, J.G.; Pickard, B.G. Wall-to-membrane linkers in onion epidermis: Some hypotheses. *Plant. Cell Environ.* **1993**, *16*, 111–123. [CrossRef]
27. Rui, Y.; Dinneny, J.R. A wall with integrity: Surveillance and maintenance of the plant cell wall under stress. *New Phytol.* **2020**, *225*, 1428–1439. [CrossRef]
28. Hoang, X.L.T.; Nguyen, Y.-N.H.; Thao, N.P.; Tran, L.-S.P. NAC transcription factors in drought and salinity tolerance. In *Salt and Drought Stress Tolerance in Plants, Signaling and Communication in Plants*; Hasanuzzaman, M., Tanveer, M., Eds.; Springer: Cham, Switzerland, 2020; pp. 351–366. ISBN 9783030402778.
29. Duan, M.; Zhang, R.; Zhu, F.; Zhang, Z.; Gou, L.; Wen, J.; Dong, J.; Wang, T. A lipid-anchored NAC transcription factor is translocated into the nucleus and activates *Glyoxalase I* expression during drought stress. *Plant Cell* **2017**, *29*, 1748–1772. [CrossRef]
30. Singh, R.K.; Deshmukh, R.; Muthamilarasan, M.; Rani, R.; Prasad, M. Versatile roles of aquaporin in physiological processes and stress tolerance in plants. *Plant Physiol. Biochem.* **2020**, *149*, 178–189. [CrossRef]
31. Kapilan, R.; Vaziri, M.; Zwiazek, J.J. Regulation of aquaporins in plants under stress. *Biol. Res.* **2018**, *51*, 4. [CrossRef]
32. Lopez, D.; Venisse, J.S.; Fumanal, B.; Chaumont, F.; Guillot, E.; Daniels, M.J.; Cochard, H.; Julien, J.L.; Gousset-Dupont, A. Aquaporins and leaf hydraulics: Poplar sheds new light. *Plant Cell Physiol.* **2013**, *54*, 1963–1975. [CrossRef] [PubMed]
33. Chu, T.T.H.; Hoang, T.G.; Trinh, D.C.; Bureau, C.; Meynard, D.; Vernet, A.; Ingouff, M.; Do, N.V.; Périn, C.; Guiderdoni, E.; et al. Sub-cellular markers highlight intracellular dynamics of membrane proteins in response to abiotic treatments in rice. *Rice* **2018**, *11*, 23. [CrossRef] [PubMed]
34. Boursiac, Y.; Chen, S.; Luu, D.T.; Sorieul, M.; Van Den Dries, N.; Maurel, C. Early effects of salinity on water transport in Arabidopsis roots. Molecular and cellular features of aquaporin expression. *Plant Physiol.* **2005**, *139*, 790–805. [CrossRef] [PubMed]
35. Leng, H.; Jiang, C.; Song, X.; Lu, M.; Wan, X. Poplar aquaporin *PIP1;1* promotes *Arabidopsis* growth and development. *BMC Plant Biol.* **2021**, *21*, 253. [CrossRef] [PubMed]
36. Bohnert, H.J.; Cushman, J.C. The ice plant cometh: Lessons in abiotic stress tolerance. *J. Plant Growth Regul.* **2000**, *19*, 334–346. [CrossRef]
37. Vera-Estrella, R.; Barkla, B.J.; Bohnert, H.J.; Pantoja, O. Novel regulation of aquaporins during osmotic stress. *Plant Physiol.* **2004**, *135*, 2318–2329. [CrossRef]
38. Ariani, A.; Barozzi, F.; Sebastiani, L.; di Toppi, L.S.; di Sansebastiano, G.P.; Andreucci, A. AQUA1 is a mercury sensitive poplar aquaporin regulated at transcriptional and post-translational levels by Zn stress. *Plant Physiol. Biochem.* **2019**, *135*, 588–600. [CrossRef]
39. Leshem, Y.; Seri, L.; Levine, A. Induction of phosphatidylinositol 3-kinase-mediated endocytosis by salt stress leads to intracellular production of reactive oxygen species and salt tolerance. *Plant J.* **2007**, *51*, 185–197. [CrossRef]

40. Luu, D.-T.; Martinière, A.; Sorieul, M.; Runions, J.; Maurel, C. Fluorescence recovery after photobleaching reveals high cycling dynamics of plasma membrane aquaporins in Arabidopsis roots under salt stress. *Plant J.* **2012**, *69*, 894–905. [CrossRef]
41. Li, X.; Wang, X.; Yang, Y.; Li, R.; He, Q.; Fang, X.; Luu, D.-T.; Maurel, C.; Lin, J. Single-molecule analysis of PIP2;1 dynamics and partitioning reveals multiple modes of *Arabidopsis* plasma membrane aquaporin regulation. *Plant Cell* **2011**, *23*, 3780–3797. [CrossRef]
42. Boursiac, Y.; Boudet, J.; Postaire, O.; Luu, D.-T.; Tournaire-Roux, C.; Maurel, C. Stimulus-induced downregulation of root water transport involves reactive oxygen species-activated cell signalling and plasma membrane intrinsic protein internalization. *Plant J.* **2008**, *56*, 207–218. [CrossRef]
43. Dhonukshe, P.; Aniento, F.; Hwang, I.; Robinson, D.G.; Mravec, J.; Stierhof, Y.-D.; Frimi, J. Clathrin-mediated constitutive endocytosis of PIN auxin efflux carriers in *Arabidopsis*. *Curr. Biol.* **2007**, *17*, 520–527. [CrossRef]
44. Ueda, M.; Tsutsumi, N.; Fujimoto, M. Salt stress induces internalization of plasma membrane aquaporin into the vacuole in *Arabidopsis thaliana*. *Biochem. Biophys. Res. Commun.* **2016**, *474*, 742–746. [CrossRef]
45. Ji, H.; Pardo, J.M.; Batelli, G.; Van Oosten, M.J.; Bressan, R.A.; Li, X. The salt overly sensitive (SOS) pathway: Established and emerging roles. *Mol. Plant* **2013**, *6*, 275–286. [CrossRef]
46. Qiu, Q.-S.; Guo, Y.; Dietrich, M.A.; Schumaker, K.S.; Zhu, J.-K. Regulation of SOS1, a plasma membrane Na^+/H^+ exchanger in *Arabidopsis thaliana*, by SOS2 and SOS3. *Proc. Natl. Acad. Sci. USA* **2002**, *99*, 8436–8441. [CrossRef]
47. Liu, J.; Ishitani, M.; Halfter, U.; Kim, C.-S.; Zhu, J.-K. The *Arabidopsis thaliana SOS2* gene encodes a protein kinase that is required for salt tolerance. *Proc. Natl. Acad. Sci. USA* **2000**, *97*, 3730–3734. [CrossRef]
48. Halfter, U.; Ishitani, M.; Zhu, J.-K. The *Arabidopsis* SOS2 protein kinase physically interacts with and is activated by the calcium-binding protein SOS3. *Proc. Natl. Acad. Sci. USA* **2000**, *97*, 3735–3740. [CrossRef]
49. Ishitani, M.; Liu, J.; Halfter, U.; Kim, C.-S.; Shi, W.; Zhu, J.-K. SOS3 function in plant salt tolerance requires N-myristoylation and calcium binding. *Plant Cell* **2000**, *12*, 1667–1677. [CrossRef]
50. Guo, Y.; Halfter, U.; Ishitani, M.; Zhu, J.-K. Molecular characterization of functional domains in the protein kinase SOS2 that is required for plant salt tolerance. *Plant Cell* **2001**, *13*, 1383–1399. [CrossRef]
51. Quan, R.; Lin, H.; Mendoza, I.; Zhang, Y.; Cao, W.; Yang, Y.; Shang, M.; Chen, S.; Pardo, J.M.; Guo, Y. SCABP8/CBL10, a putative calcium Sensor, interacts with the protein kinase SOS2 to protect *Arabidopsis* shoots from salt stress. *Plant Cell* **2007**, *19*, 1415–1431. [CrossRef]
52. Lou, L.; Yu, F.; Tian, M.; Liu, G.; Wu, Y.; Wu, Y.; Xia, R.; Pardo, J.M.; Guo, Y.; Xie, Q. ESCRT-I component VPS23A sustains salt tolerance by strengthening the SOS module in *Arabidopsis*. *Mol. Plant* **2020**, *13*, 1134–1148. [CrossRef] [PubMed]
53. Jiang, H.; Qi-Jun, M.; Zhong, M.-S.; Gao, H.-N.; Li, Y.-Y.; Hao, Y.-J. The apple palmitoyltransferase MdPAT16 influences sugar content and salt tolerance via an MdCBL1–MdCIPK13–MdSUT2.2 pathway. *Plant J.* **2021**, *106*, 689–705. [CrossRef] [PubMed]
54. Nakajima, Y.; Suzuki, S. Environmental stresses induce misfolded protein aggregation in plant cells in a microtubule-dependent manner. *Int. J. Mol. Sci.* **2013**, *14*, 7771–7783. [CrossRef] [PubMed]
55. Baniwal, S.K.; Bharti, K.; Chan, K.Y.; Fauth, M.; Arnab, G.; Kotak, S.; Mishra, S.K.; Nover, L.; Port, M.; Scharf, K.-D.; et al. Heat stress response in plants: A complex game with chaperones and more than twenty heat stress transcription factors. *J. Biosci.* **2004**, *29*, 471–487. [CrossRef] [PubMed]
56. Kotak, S.; Larkindale, J.; Lee, U.; von Koskull-Döring, P.; Vierling, E.; Scharf, K.D. Complexity of the heat stress response in plants. *Curr. Opin. Plant Biol.* **2007**, *10*, 310–316. [CrossRef]
57. Yadav, A.; Singh, J.; Ranjan, K.; Kumar, P.; Khanna, S.; Gupta, M.; Kumar, V.; Wani, S.H.; Sirohi, A. Heat shock proteins: Master players for heat-stress tolerance in plants during climate change. In *Heat Stress Tolerance in Plants: Physiological, Molecular and Genetic Perspectives*; Wani, S.H., Kumar, V., Eds.; John Wiley & Sons: Hoboken, NJ, USA, 2020; ISBN 9781119432364.
58. Zhu, J.K.; Bressan, R.A.; Hasegawa, P.M. Isoprenylation of the plant molecular chaperone ANJ1 facilitates membrane association and function at high temperature. *Proc. Natl. Acad. Sci. USA* **1993**, *90*, 8557–8561. [CrossRef]
59. Preisig-Müller, R.; Muster, G.; Kindl, H. Heat shock enhances the amount of prenylated Dnaj protein at membranes of glyoxysomes. *Eur. J. Biochem.* **1994**, *219*, 57–63. [CrossRef]
60. Wu, J.R.; Wang, T.Y.; Weng, C.P.; Duong, N.K.T.; Wu, S.J. AtJ3, a specific HSP40 protein, mediates protein farnesylation-dependent response to heat stress in Arabidopsis. *Planta* **2019**, *250*, 1449–1460. [CrossRef]
61. Barghetti, A.; Sjögren, L.; Floris, M.; Paredes, E.B.; Wenkel, S.; Brodersen, P. Heat-shock protein 40 is the key farnesylation target in meristem size control, abscisic acid signaling, and drought resistance. *Genes Dev.* **2017**, *31*, 2282–2295. [CrossRef]
62. Wang, T.Y.; Wu, J.R.; Duong, N.K.T.; Lu, C.A.; Yeh, C.H.; Wu, S.J. HSP70-4 and farnesylated AtJ3 constitute a specific HSP70/HSP40-based chaperone machinery essential for prolonged heat stress tolerance in *Arabidopsis*. *J. Plant Physiol.* **2021**, *261*, 153430. [CrossRef]
63. Wu, J.R.; Wang, L.C.; Lin, Y.R.; Weng, C.P.; Yeh, C.H.; Wu, S.J. The Arabidopsis *heat-intolerant 5 (hit5)/enhanced response to aba 1 (era1)* mutant reveals the crucial role of protein farnesylation in plant responses to heat stress. *New Phytol.* **2017**, *213*, 1181–1193. [CrossRef]
64. Li, G.L.; Chang, H.; Li, B.; Zhou, W.; Sun, D.Y.; Zhou, R.G. The roles of the atDjA2 and atDjA3 molecular chaperone proteins in improving thermotolerance of Arabidopsis thaliana seedlings. *Plant Sci.* **2007**, *173*, 408–416. [CrossRef]
65. Sjögren, L.; Floris, M.; Barghetti, A.; Völlmy, F.; Linding, R.; Brodersen, P. Farnesylated heat shock protein 40 is a component of membrane-bound RISC in *Arabidopsis*. *J. Biol. Chem.* **2019**, *293*, 16608–16622. [CrossRef]

66. Gao, H.; Brandizzi, F.; Benning, C.; Larkin, R.M. A membrane-tethered transcription factor defines a branch of the heat stress response in *Arabidopsis thaliana*. *Proc. Natl. Acad. Sci. USA* **2008**, *105*, 16397–16403. [CrossRef]
67. Sun, L.; Zhang, S.S.; Lu, S.J.; Liu, J.X. Site-1 protease cleavage site is important for the ER stress-induced activation of membrane-associated transcription factor bZIP28 in *Arabidopsis*. *Sci. China Life Sci.* **2015**, *58*, 270–275. [CrossRef]
68. Liu, X.H.; Lyu, Y.S.; Yang, W.; Yang, Z.T.; Lu, S.J.; Liu, J.X. A membrane-associated NAC transcription factor OsNTL3 is involved in thermotolerance in rice. *Plant Biotechnol. J.* **2020**, *18*, 1317–1329. [CrossRef]
69. Deng, Y.; Humbert, S.; Liu, J.X.; Srivastava, R.; Rothstein, S.J.; Howell, S.H. Heat induces the splicing by IRE1 of a mRNA encoding a transcription factor involved in the unfolded protein response in Arabidopsis. *Proc. Natl. Acad. Sci. USA* **2011**, *108*, 7247–7252. [CrossRef]
70. Lu, S.J.; Yang, Z.T.; Sun, L.; Sun, L.; Song, Z.T.; Liu, J.X. Conservation of IRE1-regulated *bZIP74* mRNA unconventional splicing in rice (*Oryza sativa* L.) involved in ER stress responses. *Mol. Plant* **2012**, *5*, 504–514. [CrossRef]
71. Lee, E.S.; Park, J.H.; Wi, S.D.; Chae, H.B.; Paeng, S.K.; Bae, S.B.; Phan, K.A.T.; Kim, M.G.; Kwak, S.S.; Kim, W.Y.; et al. Demyristoylation of the cytoplasmic redox protein trx-h2 is critical for inducing a rapid cold stress response in plants. *Antioxidants* **2021**, *10*, 1287. [CrossRef]
72. Lee, E.S.; Park, J.H.; Wi, S.D.; Kang, C.H.; Chi, Y.H.; Chae, H.B.; Paeng, S.K.; Ji, M.G.; Kim, W.Y.; Kim, M.G.; et al. Redox-dependent structural switch and CBF activation confer freezing tolerance in plants. *Nat. Plants* **2021**, *7*, 914–922. [CrossRef]
73. Ding, Y.; Lv, J.; Shi, Y.; Gao, J.; Hua, J.; Song, C.; Gong, Z.; Yang, S. EGR2 phosphatase regulates OST1 kinase activity and freezing tolerance in *Arabidopsis*. *EMBO J.* **2019**, *38*, e99819. [CrossRef] [PubMed]
74. Fiorillo, A.; Mattei, M.; Aducci, P.; Visconti, S.; Camoni, L. The salt tolerance related protein (STRP) mediates cold stress responses and abscisic acid signalling in *Arabidopsis thaliana*. *Front. Plant Sci.* **2020**, *11*, 1251. [CrossRef] [PubMed]
75. Saidi, Y.; Finka, A.; Muriset, M.; Bromberg, Z.; Weiss, Y.G.; Maathuis, F.J.M.; Goloubinoff, P. The heat shock response in moss plants is regulated by specific calcium-permeable channels in the plasma membrane. *Plant Cell* **2009**, *21*, 2829–2843. [CrossRef] [PubMed]
76. Monroy, A.F.; Sarhan, F.; Dhindsa, R.S. Cold-induced changes in freezing tolerance, protein phosphorylation, and gene expression: Evidence for a role of calcium. *Plant Physiol.* **1993**, *102*, 1227–1235. [CrossRef] [PubMed]
77. Knight, H. Calcium signaling during abiotic stress in plants. *Int. Rev. Cytol.* **1999**, *195*, 269–324.
78. Liu, Y.; Xu, C.; Zhu, Y.; Zhang, L.; Chen, T.; Zhou, F.; Chen, H.; Lin, Y. The calcium-dependent kinase OsCPK24 functions in cold stress responses in rice. *J. Integr. Plant Biol.* **2018**, *60*, 173–188. [CrossRef]
79. Wang, L.; Feng, X.; Yao, L.; Ding, C.; Lei, L.; Hao, X.; Li, N.; Zeng, J.; Yang, Y.; Wang, X. Characterization of CBL–CIPK signaling complexes and their involvement in cold response in tea plant. *Plant Physiol. Biochem.* **2020**, *154*, 195–203. [CrossRef]
80. Ludwig, A.A.; Romeis, T.; Jones, J.D.G. CDPK-mediated signalling pathways: Specificity and cross-talk. *J. Exp. Bot.* **2004**, *55*, 181–188. [CrossRef]
81. Dong, H.; Wu, C.; Luo, C.; Wei, M.; Qu, S.; Wang, S. Overexpression of *MdCPK1a* gene, a calcium dependent protein kinase in apple, increase tobacco cold tolerance via scavenging ROS accumulation. *PLoS ONE* **2020**, *15*, e0242139. [CrossRef]
82. Tao, W.; Lijuan, L.; Zeyu, L.; Lianguang, S.; Quan, W. Cloning and characterization of protein prenyltransferase alpha subunit in rice. *Rice Sci.* **2021**, *28*, 557–566. [CrossRef]
83. Tsai, T.M.; Chen, Y.R.; Kao, T.W.; Tsay, W.S.; Wu, C.P.; Huang, D.D.; Chen, W.H.; Chang, C.C.; Huang, H.J. *PaCDPK1*, a gene encoding calcium-dependent protein kinase from orchid, *Phalaenopsis amabilis*, is induced by cold, wounding, and pathogen challenge. *Plant Cell Rep.* **2007**, *26*, 1899–1908. [CrossRef]
84. Almadanim, M.C.; Alexandre, B.M.; Rosa, M.T.G.; Sapeta, H.; Leitão, A.E.; Ramalho, J.C.; Lam, T.K.T.; Negrão, S.; Abreu, I.A.; Oliveira, M.M. Rice calcium-dependent protein kinase OsCPK17 targets plasma membrane intrinsic protein and sucrose-phosphate synthase and is required for a proper cold stress response. *Plant Cell Environ.* **2017**, *40*, 1197–1213. [CrossRef] [PubMed]
85. Pawełek, A.; Duszyn, M.; Świeżawska, B.; Szmidt-Jaworska, A.; Jaworski, K. Transcriptional response of a novel *HpCDPK1* kinase gene from *Hippeastrum x hybr.* to wounding and fungal infection. *J. Plant Physiol.* **2017**, *216*, 108–117. [CrossRef] [PubMed]
86. Hettenhausen, C.; Sun, G.; He, Y.; Zhuang, H.; Sun, T.; Qi, J.; Wu, J. Genome-wide identification of calcium-dependent protein kinases in soybean and analyses of their transcriptional responses to insect herbivory and drought stress. *Sci. Rep.* **2016**, *6*, 18973. [CrossRef] [PubMed]
87. Xu, Y.; Liu, F.; Zhu, S.; Li, X. The maize NBS-LRR gene *ZmNBS25* enhances disease resistance in rice and Arabidopsis. *Front. Plant Sci.* **2018**, *9*, 1033. [CrossRef]
88. Yuan, P.; Yang, T.; Poovaiah, B.W. Calcium signaling-mediated plant response to cold stress. *Int. J. Mol. Sci.* **2018**, *19*, 3896. [CrossRef]
89. Zhao, Y.; Du, H.; Wang, Y.; Wang, H.; Yang, S.; Li, C.; Chen, N.; Yang, H.; Zhang, Y.; Zhu, Y.; et al. The calcium-dependent protein kinase ZmCDPK7 functions in heat-stress tolerance in maize. *J. Integr. Plant Biol.* **2021**, *63*, 510–527. [CrossRef]
90. Zhao, J.; Wang, S.; Qin, J.; Sun, C.; Liu, F. The lipid transfer protein OsLTPL159 is involved in cold tolerance at the early seedling stage in rice. *Plant Biotechnol. J.* **2020**, *18*, 756–769. [CrossRef]
91. Hu, X.; Wu, L.; Zhao, F.; Zhang, D.; Li, N.; Zhu, G.; Li, C.; Wang, W. Phosphoproteomic analysis of the response of maize leaves to drought, heat and their combination stress. *Front. Plant Sci.* **2015**, *6*, 298. [CrossRef]

92. Liu, H.; Che, Z.; Zeng, X.; Zhou, X.; Sitoe, H.M.; Wang, H.; Yu, D. Genome-wide analysis of calcium-dependent protein kinases and their expression patterns in response to herbivore and wounding stresses in soybean. *Funct. Integr. Genom.* **2016**, *16*, 481–493. [CrossRef]
93. Kanchiswamy, C.N.; Takahashi, H.; Quadro, S.; Maffei, M.E.; Bossi, S.; Bertea, C.; Zebelo, S.A.; Muroi, A.; Ishihama, N.; Yoshioka, H.; et al. Regulation of Arabidopsis defense responses against *Spodoptera littoralis* by CPK-mediated calcium signaling. *BMC Plant Biol.* **2010**, *10*, 97. [CrossRef]
94. Chico, J.M.; Raíces, M.; Téllez-Iñón, M.T.; Ulloa, R.M. A calcium-dependent protein kinase is systemically induced upon wounding in tomato plants. *Plant Physiol.* **2002**, *128*, 256–270. [CrossRef]
95. Szczegielniak, J.; Klimecka, M.; Liwosz, A.; Ciesielski, A.; Kaczanowski, S.; Dobrowolska, G.; Harmon, A.C.; Muszyńska, G. A wound-responsive and phospholipid-regulated maize calcium-dependent protein kinase. *Plant Physiol.* **2005**, *139*, 1970–1983. [CrossRef]
96. Romeis, T.; Piedras, P.; Jones, J.D.G. Resistance gene-dependent activation of a calcium-dependent protein kinase in the plant defense response. *Plant Cell* **2000**, *12*, 803–815. [CrossRef]
97. Cheung, M.Y.; Zeng, N.Y.; Tong, S.W.; Li, W.Y.F.; Xue, Y.; Zhao, K.J.; Wang, C.; Zhang, Q.; Fu, Y.; Sun, Z.; et al. Constitutive expression of a rice GTPase-activating protein induces defense responses. *New Phytol.* **2008**, *179*, 530–545. [CrossRef]
98. Cheung, M.Y.; Xue, Y.; Zhou, L.; Li, M.W.; Sun, S.S.M.; Lam, H.M. An ancient P-loop GTPase in rice is regulated by a higher plant-specific regulatory protein. *J. Biol. Chem.* **2010**, *285*, 37359–37369. [CrossRef]
99. Kim, C.Y.; Koo, Y.D.; Jin, J.B.; Moon, B.C.; Kang, C.H.; Kim, S.T.; Park, B.O.; Lee, S.Y.; Kim, M.L.; Hwang, I.; et al. Rice C2-domain proteins are induced and translocated to the plasma membrane in response to a fungal elicitor. *Biochemistry* **2003**, *42*, 11625–11633. [CrossRef]
100. Zhang, G.; Sun, Y.F.; Li, Y.M.; Dong, Y.L.; Huang, X.L.; Yu, Y.T.; Wang, J.M.; Wang, X.M.; Wang, X.J.; Kang, Z.S. Characterization of a wheat C2 domain protein encoding gene regulated by stripe rust and abiotic stresses. *Biol. Plant.* **2013**, *57*, 701–710. [CrossRef]
101. Medina-Puche, L.; Tan, H.; Dogra, V.; Wu, M.; Rosas-Diaz, T.; Wang, L.; Ding, X.; Zhang, D.; Fu, X.; Kim, C.; et al. A defense pathway linking plasma membrane and chloroplasts and co-opted by pathogens. *Cell* **2020**, *182*, 1109–1124.e25. [CrossRef]
102. Nomura, H.; Komori, T.; Uemura, S.; Kanda, Y.; Shimotani, K.; Nakai, K.; Furuichi, T.; Takebayashi, K.; Sugimoto, T.; Sano, S.; et al. Chloroplast-mediated activation of plant immune signalling in Arabidopsis. *Nat. Commun.* **2012**, *3*, 926. [CrossRef]
103. Arimura, G.I.; Maffei, M.E. Calcium and secondary CPK signaling in plants in response to herbivore attack. *Biochem. Biophys. Res. Commun.* **2010**, *400*, 455–460. [CrossRef] [PubMed]
104. Greco, M.; Chiappetta, A.; Bruno, L.; Bitonti, M.B. The interaction of plant biotic and abiotic stresses: From genes to the field. *J. Exp. Bot.* **2012**, *63*, 3523–3544.
105. Mansoor, S.; Wani, O.A.; Lone, J.K.; Manhas, S.; Kour, N.; Alam, P.; Ahmad, A.; Ahmad, P. Reactive oxygen species in plants: From source to sink. *Antioxidants* **2022**, *11*, 225. [CrossRef] [PubMed]
106. Wang, H.; Zhang, L.; Tao, Y.; Wang, Z.; Shen, D.; Dong, H. Transmembrane helices 2 and 3 determine the localization of plasma membrane intrinsic proteins in eukaryotic cells. *Front. Plant Sci.* **2020**, *10*, 1671. [CrossRef] [PubMed]
107. Wudick, M.M.; Li, X.; Valentini, V.; Geldner, N.; Chory, J.; Lin, J.; Maurel, C.; Luu, D.-T. Subcellular redistribution of root aquaporins induced by hydrogen peroxide. *Mol. Plant* **2015**, *8*, 1103–1114. [CrossRef]
108. Del Rodríguez-Gacio, M.C.; Matilla-Vázquez, M.A.; Matilla, A.J. Seed dormancy and ABA signaling: The breakthrough goes on. *Plant Signal. Behav.* **2009**, *4*, 1035–1048. [CrossRef]
109. Tran, L.S.P.; Urao, T.; Qin, F.; Maruyama, K.; Kakimoto, T.; Shinozaki, K.; Yamaguchi-Shinozaki, K. Functional analysis of AHK1/ATHK1 and cytokinin receptor histidine kinases in response to abscisic acid, drought, and salt stress in *Arabidopsis*. *Proc. Natl. Acad. Sci. USA* **2007**, *104*, 20623–20628. [CrossRef]
110. Máthé, C.; Garda, T.; Freytag, C.; M-Hamvas, M. The role of serine-threonine protein phosphatase PP2A in plant oxidative stress signaling-facts and hypotheses. *Int. J. Mol. Sci.* **2019**, *20*, 3028. [CrossRef]
111. Ku, Y.-S.; Sintaha, M.; Cheung, M.-Y.; Lam, H.-M. Plant hormone signaling crosstalks between biotic and abiotic stress responses. *Int. J. Mol. Sci.* **2018**, *19*, 3206. [CrossRef]
112. Park, S.-Y.; Fung, P.; Nishimura, N.; Jensen, D.R.; Fujii, H.; Zhao, Y.; Lumba, S.; Santiago, J.; Rodrigues, A.; Chow, T.-F.F.; et al. Abscisic acid inhibits type 2C protein phosphatases via the PYR/PYL family of START proteins. *Science* **2009**, *324*, 1068–1071. [CrossRef]
113. Belda-Palazon, B.; Julian, J.; Coego, A.; Wu, Q.; Zhang, X.; Batistic, O.; Alquraishi, S.A.; Kudla, J.; An, C.; Rodriguez, P.L. ABA inhibits myristoylation and induces shuttling of the RGLG1 E3 ligase to promote nuclear degradation of PP2CA. *Plant J.* **2019**, *98*, 813–825. [CrossRef]
114. Clause, S.D.; Sasse, J.M. Brassinosteroids: Essential regulators of plant growth and development. *Annu. Rev. Plant Biol.* **1998**, *49*, 427–451. [CrossRef]
115. Su, B.; Zhang, X.; Li, L.; Abbas, S.; Yu, M.; Cui, Y.; Baluška, F.; Hwang, I.; Shan, X.; Lin, J. Dynamic spatial reorganization of BSK1 complexes in the plasma membrane underpins signal-specific activation for growth and immunity. *Mol. Plant* **2021**, *14*, 588–603. [CrossRef]
116. Su, B.; Wang, A.; Shan, X. The role of N-myristoylation in homeostasis of brassinosteroid signaling kinase 1. *Planta* **2022**, *255*, 73. [CrossRef]

117. Wurzinger, B.; Nukarinen, E.; Nägele, T.; Weckwerth, W.; Teige, M. The SnRK1 kinase as central mediator of energy signaling between different organelles. *Plant Physiol.* **2018**, *176*, 1085–1094. [CrossRef]
118. Pierre, M.; Traverso, J.A.; Boisson, B.; Domenichini, S.; Bouchez, D.; Giglione, C.; Meinnel, T. N-myristoylation regulates the SnRK1 pathway in arabidopsis. *Plant Cell* **2007**, *19*, 2804–2821. [CrossRef]
119. Rodríguez-Concepción, M.; Yalovsky, S.; Zik, M.; Fromm, H.; Gruissem, W. The prenylation status of a novel plant calmodulin directs plasma membrane or nuclear localization of the protein. *EMBO J.* **1999**, *18*, 1996–2007. [CrossRef]
120. Wolff, H.; Jakoby, M.; Stephan, L.; Koebke, E.; Hülskamp, M. Heat stress-dependent association of membrane trafficking proteins with mRNPs is selective. *Front. Plant Sci.* **2021**, *12*, 670499. [CrossRef]
121. Barth, O.; Zschiesche, W.; Siersleben, S.; Humbeck, K. Isolation of a novel barley cDNA encoding a nuclear protein involved in stress response and leaf senescence. *Physiol. Plant.* **2004**, *121*, 282–293. [CrossRef]
122. Barth, O.; Vogt, S.; Uhlemann, R.; Zschiesche, W.; Humbeck, K. Stress induced and nuclear localized HIPP26 from *Arabidopsis thaliana* interacts via its heavy metal associated domain with the drought stress related zinc finger transcription factor ATHB29. *Plant Mol. Biol.* **2009**, *69*, 213–226. [CrossRef]

membranes

MDPI

Communication

ssPINE: Probabilistic Algorithm for Automated Chemical Shift Assignment of Solid-State NMR Data from Complex Protein Systems

Adilakshmi Dwarasala [1], Mehdi Rahimi [1], John L. Markley [1,2] and Woonghee Lee [1,*]

[1] Department of Chemistry, University of Colorado Denver, Denver, CO 80217, USA
[2] Department of Biochemistry, University of Wisconsin-Madison, Madison, WI 53706, USA
* Correspondence: woonghee.lee@ucdenver.edu

Abstract: The heightened dipolar interactions in solids render solid-state NMR (ssNMR) spectra more difficult to interpret than solution NMR spectra. On the other hand, ssNMR does not suffer from severe molecular weight limitations like solution NMR. In recent years, ssNMR has undergone rapid technological developments that have enabled structure–function studies of increasingly larger biomolecules, including membrane proteins. Current methodology includes stable isotope labeling schemes, non-uniform sampling with spectral reconstruction, faster magic angle spinning, and innovative pulse sequences that capture different types of interactions among spins. However, computational tools for the analysis of complex ssNMR data from membrane proteins and other challenging protein systems have lagged behind those for solution NMR. Before a structure can be determined, thousands of signals from individual types of multidimensional ssNMR spectra of samples, which may have differing isotopic composition, must be recognized, correlated, categorized, and eventually assigned to atoms in the chemical structure. To address these tedious steps, we have developed an automated algorithm for ssNMR spectra called "ssPINE". The ssPINE software accepts the sequence of the protein plus peak lists from a variety of ssNMR experiments as inputs and offers automated backbone and side-chain assignments. The alpha version of ssPINE, which we describe here, is freely available through a web submission form.

Keywords: solid-state NMR; membrane proteins; assignment; automation; ssPINE; MAS-NMR

1. Introduction

NMR spectroscopy is one of the major biophysical methods, along with X-ray crystallography [1,2] and cryo-electron microscopy [3], for determining structures of biomolecules. NMR is used to study structure–function relationships of membrane proteins and large macromolecular assemblies [4] along with their interactions with small molecules [5] as an approach to drug discovery [6].

Both solution and solid-state NMR techniques provide important information about the structures and dynamics of membrane proteins [7,8]. Solid-state NMR (ssNMR) with magic angle spinning (MAS) has advantages over solution NMR for studies of large and immobilized proteins [9,10]. Anisotropic nuclear spin interaction information from ssNMR can be extremely useful for structure determination and dynamics [11,12]. The orientation of regions of membrane proteins can be extracted from ssNMR spectra of mechanically- or magnetically-aligned membranes [13]. The broad lines and low resolution of ssNMR spectra resulting from anisotropy can be overcome in part by ultra-high MAS, cross-polarization, refined pulse sequences [14], and non-uniform sampling (NUS). Ultra-high-field NMR spectrometers operating at 1.1 GHz and 1.2 GHz are improving the resolution and sensitivity of ssNMR spectra of membrane proteins and their complexes. The above-mentioned methods are enabling the collection of improved spectral data, but manual analysis of the data to obtain chemical shift assignments and structural constraints is tedious because thousands of signals need to be analyzed, correlated, and labeled.

Software technology has reduced the burden of analyzing data from solution NMR studies of biomolecules. Available web-based resources provide automated and semi-automated algorithms for determining different parameters of biomolecules and their structure [15–18]. We recently developed an updated version of the assignment engine PINE [19], I-PINE (Integrative Probabilistic Interaction Network of Evidence) [20], which utilizes a Bayesian-based probabilistic interaction network. I-PINE supports a larger range of NMR experiments and integrates real-time statistical analysis of the PACSY database [21]. The I-PINE web server produces higher assignment coverage and accuracy than PINE and supports structure determinations based on chemical shift assignments. The POKY suite includes iPick [22], for peak picking and cross-validation of peaks from different spectra, I-PINE, and PINE-SPARKY.2 [23], a user-friendly graphical user interface (GUI) for submitting, importing, and validating the data [24].

For ssNMR data, PISA-SPARKY [25], a plugin for the assignment program, NMRFAM-SPARKY [26], supports the analysis of data from oriented samples [27]. PISA-SPARKY, along with its features, are now included in the POKY suite. Recently, the Veglia group introduced "a one-shot approach" called PHORONESIS, which generates up to ten 3D ^{1}H-detected ssNMR spectra [28]. They used the I-PINE webserver to analyze the spectra, and found that the yield of sequential assignments was similar to that for solution NMR data. The Hunter Moseley and Chad Rienstra groups developed an ssNMR version of AutoAssign and demonstrated its ability to assign ssNMR data from the small protein, GB1 [29]. The software returned 84.1% correct assignments. The ssFLYA algorithm, which was introduced by Schmidt and colleagues [30], and is currently available for only commercial users, yielded 88–87% and 77–90% correctness on protein microcrystals and amyloids.

Here, we describe ssPINE (solid-state PINE), a software package that is designed to handle the challenging features of ssNMR data from membrane proteins and other complex protein systems. ssPINE accepts, as inputs, 2D and 3D ssNMR data and gives, as an output, chemical shift assignments and their probabilistic correctness. We have evaluated the performance of ssPINE with data from GB1 and with additional protein NMR data from the BMRB database [31]. The alpha version of ssPINE is freely available through a web server utility at https://poky.clas.ucdenver.edu/ssPINE.

2. Materials and Methods

2.1. ssPINE Algorithm

As its first step, ssPINE generates spin system matrices [32], as shown in Figure 1. The main difference between the I-PINE and ssPINE algorithms is in their approach to comparing peaks from different experiments. I-PINE uses N_i and H_i in root experiments to find correlated signals ($CA/CB/CO_{i-1}$, $CA/CB/CO_i$) in different experiments and to establish di-peptide arrays {$CA/CB/CO_{i-1}$ N_i $CA/CB/CO_i$}. Then, it establishes a vector [CA_{i-1}, CB_{i-1}, CO_{i-1}, N_i, CA_i, CB_i, CO_i] and compares it to [CA_{i-1}, CB_{i-1}, CO_{i-1}] and [CA_i, CB_i, CO_i] from the other di-peptide arrays. Finally, I-PINE compares [CA_{i-1}, CB_{i-1}, CO_{i-1}] to [CA_i, CB_i, CO_i] in all vectors. By contrast, ssPINE uses CO_{i-1}, N_i, and CA_i in root experiments to find di-peptide signals (CX_{i-1}, CX_i). If an experiment is providing CO_{i-1}, N_i, and CA_i, but a single peak is not provided (e.g., CANCO), ssPINE combines information from different experiments, such as NCOCACB and NCACB, to obtain these correlations. Unlike I-PINE, ssPINE generates each spin system matrix by iterated spectral resolution steps (tolerances) using inter-residue connectivities optimized by a probability approach. This is a basic and important component of the ssNMR algorithm because it helps to overcome the variable spectral resolution of ssNMR spectra. ssPINE calculates the quality of the data at each step until it reaches the point where there is no further improvement in the spin system matrix. If the quality of the data is above a threshold value, as determined by the number of spin systems and correlations between spin systems is identified compared to the numbers expected, then the process continues to the pentapeptide generation step. Otherwise, the process terminates and informs the user that more information is required. The pentapeptide generation step, which assigns signals to atoms in sequences of five amino acid residues, finds the best marginal probabilities by using the belief propagation algorithm [33] to evaluate relations between spin systems. This step includes the identification of secondary structural elements, the evaluation of possible referencing errors, and the continued assignment of backbone spin systems until convergence is reached or, alternatively, until the specified number of iterations has occurred. The last step utilizes the Bayesian network model of PINE and I-PINE to assign side chain signals [16,20]. See the Supporting Information for a detailed description of the ssPINE algorithm.

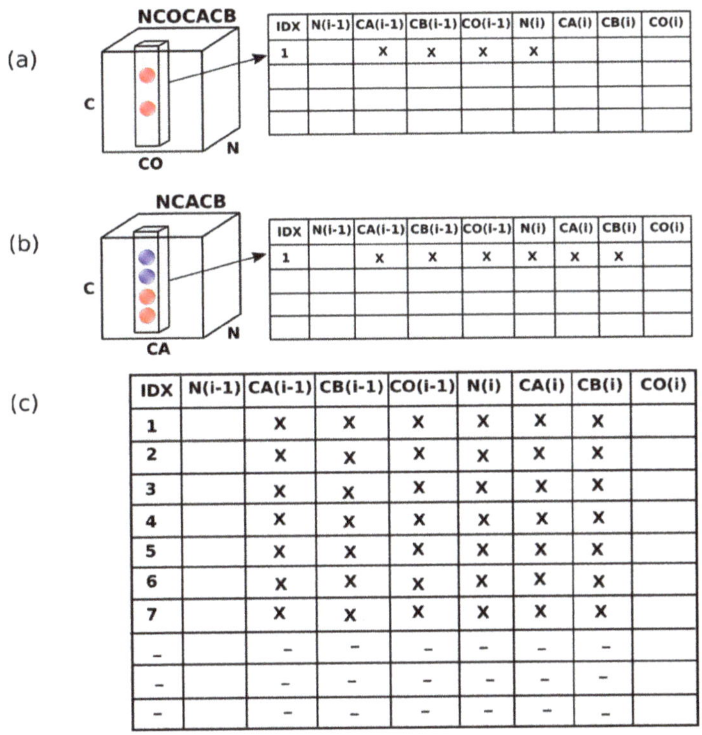

Figure 1. Spin system matrix assembly in ssPINE. (**a**) Peaks from the strip of the NCOCACB experiment containing the CA(i − 1) and CB(i − 1) resonances are inserted in the row of the table corresponding to the ith residue (IDX). Note that the peak is selected from the CO(i − 1) and N(i), in root information. (**b**) Similarly, peaks from the strip of the NCACB experiment containing the CA(i) and CB(i) resonances are inserted in the row of the table corresponding to the ith residue. N(i) and CA(i) in root information are used to select a peak from NCACB. (**c**) The process is repeated for all residues in the peptide sequence.

2.2. Input Files

As with PINE and I-PINE, peak lists (either raw or refined) and sequence files are used as inputs to ssPINE. The supported solid-state NMR experiments and their profiles are shown in Table 1. The minimum set of peak lists for assignments are those from 2D-CC, 2D-NCA, 2D-NCO, 3D-NCACX, 3D-NCOCX, and CAN(CO)CX ssNMR experiments. Data from additional ssNMR experiments can be added to improve the accuracy and completeness of the results.

Table 1. ssNMR experiments supported by ssPINE with their dimensionality and connectivity profiles. CX(i) represents carbon A, B, D, E, G, or H atoms of the ith residue; N(i) represents the nitrogen atom of the ith residue; and CO(i − 1) represents the carbon atom of the carboxyl group of the preceding residue. The minimum set of experiments needed is indicated by asterisks.

Experiment	Dimension	Profile
CC *	2D	CX/O(i)-CX/O(i)
NCA *	2D	N(i)-CA(i)
NCACB	2D	N(i)-CA/B(i)
NCO *	2D	N(i)-CO(i − 1)
NCACO	3D	N(i)-CA(i)-CO(i)
NCACB	3D	N(i)-CA(i)-CA/B(i)
NCACX *	3D	N(i)-CA(i)-CX(i)
NCOCX *	3D	N(i)-CO(i − 1)-CX/C(i − 1)
NCOCA	3D	N(i)-CO(i − 1)-CA(i − 1)
NCOCACB	3D	N(i)-CO(i − 1)-CA/B(i − 1)
CANCO	3D	CA(i)-N(i)-CO(i − 1)
CANCOCX *	3D	CA(i)-N(i)-CX/O(i − 1)
CANCOCA	3D	CA(i)-N(i)-CA/O(i − 1)
CANCOCACB	3D	CA(i)-N(i)-CO/A/B(i − 1)

* Minimum experiments to run ssPINE.

2.2.1. Preparation of Peak Lists

Several peak list formats are accepted: Sparky (UCSF-/NMRFAM-SPARKY or POKY) with the *.list* file extension prepared in the peak list window (two-letter-code "lt" with the *Data Heights* option turned on), XEASY with the *.peaks* file extension [34], nmrDraw with the *.ft2* file extension, NMRView with the *.xpk* file extension, and I-PINE with the *.txt* file extension. The file extension in the file name should match its actual format. Other programs can generate the Sparky format, which is one of the most common file formats in the field. For example, CARA has the *WriteSparkyPeakList.lua* script, and CCPNMR v2 has the *Format Converter* program [35,36]. The POKY suite contains multiple options for generating peak lists; of these, one of the easiest approaches is iPick. With iPick, the user simply selects one or more spectra from the session and clicks on the "Run iPick" button. After peak lists have been generated for each spectrum, the "Peak List" window opens, and by clicking on the "Save" button, the user can designate the names for the peak lists. Peak lists can be refined by hand or by software to remove noise or other spurious peaks.

2.2.2. Protein Sequence

ssPINE accepts peptide sequences in either one- or three-letter amino acid codes as ASCII text files. Sequences submitted in RTF (Rich Text Format; *.rtf*), ODT (OpenDocument Text; *.odt*), or DOCX (Office Open XML; *.docx*) are automatically converted to ASCII.

2.3. *Output Files*

The ssPINE output consists of several files: (1) The list of ssNMR experiments used. (2) A bar graph indicating the assignment probabilities of each residue in the protein (Figure 2). (3) Separate files with the backbone and sidechain chemical shift assignments of each residue of the protein in NMR-STAR 2.1 and 3.1 formats. (4) Sparky format assignment labels and frequencies. (5) Protein secondary structure prediction by PECAN (Protein Energetic Conformational Analysis from NMR chemical shifts) [37]. (6) Chemical shift referencing errors in each experiment, as detected by LACS (Linear Analysis of Chemical Shifts) [38], which is used in redefining offsets during the assignment iteration.

(It is recommended that the user use these values to correct the offset for each peak list when a job is resubmitted. This will reduce the computational time and improve the assignment accuracy).

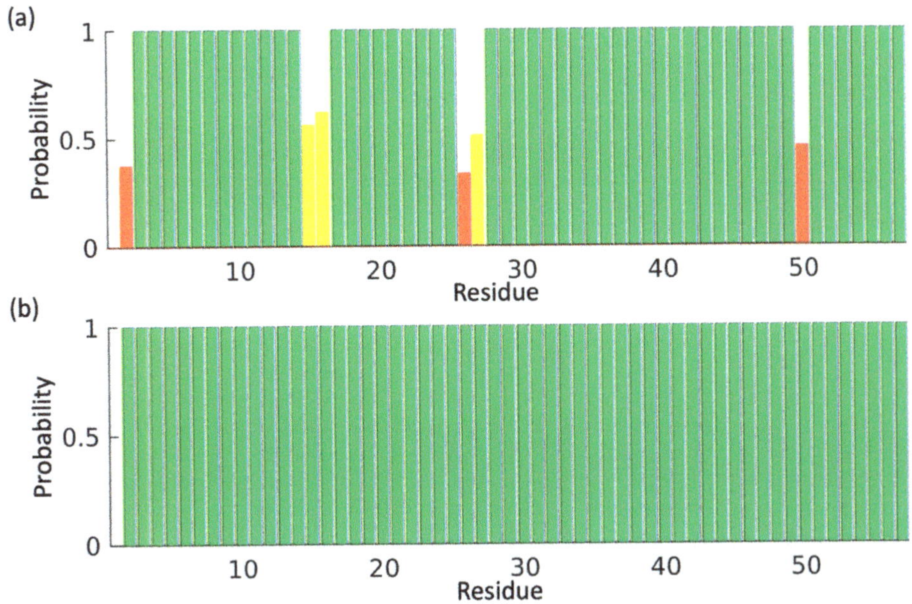

Figure 2. Bar graphs indicating the correct assignment probability (p) for each residue of GBI resulting from ssPINE analysis. Green indicates p *greater than* 0.99; cyan indicates p = 0.85–0.99; yellow indicates p = 0.5–0.84; red indicates p *less than* 0.5; and gray indicates no assignment (not seen with these test sets). (**a**) Unrefined GBI data as input. (**b**) Refined GBI data as input.

2.4. Data Used in Developing ssPINE

2.4.1. Data from GB1

In the early stages of developing ssPINE, we used unpublished ssNMR data from the uniformly $^{13}C/^{15}N$-labeled small (56 residue, 6.2 kDa) protein GB1 that was generously provided by Chad Rienstra's group. GB1, which is the streptococcal B1 immunoglobulin-binding domain of protein G20, has been used frequently as a standard sample in the development of NMR technology. We prepared both unrefined and refined peak lists from raw data from the following ssNMR experiments: 2D-CC, 2D-NCA, 2D-NCACB, 2D-NCO, 3D-NCACB, 3D-NCACX, 3D-NCACO, 3D-CANCO, 3D-CANCOCX, 3D-NCOCA, 3D-NCOCACB, and 3D-NCOCX. We prepared unrefined peak lists automatically with the iPick peak picking tool of POKY (two-letter-code *iP*). Subsequently, we created refined peak lists by using the cross-validation tool of iPick to weed out noise and non-sequential signals.

2.4.2. Other Protein NMR Data

Additional data from the PACSY database [21] were used in refining and optimizing the ssPINE algorithm. PACSY is a relational database that contains post-processed information from BMRB [39] and PDB [40]. Data from 82 proteins, including both large (181 residues) and small (26 residues) proteins, were included (SI Table S1). Most datasets were from solution NMR, because few ssNMR entries in the BMRB contain complete assignments. We created synthetic peak lists for 2D-CC, 2D-NCA, 2D-NCACB, 2D-NCO, 3D-NCACB, 3D-NCACX, 3D-NCACO, 3D-CANCO, 3D-CANCOCX, 3D-NCOCA, 3D-NCOCACB, and 3D-NCOCX ssNMR spectra of these proteins. For a more controlled evaluation, we only regarded sequential cross-peaks.

2.5. ssPINE Web Server

We utilized multiple technologies in implementing the ssPINE algorithm as a web server. Programs written in Perl, Python, and shell scripting handle various parts of the task. A web-facing server hosts a form that the user can fill out with their information: the amino acid sequence file and the peak lists from specified 2D and 3D solid-state NMR experiments. By clicking the "Submit" button, this information is validated and sent to a processing server. After the automated backbone and sidechain assignments are completed, the result is sent back to the user's email address. From there, the user can download all the result files. The actual running time is determined by the size of the protein and the complexity of the problem, including peak list quality provided by the user, but jobs usually require less than one hour. The ssPINE web server is hosted at the University of Colorado, Denver and is accessible at: https://poky.clas.ucdenver.edu/ssPINE. No login or signup is required, and the server is open to all researchers at no cost and processes submissions in the order in which they are received.

3. Results

We evaluated the results with GB1 in terms of their completeness and correctness. "Completeness" is the number of automatically-assigned chemical shifts by ssPINE divided by the number of assignments for GB1 derived from our manual assignment of the ssNMR data. "Correctness" is the number of correct assignments made by ssPINE divided by the number of manual assignments. Given that ssPINE provides multiple assignment candidates with associated probabilities, only the assignment candidate with the highest probability is used in the evaluation of completeness and correctness.

With the unrefined peak lists of GB1 as inputs, ssPINE yielded 100% (219/219) completeness and an average of 97.26% (213/219) correctness for the backbone chemical shift assignments (Figure 2a). With the refined peak lists of GB1, ssPINE yielded 100% (219/219) completeness and 100% (219/219) accuracy (Figure 2b).

We also tested ssPINE algorithm with synthetic peak lists from other proteins whose assigned chemical shifts had been deposited in BMRB (see Section 2.4.2). These BMRB assignments are assumed to be correct and were used in evaluating the correctness of the ssPINE results. The numbers of BMRB and ssPINE assignments were used, respectively, as the denominator and numerator in the completeness calculation. The number of valid ssPINE assignments ("given" assignments) at the different probability cutoffs were used as the denominator in the correctness calculation.

The total number of assignment candidates returned by ssPINE are plotted as a function of their probability scores in Figure 3a. They are shown as "correct", "incorrect", "given" (sum of correct and incorrect), and "all". The "all" category includes "given" plus invalid assignments, namely those with scores below the probability cutoff.

The correctness and completeness parameters for all assignment candidates with the highest probability for each protein are plotted with respect to their probability in Figure 3b. The correctness decreased moderately as a function of lower probability. The fact that it remained above 85% means that more than 85% of the given chemical shift values were assigned correctly. Overall, the completeness ranged between 85% and 97%. The completeness increased abruptly between 1.0 and 0.9 probability, and then more gradually to 0.0 probability. Plots of percentages of completeness versus correctness for each BMRB entry at each probability are given in SI Figure S2.

The unrefined GB1 peak lists led to a few incorrect $^{13}C^{\alpha}$ assignments (Figure 2a) because false signals picked by automated peak picking algorithm were close to the BMRB average chemical shift value. Manual refinement of the peak lists alleviated this problem by removing false-positive peaks, adding unpicked peaks, and resolving overlaps.

Of the 82 synthetic sets of peak lists analyzed by ssPINE, only three yielded assignment correctness below 70% with a probability cutoff of 0.5. These are denoted by red circles in SI Figure S2 and by red text in SI Table S1. One of the poorest scoring datasets (completeness = 84.5% (474/561); correctness = 66.9% (317/474)) corresponded to BMRB entry 15,716 (the AlgE6R1 subunit from the *Azotobacter vinelandii* Mannuronan C5-epimerase), a 153 amino acid protein containing 27 glycine residues with many overlapping peaks in the carbon alpha region (~45 ppm).

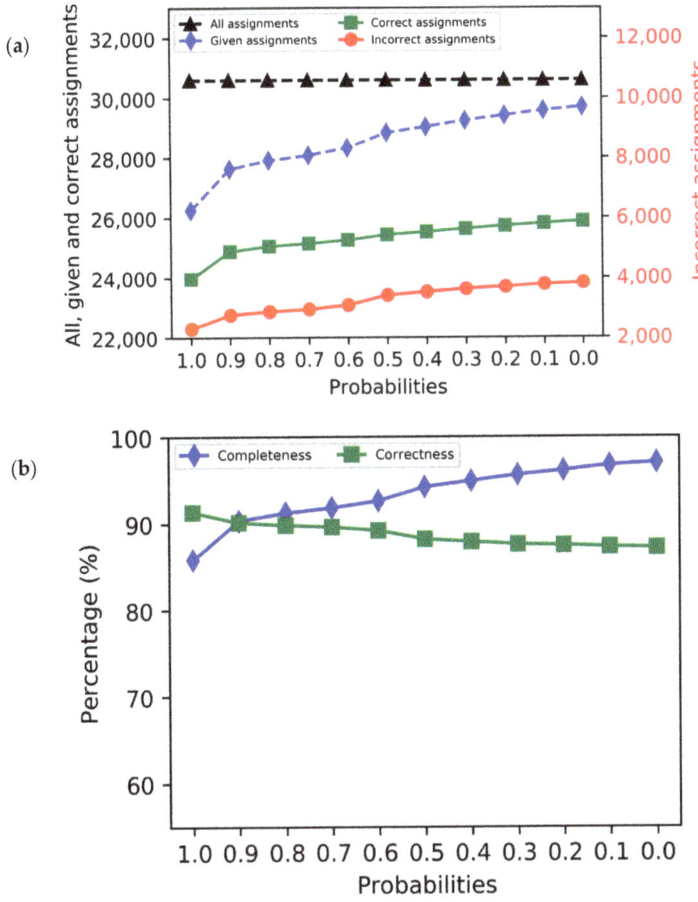

Figure 3. Results from ssPINE analysis of synthetic ssNMR data as averages for the 82 proteins studied. (**a**) Chemical shift assignment probabilities returned by ssPINE for all assignment candidates (x-axis) versus assignment type (y-axis). All (dashed black), given (dashed blue), and correct (solid green) assignments are represented by the numbers on the left side, whereas the incorrect assignments (solid red) are represented by the numbers on the right side. (**b**) Data from the assignment candidate for each protein with the highest assignment probability. Completeness (solid blue) and correctness (solid green) are plotted as a function of that assignment probability.

4. Discussion

In this report, we have introduced the ssPINE algorithm for the automated analysis and assignment of solid-state

NMR data from membrane proteins and other difficult protein systems. ssPINE builds on the technology of our I-PINE web server for solution NMR data, which serves several thousand jobs annually. We have adapted the I-PINE algorithm to account for the challenging features of ssNMR data from these systems. These include broader lines, extensive inter-residue dipolar interactions, and 2/3D ssNMR experiments that yield a variety of connectivities. As with I-PINE, ssPINE accepts the amino acid sequence of the protein and raw or refined peak lists as an input from a variety of NMR experiments (Table 1). The output of ssPINE includes peak assignments and their probabilities. We have tested and refined the implementation of the ssPINE algorithm with the excellent set of ssNMR data from the small protein, GB1. We also used ssPINE as an input for a set of synthetic peak lists that simulated ssNMR data from 82 other proteins of various sizes that were generated from solution NMR data deposited in BMRB. As shown above, the choice of probability cutoff is an important factor in maximizing correct assignments. In solution NMR, the recommended probability cutoff for I-PINE is 0.5 because it leads to a higher probability of correct assignments [20]. With ssNMR data, a cutoff of 0.6 appears to provide optimal completeness and assignment correctness. Glycine residues are harder to assign because they lack the CB signals that ssPINE uses to evaluate connectivities. Proteins that contain a high glycine content (e.g., BMRB entry 15,716) are particularly problematic because ssPINE has difficulty distinguishing among the several glycine candidates.

Currently, the user can use the ssPINE extension in POKY (two-letter-code *EP*) to generate and submit peak lists from the web browser to the ssPINE webserver. The user can use the *Convert (ss)I-PINE outputs to POKY plugin* in POKY (two-letter-code *ip*) to convert the assigned chemical shift table file from ssPINE to the POKY resonance list file with the chosen probability cutoff. Finally, the POKY Notepad (two-letter-code *Pn*) can be used to propagate assigned peaks onto ssNMR spectra: this is enabled by the script, *Simulate SSNMR peaks with assignments labels* (*predict-and-confirm*).

The analysis of ssNMR data from membrane proteins is highly challenging. ssPINE offers a promising approach for resolving the chemical, structural, and dynamic information contained in these spectra. Information of this kind is crucial for understanding the mechanisms underlying membrane transport, energy transfers, and signaling. We encourage feedback from users of ssPINE, particularly those analyzing ssNMR spectra of membrane proteins, as a means for guiding its further development. Our immediate goals with ssPINE are to incorporate information from strategies commonly used in NMR spectroscopy of membrane proteins, including mutational analysis, ^{19}F labeling, and/or selective isotopic labeling.

Longer-term plans are to develop and release a program (ssPINE-POKY) that will include a graphical user interface analogous to that in PINE-SPARKY.2 for solution NMR. In addition, we envision an "integrative" version of ssPINE that will increase assignment correctness and completeness by implementing adaptive probability density functions that incorporate machine learning (ML)-based chemical shift and structure prediction methods, and will provide a comprehensive visualization of structural and dynamic information from ssNMR data, which is analogous to that afforded by I-PINE for solution NMR data.

5. Web Server Availability

The usage of the webserver is described in Section 2.5. The web server for ssPINE is freely accessible at https://poky.clas.ucdenver.edu/ssPINE.

Supplementary Materials: The following are available online at https://www.mdpi.com/article/10.3390/membranes12090834/s1. Figure S1: Flowchart of the ssPINE algorithm; Figure S2: Scatter plots of percentages of completeness verses correctness of each BMRB entry at different probabilities. Red circle indicates the three poorly performing entries: 15716, 15797 and 19755. These are indicated by red text in SI Table S1; Table S1: List of BMRB entries used for the ssPINE performance benchmark. Poorly performing entries (15716, 15797 and 19755) are shown in red.

Author Contributions: Conceptualization, W.L.; methodology, W.L.; software, A.D., M.R. and W.L.; validation, A.D.; formal analysis, A.D.; resources, M.R.; data curation, A.D.; writing—original draft preparation, A.D., M.R. and W.L.; writing—review and editing, J.L.M.; supervision, W.L.; funding acquisition, W.L. and J.L.M. All authors have read and agreed to the published version of the manuscript.

Funding: This work has been supported by the NIH P41GM103399 (to J.L.M.), the NSF DBI-1902076 and DBI-2051595 (to W.L.), and new faculty research support from the University of Colorado, Denver (to W.L.).

Institutional Review Board Statement: Not applicable.

Informed Consent Statement: Not applicable.

Data Availability Statement: Not applicable.

Acknowledgments: We thank Chad Rienstra's group, Biochemistry Department, University of Wisconsin-Madison for providing the GB1 experimental data.

Conflicts of Interest: The authors declare no conflict of interest.

References

1. Ilari, A.; Savino, C. Protein Structure Determination by X-ray Crystallography. *Bioinformatics* **2008**, 63–87.
2. Fujiyoshi, Y. The Structural Study of Membrane Proteins by Electron Crystallography. *Adv. Biophys.* **1998**, *35*, 25–80. [CrossRef]
3. Allen O áHill, H. Immobilization of Small Proteins in Carbon Nanotubes: High-Resolution Transmission Electron Microscopy Study and Catalytic Activity. *J. Chem. Soc. Chem. Commun.* **1995**, *17*, 1803–1804.
4. Wüthrich, K. The Way to NMR Structures of Proteins. *Nat. Struct. Biol.* **2001**, *8*, 923–925. [CrossRef] [PubMed]
5. Shuker, S.B.; Hajduk, P.J.; Meadows, R.P.; Fesik, S.W. Discovering High-Affinity Ligands for Proteins: SAR by NMR. *Science* **1996**, *274*, 1531–1534. [CrossRef]
6. Karsisiotis, A.I. NMR Studies of Inhibitor Binding to Metallo-B-Lactamases. Ph.D. Thesis, University of Leicester, Leicester, UK, 2008.
7. Kempf, J.G.; Loria, J.P. Protein Dynamics from Solution NMR. *Cell Biochem. Biophys.* **2002**, *37*, 187–211. [CrossRef]
8. Kovermann, M.; Rogne, P.; Wolf-Watz, M. Protein Dynamics and Function from Solution State NMR Spectroscopy. *Q. Rev. Biophys.* **2016**, *49*, e6. [CrossRef]
9. Castellani, F.; Van Rossum, B.; Diehl, A.; Schubert, M.; Rehbein, K.; Oschkinat, H. Structure of a Protein Determined by Solid-State Magic-Angle-Spinning NMR Spectroscopy. *Nature* **2002**, *420*, 99–102. [CrossRef]
10. Shahid, S.A.; Bardiaux, B.; Franks, W.T.; Krabben, L.; Habeck, M.; van Rossum, B.-J.; Linke, D. Membrane-Protein Structure Determination by Solid-State NMR Spectroscopy of Microcrystals. *Nat. Methods* **2012**, *9*, 1212. [CrossRef]
11. Hong, M. Solid-State NMR Determination of 13Cα Chemical Shift Anisotropies for the Identification of Protein Secondary Structure. *J. Am. Chem. Soc.* **2000**, *122*, 3762–3770. [CrossRef]
12. Reif, B. Ultra-High Resolution in MAS Solid-State NMR of Perdeuterated Proteins: Implications for Structure and Dynamics. *J. Magn. Reson.* **2012**, *216*, 1–12. [CrossRef] [PubMed]
13. De Angelis, A.A.; Jones, D.H.; Grant, C.V.; Park, S.H.; Mesleh, M.F.; Opella, S.J. NMR Experiments on Aligned Samples of Membrane Proteins. *Methods Enzymol.* **2005**, *394*, 350–382. [PubMed]
14. Wi, S.; Frydman, L. An Efficient, Robust New Scheme for Establishing Broadband Homonuclear Correlations in Biomolecular Solid State NMR. *ChemPhysChem* **2020**, *21*, 284–294. [CrossRef] [PubMed]
15. Kobayashi, N.; Iwahara, J.; Koshiba, S.; Tomizawa, T.; Tochio, N.; Güntert, P.; Kigawa, T.; Yokoyama, S. KUJIRA, a Package of Integrated Modules for Systematic and Interactive Analysis of NMR Data Directed to High-Throughput NMR Structure Studies. *J. Biomol. NMR* **2007**, *39*, 31–52. [CrossRef]
16. Bahrami, A.; Assadi, A.H.; Markley, J.L.; Eghbalnia, H.R. Probabilistic Interaction Network of Evidence Algorithm and Its Application to Complete Labeling of Peak Lists from Protein NMR Spectroscopy. *PLoS Comput. Biol.* **2009**, *5*, e1000307. [CrossRef]
17. Lee, W.; Kim, J.H.; Westler, W.M.; Markley, J.L. PONDEROSA, an Automated 3D-NOESY Peak Picking Program, Enables Automated Protein Structure Determination. *Bioinformatics* **2011**, *27*, 1727–1728. [CrossRef]
18. Lee, W.; Stark, J.L.; Markley, J.L. PONDEROSA-C/S: Client–Server Based Software Package for Automated Protein 3D Structure Determination. *J. Biomol. NMR* **2014**, *60*, 73–75. [CrossRef]
19. Lee, W.; Westler, W.M.; Bahrami, A.; Eghbalnia, H.R.; Markley, J.L. PINE-SPARKY: Graphical Interface for Evaluating Automated Probabilistic Peak Assignments in Protein NMR Spectroscopy. *Bioinformatics* **2009**, *25*, 2085–2087. [CrossRef]
20. Lee, W.; Bahrami, A.; Dashti, H.T.; Eghbalnia, H.R.; Tonelli, M.; Westler, W.M.; Markley, J.L. I-PINE Web Server: An Integrative Probabilistic NMR Assignment System for Proteins. *J. Biomol. NMR* **2019**, *73*, 213–222. [CrossRef]
21. Lee, W.; Yu, W.; Kim, S.; Chang, I.; Lee, W.; Markley, J.L. PACSY, a Relational Database Management System for Protein Structure and Chemical Shift Analysis. *J. Biomol. NMR* **2012**, *54*, 169–179. [CrossRef]
22. Rahimi, M.; Lee, Y.; Markley, J.L.; Lee, W. iPick: Multiprocessing Software for Integrated NMR Signal Detection and Validation. *J. Magn. Reson.* **2021**, *328*, 106995. [CrossRef] [PubMed]
23. Lee, W.; Markley, J.L. PINE-SPARKY. 2 for Automated NMR-Based Protein Structure Research. *Bioinformatics* **2018**, *34*, 1586–1588. [CrossRef]
24. Lee, W.; Rahimi, M.; Lee, Y.; Chiu, A. POKY: A Software Suite for Multidimensional NMR and 3D Structure Calculation of Biomolecules. *Bioinformatics* **2021**, *37*, 3041–3042. [CrossRef] [PubMed]

25. Weber, D.K.; Wang, S.; Markley, J.L.; Veglia, G.; Lee, W. PISA-SPARKY: An Interactive SPARKY Plugin to Analyze Oriented Solid-State NMR Spectra of Helical Membrane Proteins. *Bioinformatics* **2020**, *36*, 2915–2916. [CrossRef] [PubMed]
26. Lee, W.; Tonelli, M.; Markley, J.L. NMRFAM-SPARKY: Enhanced Software for Biomolecular NMR Spectroscopy. *Bioinformatics* **2015**, *31*, 1325–1327. [CrossRef]
27. Opella, S.J.; Marassi, F.M. Structure Determination of Membrane Proteins by NMR Spectroscopy. *Chem. Rev.* **2004**, *104*, 3587–3606. [CrossRef]
28. Gopinath, T.; Manu, V.S.; Weber, D.K.; Veglia, G. PHRONESIS: A One-Shot Approach for Sequential Assignment of Protein Resonances by Ultrafast MAS Solid-State NMR Spectroscopy. *ChemPhysChem* **2022**, *23*, e202200127. [CrossRef]
29. Moseley, H.N.; Sperling, L.J.; Rienstra, C.M. Automated Protein Resonance Assignments of Magic Angle Spinning Solid-State NMR Spectra of B1 Immunoglobulin Binding Domain of Protein G (GB1). *J. Biomol. NMR* **2010**, *48*, 123–128. [CrossRef]
30. Schmidt, E.; Gath, J.; Habenstein, B.; Ravotti, F.; Székely, K.; Huber, M.; Buchner, L.; Böckmann, A.; Meier, B.H.; Güntert, P. Automated Solid-State NMR Resonance Assignment of Protein Microcrystals and Amyloids. *J. Biomol. NMR* **2013**, *56*, 243–254. [CrossRef]
31. Zhou, D.H.; Nieuwkoop, A.J.; Berthold, D.A.; Comellas, G.; Sperling, L.J.; Tang, M.; Shah, G.J.; Brea, E.J.; Lemkau, L.R.; Rienstra, C.M. Solid-State NMR Analysis of Membrane Proteins and Protein Aggregates by Proton Detected Spectroscopy. *J. Biomol. NMR* **2012**, *54*, 291–305. [CrossRef]
32. Gopinath, T.; Veglia, G. Multiple Acquisition of Magic Angle Spinning Solid-State NMR Experiments Using One Receiver: Application to Microcrystalline and Membrane Protein Preparations. *J. Magn. Reson.* **2015**, *253*, 143–153. [CrossRef] [PubMed]
33. Yedidia, J.S.; Freeman, W.T.; Weiss, Y. Constructing Free-Energy Approximations and Generalized Belief Propagation Algorithms. *IEEE Trans. Inf. Theory* **2005**, *51*, 2282–2312. [CrossRef]
34. Bartels, C.; Xia, T.; Billeter, M.; Güntert, P.; Wüthrich, K. The Program XEASY for Computer-Supported NMR Spectral Analysis of Biological Macromolecules. *J. Biomol. NMR* **1995**, *6*, 1–10. [CrossRef] [PubMed]
35. Keller, R.L.J. Optimizing the Process of Nuclear Magnetic Resonance Spectrum Analysis and Computer Aided Resonance Assignment. Ph.D. Thesis, ETH Zurich, Zürich, Switzerland, 2005.
36. Vranken, W.F.; Boucher, W.; Stevens, T.J.; Fogh, R.H.; Pajon, A.; Llinas, M.; Ulrich, E.L.; Markley, J.L.; Ionides, J.; Laue, E.D. The CCPN Data Model for NMR Spectroscopy: Development of a Software Pipeline. *Proteins Struct. Funct. Bioinforma.* **2005**, *59*, 687–696. [CrossRef] [PubMed]
37. Eghbalnia, H.R.; Wang, L.; Bahrami, A.; Assadi, A.; Markley, J.L. Protein Energetic Conformational Analysis from NMR Chemical Shifts (PECAN) and Its Use in Determining Secondary Structural Elements. *J. Biomol. NMR* **2005**, *32*, 71–81. [CrossRef]
38. Wang, L.; Eghbalnia, H.R.; Bahrami, A.; Markley, J.L. Linear Analysis of Carbon-13 Chemical Shift Differences and Its Application to the Detection and Correction of Errors in Referencing and Spin System Identifications. *J. Biomol. NMR* **2005**, *32*, 13–22. [CrossRef] [PubMed]
39. Ulrich, E.L.; Akutsu, H.; Doreleijers, J.F.; Harano, Y.; Ioannidis, Y.E.; Lin, J.; Livny, M.; Mading, S.; Maziuk, D.; Miller, Z. BioMagResBank. *Nucleic Acids Res.* **2008**, *36*, D402–D408. [CrossRef]
40. Berman, H.; Henrick, K.; Nakamura, H.; Markley, J.L. The Worldwide Protein Data Bank (WwPDB): Ensuring a Single, Uniform Archive of PDB Data. *Nucleic Acids Res.* **2007**, *35*, D301–D303. [CrossRef]

membranes

MDPI

Article

Dynamin-Related Proteins Enhance Tomato Immunity by Mediating Pattern Recognition Receptor Trafficking

Meirav Leibman-Markus [1,2], Silvia Schuster [1], Beatriz Vasquez-Soto [3], Maya Bar [2], Adi Avni [1] and Lorena Pizarro [1,2,3,*]

1 School of Plant Sciences and Food Security, Faculty of Life Sciences, Tel Aviv University, Tel Aviv 6997801, Israel; meiravleibman@gmail.com (M.L.-M.); schus@tauex.tau.ac.il (S.S.); lpavni@tauex.tau.ac.il (A.A.)
2 Department of Plant Pathology and Weed Research, Institute of Plant Protection, ARO, Volcani Institute, Rishon LeZion 7505101, Israel; mayabar@volcani.agri.gov.il
3 Institute of Agri-Food, Animal and Environmental Sciences, Universidad de O'Higgins, Rancagua 2820000, Chile; bgvasquez@uc.cl
* Correspondence: lorena.pizarro@uoh.cl; Tel.: +56-233-286-050

Abstract: Pattern recognition receptor (PRR) trafficking to the plasma membrane and endocytosis plays a crucial role in pattern triggered immunity (PTI). Dynamin-related proteins (DRPs) participate in endocytosis and recycling. In Arabidopsis, DRP1 and DRP2 are involved in plasma membrane scission during endocytosis. They are required for the PRR FLS2 endocytosis induction and PTI activation after elicitation with flg22, the MAMP recognized by FLS2. In tomato, SlDRP2A regulates the PRR LeEIX2 endocytosis and PTI activation in response to EIX, the MAMP recognized by LeEIX2. However, it is unknown if other DRPs participate in these processes. Taking advantage of bioinformatics tools, we selected SlDRP2B among the eight DRP2 tomato orthologues to study its functionality in trafficking and plant immunity. Through transient expression of SlDRP1B and its dominant-negative mutant on *Nicotiana benthamiana* and *Nicotiana tabacum*, we analyzed SlDRP1B function. We observed that SlDRP1B is physically associated with the LeEIX2 and modifies LeEIX2 trafficking, increasing its presence in endosomes. An enhancement of EIX-elicitated defense responses accompanies the role of SlDRP1B on LeEIX endocytosis. In addition, SlDRP1B overexpression enhanced flg22-elicited defense response. With these results, we conclude that SlDRP1B regulates PRR trafficking and, therefore, plant immunity, similarly to the SlDRP2A role.

Keywords: dynamin-related protein; DRP1; endomembrane trafficking; FLS2; LeEIX2; tomato; defense responses

1. Introduction

During pathogen colonization, plants sense an intruder by recognizing exogenous molecules through the plasma membrane (PM) and intracellular receptors. Pattern recognition receptors (PRRs) are the PM receptors that recognize microbial-associated molecular patterns (MAMPs) [1]. PRRs are part of the first layer of plant immunity, sensing pathogens and triggering defense responses to halt pathogen invasion. There are two main groups of PRRs: receptor-like kinases (RLKs) and receptor-like proteins (RLPs). RLKs are generally composed of an extracellular recognizing domain, such as a leucine-rich repeat (LRR) domain thought to be involved in ligand binding; a transmembrane domain; and an intracellular kinase domain involved in signal transduction [2]. RLPs have an extracellular recognizing domain and a transmembrane domain, but lack any obvious domains in charge of intracellular signaling. RLPs are assumed to form receptor complexes with RLKs or receptor-like cytoplasmic kinases to transduce defense signals to target proteins [3–5]. PRRs are integral membrane proteins synthesized in the endoplasmic reticulum (ER), and carried to the PM through the Golgi, early endosomes/trans-Golgi network (EE/TGN). PRRs must be adequately delivered and localized at the PM for sensing MAMPs from the apoplast [6].

Roles for endocytosis and cellular trafficking in plant cellular responses to invading pathogens are well documented [6,7]. Receptor-mediated endocytosis (RME) serves as a mechanism for both signaling initiation and termination through the degradation of activated PRR complexes after internalization [7,8]. Plant cells require PRR presence at the PM to mount effective defenses. It has been suggested that, from an evolutionary perspective, targeting plant vesicular cellular trafficking is a strategy adopted by many pathogens through their effector molecules [9–11]. Plant endocytic trafficking components are diverted towards pathogen infection sites when required, reinforcing the importance of cellular trafficking in immune responses [12–15]. The LRR-RLK FLS2 recognizes the epitope flg22 from flagellin [16]. FLS2 undergoes constitutive endocytosis and recycling between the PM and the trans-Golgi network as a mechanism of maintaining its steady-state level at the PM [17,18]. Upon ligand stimulation, FLS2 transmits from the PM and localizes to endomembrane vesicles in a BAK1-dependent manner [18–20], and then traffics toward the vacuole for degradation [21]. FLS2 thus traffics through different endocytic pathways, aimed both at maintaining cellular readiness to sense pathogens and mounting an efficient response once exposed to an actual pathogen.

The LRR-RLP LeEIX2 traffics on endomembrane compartments. The binding of the fungal Xylanase-11 ligand (EIX) to LeEIX2 induces receptor-mediated endocytosis [22]. Inhibition of endosome formation reduces EIX-mediated responses, while arresting post internalization cellular trafficking increases these responses [23–25]. This suggests that LeEIX2 transmits defense signals from endosomes. Interestingly, although much of the defense signals and signal-propagation mechanisms differ between LeEIX2 and FLS2 [20,23], the involvement of the endomembrane system as a tool for both signal propagation and maintenance of immune system readiness is conserved. This testifies to the importance of cellular trafficking in plant response to attacking pathogens as a general mechanism. Indeed, it was recently reported that overexpression of PRRs is sufficient to induce localized immune responses [26], and furthermore, plants that are constitutively immuno- activated have increased expression of a variety of PRRs from both the RLK and RLP classes in steady-state [27,28].

Dynamin-related proteins (DRPs) are mechanochemical GTPases that remodel membranes in diverse cellular processes [29,30]. In plants, DRPs are classified into six families, and are involved in vesicle formation and scission during endocytosis and post-Golgi traffic [7,31,32]. They are also critical for homotypic membrane fission and fusion in different organelles [31,32]. DRP1 and DRP2 have been implicated in PM scission during endocytosis [33,34]. In plant immunity, AtDRP1 and AtDRP2 are required for flg-22-induced FLS2 endocytosis, and are involved in PTI in response to flg22 and *Pseudomonas syringe* in *Arabidopsis thaliana* [25,35,36]. As such, DRP1s and DRP2s are involved in pathogen resistance; they are targeted by pathogen effectors in several cases [34–39]. We previously reported that DRP2A of tomato associates with LeEIX2, regulating its trafficking, defense signaling, and defense response [25]. As both AtDRP1 and AtDRP2 are required for endocytosis of PRRs, we aim to elucidate if the same occurs in tomato. Tomato has eight orthologs in the AtDRP1 sub-family [25,40]. Here, we identified a drp1 family member protein based on homology and characterized its role in PRR endocytosis and plant immunity.

2. Materials and Methods

2.1. Plant Materials and Growth Conditions

Nicotiana tabacum cv samsun NN, *N. benthamiana*, and *S. lycopersicum* cv M82 were grown from seeds in soil (Green Mix; Even-Ari, Ashdod, Israel) in a growth chamber, under long-day conditions (16 h:8 h, light:dark) at 24 °C.

2.2. Bioinformatic Analysis

Protein domain analysis of tomato DRP1 orthologues [25] was performed using HMMER: biosequence analysis using profile hidden Markov models [41], and a domain representation was drawn. Expression analysis of each of the DRP1 orthologues, and co-expression analysis of SlDRP1A, SlDRP1B, and SlDRP2A, were made using Genevestigator® software (all samples database, Zurich, Switzerland) [42]. A protein–protein interacting (PPI) network was built using the physical PPI analysis with the PTIR database [43] and plotted using the Cytoscape program.

Gene ontology overrepresentation analysis was performed by PANTHER GO over-representation test. A set of 180 out of 191 SlDRP1B co-expressed genes were found and analyzed, using the complete *S. lycopersicum* genome as a reference list (34,652 genes), and Fisher's test and false discovery rate (FDR) as statistical methods [44]. Data visualization was performed using the Python library plotline (based on ggplot2) and considering fold-enrichment, overrepresentation significance (p-value), and number of genes related to each GO term (count). For corroborating the expression level of Solyc01g095970, Solyc05g050600, Solyc01g005310, and Solyc08g077360, specific real-time PCR primers were designed (Table S5). RNA was extracted from M82 tomato leaves (2 months old). M-MLV reverse transcriptase (Promega, Madison, WI, USA) and oligo-d(T) primers were used for cDNA synthesis according to manufacturer's instructions. qRT-PCR was performed according to the Fast SYBR qPCR Master Mix (BioGate, Yad HaShmona, Israel), using a Rotor-Gene Q machine (Qiagen, Valencia, CA, USA). Relative expression was determined using the following formula: Relative expression = $2^{\text{Ct Hkp–Ct SlDRP1X}}$. For determining relative expression, we used RPL8 (Solyc10g006580) housekeeping gene [45,46]. Three RNA biological replicates were used; each reaction was performed in triplicate.

2.3. Transient Expression

Binary vector clones were introduced by electroporation into *Agrobacterium tumefaciens* strain GV3101. *Agrobacterium* cells were grown in LB medium containing 50 mg/L Kanamycin, 40 mg/L Gentamycin, and 100 mg/L Rifampicin overnight at 28 OC, diluted into VIR induction medium (50 mM MES pH 5.6, 0.5% (w/v) glucose, 1.7 mM NaH_2PO_4, 20 mM NH_4Cl, 1.2 mM $MgSO_4$, 2 mM KCl, 17 µM $FeSO_4$, 70 µM $CaCl_2$, and 200 µM acetosyringone), and grown for 6 additional hours until OD600 reached 0.4–0.6. Suspensions containing single or mixed *Agrobacterium* cultures were diluted to a final OD600 of 0.15–0.2 in VIR induction medium. Cultures were infiltrated abaxially with a needless syringe into leaves of *N. tabacum* cv samsun NN or *N. benthamiana*. Leaves were harvested 40 h after injection for ethylene measurement, ROS measurement, CoIP, BiFC, or confocal microscopy analysis.

2.4. Cloning of SlDRP1B and Site-Directed Mutagenesis

SlDRP1B cDNA C-terminally tagged with GFP or mCherry were amplified from *S. lycopersicum* cv M82 and cloned into the SalI, XbaI sites of pBINPLUS, following the strategy used by Pizarro et al., 2019 [25] using the following primers: SlDRP1B forward primer 5′-CCGTCGACATGGGAATCTCATAGCATTAG-3′ and SlDRP1B reverse primer 5′-CTCTAGACTTAGACCATGCTACTGAATC-3′. Amplified fragments were cloned between the CAM35SΩ promoter containing the Ω translation enhancer signal and the Nos terminator.

For site-directed mutagenesis, GTPase domain and the catalytic site were predicted using Conserved Domain search from National Center of Biotechnology Information website2, following the strategy used by Pizarro et al, 2019 [25]. Site-directed mutagenesis on the GTPase catalytic site of SlDRP1B was created using the Q5 site-directed mutagenesis kit (NEB #E0554) and the following primers: forward 5′-RTGTCCACCGACAACAGCGU-3′ and reverse 5′-FGAGTTCTGGGGCATCTTCGGTGCTTG-3′.

2.5. Co-Immunoprecipitation

Co-immunoprecipitation assays were performed as described by Leibman-Markus [47]. *N. benthamiana* leaves transiently co-expressing LeEIX2-HA and SlDRP1B-GFP were harvested 40 h after infiltration. Leaf petioles were immersed in EIX 3 μg/mL (or water as mock) for seven minutes and then transferred to the water for an additional seven minutes. A total of 500 mg leaf tissue was used for co-immunoprecipitation, with 13 μL α-HA Affinity Matrix (Roche, Indianapolis, ID, USA). Samples were run in SDS-PAGE, blotted onto nitrocellulose membranes, and incubated with antibodies as required: rat α-GFP (Chromotek, Planegg-Martinsried, Germany) and mouse α-HA (Biolegend, San Diego, CA, USA).

2.6. BiFluorescence Complementation

LeEIX2 cytoplasmic domain and SlDRP1B cDNA were cloned into the Spe1 site of pSY751 and pSY752, downstream of the N-terminal fragment of YFP (YN) and the C-terminal fragment (YC), respectively, using the following primers: LeEIX2 forward primer 5′-GGGGCCTTTTAGGCTG-3′ and LeEIX2 reverse primer 5′-CTGGCGGCCGCTCAGTTC-CTTAGCTTTCCC-3′; SlDRP1B forward primer 5′-CCACTAGTATGGAGAATCTCATAGC-ATTAG-3′ and SlDRP1B reverse primer 5′-CCACTAGTTTACTTAGACCATGCTACTG-3′. The resulting plasmids were used for transient expression assay in *N. benthamiana* leaves.

2.7. ROS Burst Assay

ROS burst was measured as previously described by Leibman-Markus [47]. Leaf disks (0.5 cm diameter) were taken from tobacco plants transiently expressing SlDRP1B tagged with GFP, mCherry, or HA, and free GFP was used as control. Disks were floated in 250 μL ddH_2O in a white 96-well plate for 4–6 h at room temperature. After incubation, water was removed entirely and ROS measurement reaction containing EIX 1 μg/mL or Flg22 1 mM was added, and light emission was immediately measured using a micro-plate luminometer (Turnerbiosystems Veritas, Santa Clara, CA, USA).

2.8. Ethylene Production Assay

Ethylene biosynthesis was measured as previously described by Leibman-Markus [47]. Leaf disks (0.9 cm diameter) were taken from transiently expressing tobacco plants. Five disks were sealed in each 10 mL flask containing 1 mL assay medium (with or without 1 μg/mL EIX) and incubated with shaking for four hours at room temperature. Ethylene production was measured by gas chromatography (Varian 3350, Varian, Palo Alto, CA, USA).

2.9. Live-Cell Imaging

Confocal microscopy images were acquired using a Zeiss LSM780 confocal microscope system using Objective C-Apochromat 40x/1.2 W Corr M27 (Zeiss, Oberkochen, Germany). Microscopy configuration includes two tracks: Track 1 for collecting chlorophyll autofluorescence using, excitation laser wavelength of 633 nm (2% power), and emission collection range of 652–721 nm; Track 2 for collecting GFP and dsRed/mCherry fluorescence using excitation laser of 488 nm (5% power) and 561 nm (2% power), respectively. GFP and dsRed/mCherry emission was collected in the range of 493–535 nm and 588–641 nm, respectively. For YFP image acquisition, an excitation laser of 514 nm (5% power) was used, and emission was collected in the 522–530 nm range. 1024X1024 pixels images were acquired using a pixel dwell time of 1.27, pixel averaging of 4, and pinhole of 1 airy unit.

2.10. Confocal Image Analysis

Image analysis was conducted using Fiji-ImageJ on the original files [48]. The analysis was performed on a single epidermal cell region of interest (ROI) where both proteins expressed with similar intensity. Colocalization of SlDRP1B and the endosome markers was determined using Coloc2 function. The 3D object counter function was used for quantifying compartment density and size. Fluorescence ratio was performed by quantifying the integrated pixel intensity in the PM divided by the integrated pixel intensity in the whole cell.

3. Results

3.1. Searching for a Functional Tomato DRP1 Involved in PRR Endocytic Trafficking

DRP1 proteins contain three dynamin core domains: an N-terminal GTPase domain, a middle domain, and a GTPase effector domain (GED) [41–44]. To identify a functional tomato DRP1, we searched for the DRP1 orthologs in the tomato genome, filtering in by homology to AtDRP1A (At5g42080) and AtDRP1B (At3g61760), using the Sol Genomics Network [47]. Then, we analyzed the putative orthologues using three available bioinformatics platforms: HMMR EMBL-EBI, GENEVESTIGATOR, and PTIR [41–44]. We found eight DRP1s gene orthologues to DRP1s; among them, four possessed the three canonical domains of a DRP1: Solyc01g095970, Solyc05g050600, Solyc01g005310, and Solyc08g077360 (Figure 1A, Supplementary Figure S1). Expression levels of the DRP1 orthologues were analyzed using GENEVESTIGATOR®bioinformatic software. When examining a dataset of 877 samples, Solyc01g095970, Solyc05g050600, and Solyc01g005310 have the highest average expression among the eight DRP1 orthologues (Figure 1B). Expression analysis of Solyc01g095970, Solyc05g050600, Solyc01g005310, and Solyc08g077360 was corroborated via real-time PCR using RNA from tomato leaves (Supplementary Figure S2).

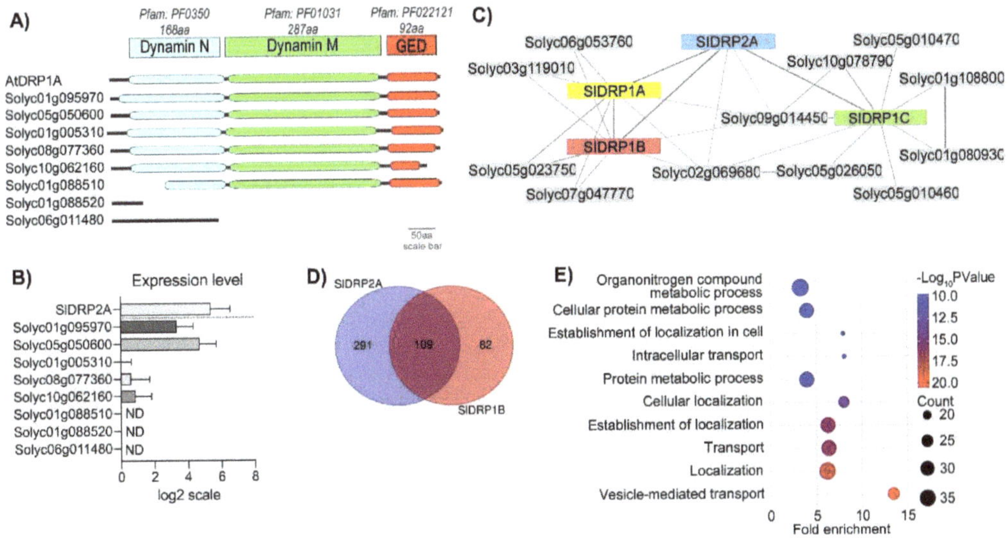

Figure 1. Bioinformatic analysis of tomato DRP1 orthologues. (**A**) Protein domain analysis was per this figure. Orthologues: Solyc01g095970, Solyc05g050600, Solyc01g005310, Solyc08g077360, Solyc10g062160, Solyc01g088510, Solyc01g088520, and Solyc06g011480. (**B**) Gene expression analysis of tomato DRP1 orthologues was performed using GENEVETIGATOR®. (**C**). Interactome network of SlDRP1A, SlDRP1B, and SlDRP1C based on PTIR database (Supplementary Table S2). (**D**) Venn diagram of SlDRP1B and SlDRP2A co-expressed genes based on GENEVETIGATOR® database. (**E**) GO overrepresentation analysis of SlDRP1B co-expressed genes.

To collect information about the possible interactors of these genes, we used the PTIR database. This analysis predicted interactors for three tomato putative DRP1 proteins: Solyc01g095970 (SlDRP1A), Solyc05g050600 (SlDRP1B), and Solyc01g005310 (SlDRP1C). SlDRP1A and SlDRP1B were predicted to share the same interactors, while SlDRP1C was found to differ in some putative interactors (Figure 1C, Supplementary Table S1). Interestingly, all of them interact with SlDRP2A (Figure 1C, Supplementary Table S2). In addition, SlDRP1A and B share four out of five interactors with SlDRP2A (Figure 1C, Supplementary Tables S1 and S2). The interaction between DRP1 and DRP2 during the formation of clathrin-coated vesicles was previously described in Arabidopsis [35]. We searched for the genes that share similar expression profiles with SlDRP1A, SlDRP1B, and SlDRP1C. We set a cut-off = 0.8 in Pearson's correlation factor for co-expression. The highest Pearson's correlation value obtained for SlDRP1A was 0.74; therefore, under our criterion, SlDRP1A does not co-express with any gene (Supplementary Table S3). SlDRP1B was found to be co-expressed with 191 genes, including SlDRP2A (Supplementary Table S3). Interestingly, 109 of the genes found to be co-expressed with SlDRP1B were also co-expressed with SlDRP2A (Figure 1C, Supplementary Table S3). The list of genes correlatively expressed with both SlDRP1B and SlDRP2A (Supplementary Table S3) includes genes such as two Coatomers (Solyc04g080980 and Solyc12g043000), a DRP3 (Solyc11g043210), a Myosin (Solyc07g041150), and a Clathrin Heavy Chain (Solyc05g052510). These results suggest that SlDRP1B could function in the processes described for SlDRP2A. In contrast, SlDRP1C was found to have 382 co-expressed genes; however, none of them were co-expressed with SlDRP2A.

Additionally, gene ontology enrichment analysis for molecular function classification of SlDRP1B co-expressed genes showed overrepresentation of genes belonging to the classification terms "Vesicle-mediated transport", "Localization", "Transport establishment of localization", and "Cellular localization" (Supplementary Table S4). Given the results of the bioinformatic analysis, we determined that SlDRP1B showed the most promise to be a functional DRP1 family protein and proceeded to characterize the roles of SlDRP1B in trafficking and defense.

3.2. *SlDRP1B Subcellular Localization*

To characterize SlDRP1B and determine its subcellular localization, we conducted confocal microscopy experiments with fluorescently labeled SlDRP1B. Confocal images of *N. benthamiana* epidermal cells reveal that SlDRP1B has a dual localization (Figure 2), as reported for Arabidopsis members of this sub-family [33,49]. SlDRP1B forms discrete foci at the PM (marked in white circles), ranging in diameter between 200–550 nm (Figure 2A,B) [33,50], and colocalizing with a PM marker such as Flot1 (Pearson correlation coefficient of 0.83) [51]. This result suggests that SlDRP1B, similarly to AtDRP1, is involved in the formation of endocytic vesicles at the PM [33]. In addition, SlDRP1B also appears at punctuated structures (marked with white arrowheads), resembling endosomes (Figure 2A). Comparing the subcellular localization of SlDRP1B and SlDRP2A, we observed that SlDRP2A distribution is enriched at the PM (Supplementary Figure S3) [25].

To better examine the subcellular localization of SlDRP1B and its punctuated pattern, we expressed SlDRP1B together with several endomembrane markers. SlDRP1B c-localizes with late endosomes/multivesicular body (LE/MVB) marker, RabG3f [52,53], having a Pearson correlation coefficient of 0.69 (Figure 2A,B). No colocalization was observed when co-expressing SlDRP1B with the EE/TGN marker, RabA1e [54–56], or with the Golgi marker, ST [56,57], having Pearson correlation coefficients of 0.54 and 0.22, respectively (Figure 2A,B). Surprisingly, although it colocalized with RabG3f, SlDRP1B did not colocalize with the LE/MVB/PVC marker FYVE [54], having a negative Pearson correlation coefficient of-0.04 (Figure 2A,B). This suggests that SlDRP1B localizes to a sub-population of LE/MVB. Interestingly, RabG3f affects SlDRP1B sub-cellular distribution, causing it to shift towards punctuated compartments, leading to an increase in SlDRP1B compartment number and size (marked with yellow arrowheads, Figure 2A,C,D). These changes might be a result of the RabG3f function or due to its overexpression.

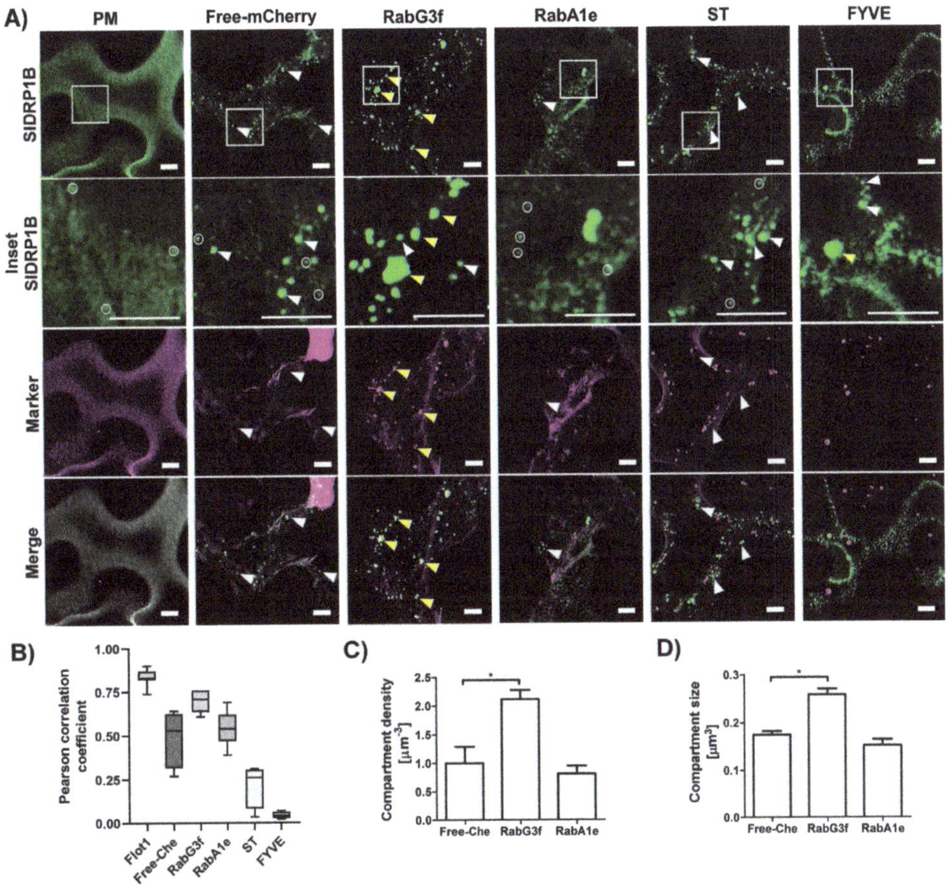

Figure 2. SlDRP1B localizes to a subset of MVB/LE/PVC compartments. Localization of SlDRP1B was analyzed by confocal microscopy of *N. benthamiana* epidermal cells transiently expressing SlDRP1B-GFP and different endomembrane compartment markers as indicated. Images were acquired using a Zeiss LSM 780 microscope. (**A**) Representative images of SlDRP1B-GFP colocalization with Flot1-mCherry, free mCherry, RabG3f-mCherry, RabA1e-mCherry, ST-mCherry, and FYVE-dsRED are shown. White and yellow arrowheads point to SlDRP1B compartments and SlDRP1B compartments colocalizing with the markers, respectively. Circles indicate SlDRP1B foci. Squares indicate the inset region. Scale bar 7.5 μm. (**B**) Colocalization analysis between SlDRP1B and the markers. (**C**,**D**) SlDRP1B endosomes density Endosome density was normalized to control (Free mCherry). Control = 0.013 endosomes per μm^2. (**B**–**D**) Twelve images each were analyzed. (**C**,**D**). Error bars represent mean ± SEM. Asterisks indicate significant differences with free mCherry control (*t*-test, * $p < 0.05$).

3.3. *SlDRP1B Associates with LeEIX2 and Enhances Defense Responses Mediated by EIX*

We previously observed that SlDRP2A enhances defense responses mediated by LeEIX2 and FLS2 [25]. In order to examine a possible physical association between LeEIX2 and the tomato SlDRP1B ortholog, we performed co-immunoprecipitation (CoIP) and bi-fluorescence complementation (BiFC) experiments. We transiently co-expressed tomato SlDRP1B-GFP and LeEIX2-HA in *N. benthamiana* (an EIX nonresponsive plant species), and pulled down LeEIX2-HA using HA affinity beads. SlDRP1B was successfully pulled down in the presence of LeEIX2. The association between the two proteins was dramatically enhanced when pretreated with the ligand EIX (Figure 3A). We performed BiFC analysis to further validate the association. When co-expressing YN-LeEIX2 (fused to the N-terminal portion of YFP) and YC-SlDRP1B (fused to the C-terminal portion of YFP), we observed reconstitution of YFP fluorescence, indicating protein association (Figure 3B).

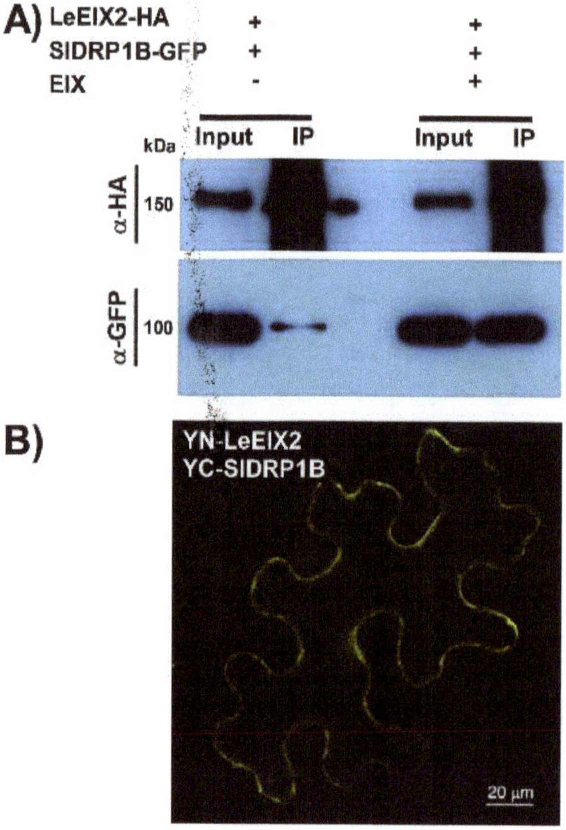

Figure 3. SlDRP1B associates with LeEIX2. (**A**) *N. benthamiana* was transiently transformed with LeEIX2-HA and SlDRP1B-GFP. Leaves were harvested and treated with EIX or water (mock). Input and immunoprecipitated samples were subjected to SDS-PAGE and immunoblot with anti-HA antibodies to detect LeEIX2-HA, and anti-GFP antibodies to detect SlDRP1B-GFP. (**B**) BiFC visualization of the association between LeEIX2 and SlDRP1B. *N. benthamiana* leaves transiently expressing YN-LeEIX2 cytoplasmatic domain and YC-SlDRP1B as indicated. Leaf sections were visualized using a laser-scanning-meta confocal microscope. Scale bar 20 μm.

To examine the effect of SlDRP1B overexpression on ROS burst and ethylene production elicited by EIX, we transiently expressed SlDRP1B in *N. tabacum* cv Samsun NN (EIX responsive cultivar) followed by EIX application. SlDRP1B enhanced LeEIX2-mediated defense responses compared to control, leading to an increase in ROS burst and ethylene production (Figure 4). Based on bioinformatic analysis, we mutated the GTPase active site (Lysine 47 to Alanine, SlDRP1BK47A), creating a predicted loss of function. The mutated form lost the ability to enhance EIX-mediated defense responses, effecting a decrease in EIX-dependent ROS burst (Figure 4).

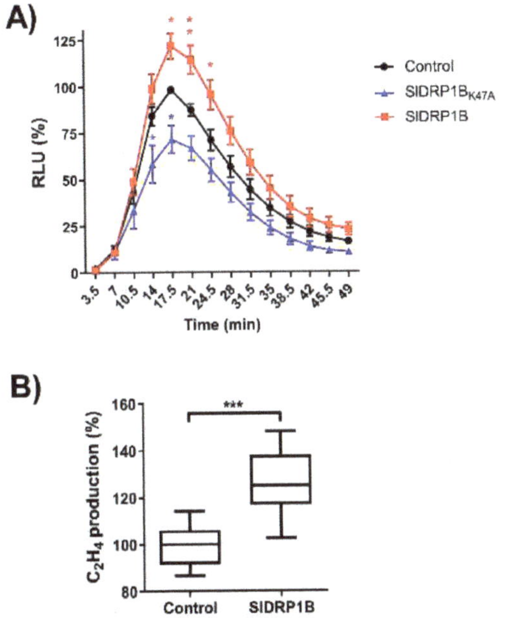

Figure 4. Effect of overexpression of SlDRP1B on LeEIX2-mediated defense responses. Leaf disks of *N. tabacum* transiently expressing SlDRP1B-tagged, SlDRP1B$_{K47A}$-tagged, or free-tag (control) were harvested 48 h after transformation. (**A**) Luminescence (RLU) was measured immediately after EIX application. Error bars represent the average +SEM of 7 independent experiments, N = 12. The average value of control peak was defined as 100%. Time points represent the average ± SEM. Asterisks indicate significant differences with control treatment (two-way ANOVA, * $p < 0.05$, ** $p < 0.01$). (**B**) Ethylene (C_2H_4) biosynthesis was measured 4 h after EIX application. Average value of control was defined as 100%. Boxplots represent minimum to maximum values, with boxes representing the inner quartile ranges, whiskers representing the outer quartile ranges, and the line in the box representing the median, of 3 independent experiments, with asterisks denoting significant differences to control treatment (*t*-test, *** $p < 0.001$).

We next tested the possibility of SlDRP1B affecting the signal transduction cascades of additional PRR proteins. We turned to the hallmark PTI receptor FLS2, and examined a possible role of SlDRP1B in flagellin-mediated defense responses. Overexpression of SlDRP1B enhanced ROS burst upon flg22 elicitation by more than 60% (Supplementary Figure S4), suggesting that SlDRP1B's role as a positive immunity regulator is not restricted to LeEIX2-mediated defense responses.

In our bioinformatics analyses (Figure 1), SlDRP1B and SlDRP2A were also likely co-expressed with a DRP3 family protein, Solyc11g043210 (Supplementary Tables S3 and S4), here named SlDRP3A. We cloned SlDRP3A and conducted similar experiments to those detailed here for SlDRP1B, finding that SlDRP3A also enhances LeEIX2-mediated defense (Supplementary Figure S5).

3.4. SlDRP1B Affects LeEIX2 Endosomal Distribution

DRPs are essential modulators of endocytosis, and endocytosis affects LeEIX2 signaling [23,25,31]; therefore, we studied the effect of SlDRP1B on LeEIX2 subcellular localization. As previously published, LeEIX2 endosomes increase in both size and density upon EIX treatment (Figure 5), reflecting endocytosis involvement in EIX-induced signaling [24,25]. When overexpressing SlDRP1B, we observed a threefold increase in LeEIX2 endosome size and density in basal conditions (without EIX elicitation, Figure 5). As expected,

SlDRP1B$_{K47A}$ mutant the loss of function of SlDRP1A, behaves similarly to the control, and does not exhibit any effect on LeEIX2 endosomes (Figure 5). When overexpressing SlDRP1B, EIX elicitation led to a mild increase in LeEIX2 endosome size and density, similar to both the basal control and the control under elicited conditions (Figure 5). When comparing LeEIX2 endosome size and density in both basal and ligand-elicited conditions in the background of SlDRP1B overexpression, we observed a decrease in LeEIX2 endosome size and density, rather than an increase. In samples overexpressing the mutated SlDRP1B$_{K47A}$, we see an increase in LeEIX2 endosome size upon elicitation, but a decrease in LeEIX2 endosome density (Figure 5). Interestingly, LeEIX2 fluorescence at the PM does not change significantly in any of the conditions analyzed (Supplementary Figure S6), except for SlDRP1B. When SlDRP1B is overexpressed, LeEIX2 PM fluorescence is lower than in the control. However, after elicitation, LeEIX2 level at the PM increases, suggesting a possible activation of a recycling pathway.

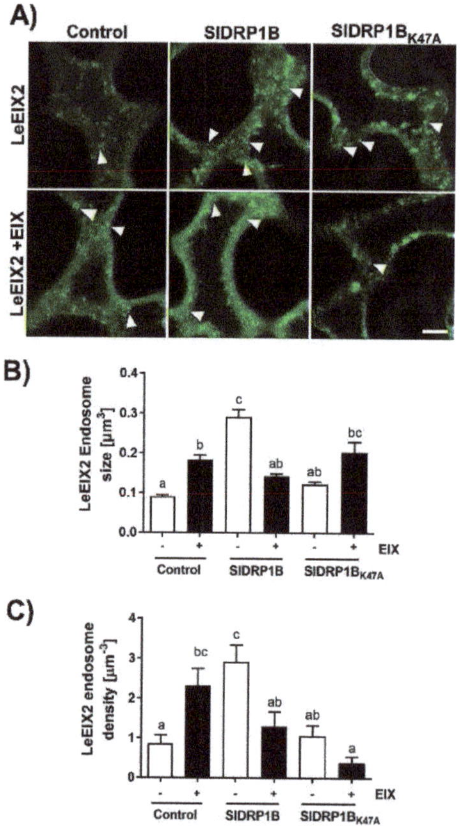

Figure 5. Effect of overexpression of SlDRP1B on LeEIX2 endosomes. *N. benthamiana* leaves transiently expressing LeEIX2-GFP and free mCherry (control), SlDRP1B-mCherry, and SlDRP1B$_{K47A}$, as indicated, were treated with EIX (1 µg mL^{-1} tissue) or water (mock) at the petiole 40 h after transformation and visualized after 15 min. (**A**) Images were acquired using Zeiss LSM 780 confocal microscope (40× objective); arrowheads indicate LeEIX2 endosomes. Scale bar 10 µm. (**B**,**C**) Size and density of LeEIX2 endosomes were quantified using 3D Object Counter (Fiji-ImageJ). LeEIX2 endosomes density was normalized to the control average (mock). Four independent replicates of four images were measured (N = 16). Different letters indicate statistically significant differences between samples in a one-way ANOVA, with Tukey's multiple comparison test; $p < 0.05$.

4. Discussion

PRRs are PM resident proteins. At the PM, they are available to detect extracellular elicitors. PRRs undergo a constant flux of endocytosis; from endosomes, they can be sorted back to PM or towards the vacuole. The fate of the PRR-trafficking directly impacts immune signaling; recycling to the PM enables PRR availability for the perception of newly arriving elicitors/MAMPs, while trafficking towards vacuole leads to PRR degradation, reducing PRR availability at the PM, resulting in signal attenuation [6,8]. Using various chemical modulators that affect different stages of endomembrane trafficking demonstrated that LeEIX2 signals from within early endosomes [23]. Upon EIX elicitation, we see an increase in both size and density of LeEIX2 endosomes (Figure 5) [24,25]. Interestingly, in previous work, we observed that overexpression of Prenylated Rab acceptor1 (PRA1) in tomato results in increased vacuolar degradation of PRRs and attenuated levels of PRRs at the PM, coupled with reduced defense signaling [57]. This and other works have demonstrated that PRR levels at the PM directly impact the amplitude of defense responses mounted by the plant cell. Furthermore, in another work, we found that overexpression of PRA1 leads to disease susceptibility [58], confirming that this attenuation in defense signaling caused by PRR insufficiency has "real" implications for plant health. Highlighting PRR relevance in plant defense, a functional interaction between PRR and pathogen effector intracellular receptors from the type NLR has been shown, indicating that PRRs are also components of effector-triggered immunity [28].

PRRs' signal recognition occurs at the PM, where they form signaling complexes. Both LeEIX2 and FLS2 undergo constant recycling to the PM for further availability [18,59]. We previously reported the involvement of SlDRP2A in LeEIX2 and FLS2 signaling [25]. In that work, we observed significant changes to the size and cellular distribution of endomembrane compartments upon overexpression of SlDRP2A, as well as enhanced defense responses following both EIX and flg-22 elicitation. Here, we observed similar results for SlDRP1B, and enhanced defense also with SlDRP3A, indicating that multiple DRPs can have similar or complementary functions. In *A. thaliana,* it was proven that DRP1 and DRP2 work together in clathrin-dependent endocytosis, plant growth, and flg22-driven immune responses [33,34]. Remarkably, in tomato, SlDRP1B, SlDRP2A, and SlDRP3A are co-expressed, and overexpression of all three DRPs led to an increase of 50–100% in defense signaling output, depending on the parameter being measured (Figures 3 and 4; Supplementary Figure S1) [25]. This suggests that different DRPs from different families could have complementary functions in modulating PRR subcellular distribution. Notably, DRP3s are related to membrane fission and fusion of peroxisomes and mitochondria [60–62], possibly indicating that SlDRP3A's effect on regulating PRR function could be indirect.

The increase in endomembrane compartment density and size, coupled with the increase in defense signaling, reinforces the idea that some of the PRR signals are emanating directly from endomembrane compartments and suggests that the altered endomembrane structures observed upon DRP overexpression do not hinder signal propagation, at least in the short-term. Interestingly, LeEIX2 endosomes are significantly bigger when SlDRP1B is overexpressed in basal conditions (without EIX elicitation). LeEIX2 endosomes return to their original size after EIX treatment (Figure 2A,C). These results suggest that SlDRP1B expression dramatically affects LeEIX2 endosomes, causing an early onset of LeEIX2 endocytosis induction, independently of EIX elicitation. EIX application seems to reverse the SlDRP1B effect partially, perhaps by facilitating the downstream trafficking of the LeEIX2 endosomes, or recycling to the PM, and thus releasing a possible bottleneck created by SlDRP1B overexpression. These results are similar to those previously observed for SlDRP2A [25]. While DRP2A and DRP1B share similar domains, DRP1B lacks the PH and PRM motifs present in DRP2A (Supplementary Figures S1 and S2), suggesting that these motifs may not be required for the interactions of DRPs with PRRs, or for defense signaling enhancement. Both DRPs are localized to endomembrane compartments. DRP2A has a significant PM presence [25], while DRP1B is more specifically localized to a subset of late endosomes (Figure 2). This may suggest that these different DRPs interact with different PRRs in different compartments, or with PRRs that are in a different "signaling state", possibly providing specificity needed to discern between "activated" receptors and "recycling-bound" receptors that are targeted either for degradation or for return to the PM, though further work is needed to elucidate this point.

Here, we show the involvement of DRPs from different families in PRR-mediated defense responses. DRP1B (Figure 3, Supplementary Figure S1) and DRP2A [25] both enhanced PRR-mediated defense signaling of the RLP LeEIX2 and the RLK FLS2, demonstrating the generality of DRP-mediated cellular processes in PRR function. Overexpression of DRPs from different families [25] (Figures 3 and 4) resulted in alterations to cellular signaling and PRR trafficking, likely underlying the observed increases in defense output and adding to the evidence of the roles of cellular trafficking in plant immune signal propagation. The similar function of the different DRPs in defense, coupled with their slightly different subcellular localization and protein domains, suggests that they may have similar functions in different cellular locales, or similar functions within complexes with different proteins (Figure 1C). Future work will tease out the different specificities of different DRPs, and elucidate general mechanisms underlying the functionalities of dynamin-family proteins in different immunity and pathogenicity contexts.

Supplementary Materials: The following supporting information can be downloaded at: https://www.mdpi.com/article/10.3390/membranes12080760/s1, Figure S1: Alignment of tomato DRP1 orthologues; Figure S2. SlDRP1s expression in tomato leaves; Figure S3: SlDRP1B and SlDRP2A comparison; Figure S4: Effect of overexpression of SlDRP1B on FLS2-mediated oxidative burs; Figure S5: Effect of overexpression of SlDRP3A on LeEIX2-mediated defense; Figure S6: LeEIX2 localization ratio at PM; Table S1. Description of interacting proteins of SlDRPs; Table S2. Interactome interaction scores; Table S3: Co-expressed genes list of SlDRP1B, SlDRP1C, and SlDRP2A; Table S4. Gene ontology analysis; Table S5. List of specific real-time PCR primers used for SlDRP1s gene expression analysis.

Author Contributions: Conceptualization, M.L.-M., L.P., M.B., and A.A.; methodology, M.L.-M., L.P., and S.S.; formal analysis, M.L.-M., L.P., and B.V.-S.; investigation, M.L.-M., L.P., S.S., and B.V.-S.; writing—original draft preparation, M.L.-M. and L.P.; writing—review and editing, M.B. and A.A.; funding acquisition, L.P., M.B., and A.A. All authors have read and agreed to the published version of the manuscript.

Funding: This research was partially supported by the Israel Science Foundation administered by the Israel Academy of Science and Humanities (550/18) (A.A.), Maïmonide-Israel 2018 (A.A. and M.B.) and by CONICYT (Chile), PAI/CONVOCATORIA NACIONAL SUBVENCIÓN A LA INSTALACIÓN EN LA ACADEMIA (CONVOCATORIA 2019) PAI 77190027 (L.P.).

Institutional Review Board Statement: Not applicable.

Informed Consent Statement: Not applicable.

Conflicts of Interest: The authors declare no conflict of interest.

References

1. Zipfel, C. Plant pattern-recognition receptors. *Trends Immunol.* **2014**, *35*, 345–351. [CrossRef] [PubMed]
2. Boutrot, F.; Zipfel, C. Function, Discovery, and Exploitation of Plant Pattern Recognition Receptors for Broad-Spectrum Disease Resistance. *Annu. Rev. Phytopathol.* **2017**, *55*, 257–286. [CrossRef] [PubMed]
3. Tor, M.; Lotze, M.T.; Holton, N. Receptor-mediated signalling in plants: Molecular patterns and programmes. *J. Exp. Bot.* **2009**, *60*, 3645–3654. [CrossRef] [PubMed]
4. Gust, A.A.; Felix, G. Receptor like proteins associate with SOBIR1-type of adaptors to form bimolecular receptor kinases. *Curr. Opin. Plant Biol.* **2014**, *21*, 104–111. [CrossRef]
5. Liang, X.; Zhou, J.-M. Receptor-like Cytoplasmic Kinases: Central Players in Plant Receptor Kinase–Mediated Signaling. *Annu. Rev. Plant Biol.* **2018**, *69*, 267–299. [CrossRef] [PubMed]
6. Khaled, S.B.; Postma, J.; Robatzek, S. A Moving View: Subcellular Trafficking Processes in Pattern Recognition Receptor–Triggered Plant Immunity. *Annu. Rev. Phytopathol.* **2015**, *53*, 379–402. [CrossRef] [PubMed]
7. Ekanayake, G.; LaMontagne, E.D.; Heese, A. Never Walk Alone: Clathrin-Coated Vesicle (CCV) Components in Plant Immunity. *Annu. Rev. Phytopathol.* **2019**, *57*, 387–409. [CrossRef] [PubMed]

8. Claus, L.A.N.; Savatin, D.V.; Russinova, E. The crossroads of receptor-mediated signaling and endocytosis in plants. *J. Integr. Plant Biol.* **2018**, *60*, 827–840. [CrossRef] [PubMed]
9. Gu, Y.; Zavaliev, R.; Dong, X. Membrane Trafficking in Plant Immunity. *Mol. Plant* **2017**, *10*, 1026–1034. [CrossRef] [PubMed]
10. Bhandari, D.D.; Brandizzi, F. Plant endomembranes and cytoskeleton: Moving targets in immunity. *Curr. Opin. Plant Biol.* **2020**, *58*, 8–16. [CrossRef] [PubMed]
11. Agaoua, A.; Bendahmane, A.; Moquet, F.; Dogimont, C. Membrane Trafficking Proteins: A New Target to Identify Resistance to Viruses in Plants. *Plants* **2021**, *10*, 2139. [CrossRef] [PubMed]
12. Matern, A.; Böttcher, C.; Eschen-Lippold, L.; Westermann, B.; Smolka, U.; Döll, S.; Trempel, F.; Aryal, B.; Scheel, D.; Geisler, M.; et al. A substrate of the ABC transporter PEN3 stimulates bacterial flagellin (flg22)-induced callose deposition in Arabidopsis thaliana. *J. Biol. Chem.* **2019**, *294*, 6857–6870. [CrossRef] [PubMed]
13. Li, X.; Bao, H.; Wang, Z.; Wang, M.; Fan, B.; Zhu, C.; Chen, Z. Biogenesis and Function of Multivesicular Bodies in Plant Immunity. *Front. Plant Sci.* **2018**, *9*, 979. [CrossRef] [PubMed]
14. Lu, Y.-J.; Schornack, S.; Spallek, T.; Geldner, N.; Chory, J.; Schellmann, S.; Schumacher, K.; Kamoun, S.; Robatzek, S. Patterns of plant subcellular responses to successful oomycete infections reveal differences in host cell reprogramming and endocytic trafficking. *Cell. Microbiol.* **2012**, *14*, 682–697. [CrossRef] [PubMed]
15. Bozkurt, T.O.; Belhaj, K.; Dagdas, Y.F.; Chaparro-Garcia, A.; Wu, C.H.; Cano, L.M.; Kamoun, S. Rerouting of Plant Late Endocytic Trafficking Toward a Pathogen Interface. *Traffic* **2015**, *16*, 204–226. [CrossRef] [PubMed]
16. Chinchilla, D.; Bauer, Z.; Regenass, M.; Boller, T.; Felix, G. The Arabidopsis receptor kinase FLS2 binds flg22 and determines the specificity of flagellin perception. *Plant Cell* **2006**, *18*, 465–476. [CrossRef]
17. Chinchilla, D.; Zipfel, C.; Robatzek, S.; Kemmerling, B.; Nurnberger, T.; Jones, J.D.; Felix, G.; Boller, T. A flagellin-induced complex of the receptor FLS2 and BAK1 initiates plant defence. *Nature* **2007**, *448*, 497–500. [CrossRef]
18. Beck, M.; Zhou, J.; Faulkner, C.; MacLean, D.; Robatzek, S. Spatio-Temporal Cellular Dynamics of the Arabidopsis Flagellin Receptor Reveal Activation Status-Dependent Endosomal Sorting. *Plant Cell* **2012**, *24*, 4205–4219. [CrossRef]
19. Choi, S.-W.; Tamaki, T.; Ebine, K.; Uemura, T.; Ueda, T.; Nakano, A. RABA Members Act in Distinct Steps of Subcellular Trafficking of the FLAGELLIN SENSING2 Receptor. *Plant Cell* **2013**, *25*, 1174–1187. [CrossRef]
20. Mbengue, M.; Bourdais, G.; Gervasi, F.; Beck, M.; Zhou, J.; Spallek, T.; Bartels, S.; Boller, T.; Ueda, T.; Kuhn, H.; et al. Clathrin-dependent endocytosis is required for immunity mediated by pattern recognition receptor kinases. *Proc. Natl. Acad. Sci. USA* **2016**, *113*, 11034–11039. [CrossRef] [PubMed]
21. Smith, J.M.; Salamango, D.J.; Leslie, M.E.; Collins, C.A.; Heese, A. Sensitivity to Flg22 is modulated by ligand-induced degradation and de novo synthesis of the endogenous flagellin-receptor FLAGELLIN-SENSING2. *Plant Physiol.* **2014**, *164*, 440–454. [CrossRef] [PubMed]
22. Ron, M.; Avni, A. The receptor for the fungal elicitor ethylene-inducing xylanase is a member of a resistance-like gene family in tomato. *Plant Cell* **2004**, *16*, 1604–1615. [CrossRef] [PubMed]
23. Sharfman, M.; Bar, M.; Ehrlich, M.; Schuster, S.; Melech-Bonfil, S.; Ezer, R.; Sessa, G.; Avni, A. Endosomal signaling of the tomato leucine-rich repeat receptor-like protein LeEix2. *Plant J.* **2011**, *68*, 413–423. [CrossRef] [PubMed]
24. Leibman-Markus, M.; Pizarro, L.; Schuster, S.; Lin, Z.J.D.; Gershony, O.; Bar, M.; Coaker, G.; Avni, A. The intracellular nucleotide-binding leucine-rich repeat receptor (SlNRC4a) enhances immune signalling elicited by extracellular perception. *Plant Cell Environ.* **2018**, *41*, 2313–2327. [CrossRef] [PubMed]
25. Pizarro, L.; Leibman-Markus, M.; Schuster, S.; Bar, M.; Avni, A. Tomato Dynamin Related Protein 2A Associates With LeEIX2 and Enhances PRR Mediated Defense by Modulating Receptor Trafficking. *Front. Plant Sci.* **2019**, *10*, 936. [CrossRef] [PubMed]
26. Zhou, F.; Emonet, A.; Dénervaud Tendon, V.; Marhavy, P.; Wu, D.; Lahaye, T.; Geldner, N. Co-incidence of Damage and Microbial Patterns Controls Localized Immune Responses in Roots. *Cell* **2020**, *180*, 440–453.e18. [CrossRef] [PubMed]
27. Pizarro, L.; Leibman-Markus, M.; Gupta, R.; Kovetz, N.; Shtein, I.; Bar, E.; Davidovich-Rikanati, R.; Zarivach, R.; Lewinsohn, E.; Avni, A.; et al. A gain of function mutation in SlNRC4a enhances basal immunity resulting in broad-spectrum disease resistance. *Commun. Biol.* **2020**, *3*, 404. [CrossRef] [PubMed]
28. Yuan, M.; Jiang, Z.; Bi, G.; Nomura, K.; Liu, M.; Wang, Y.; Cai, B.; Zhou, J.M.; He, S.Y.; Xin, X.F. Pattern-recognition receptors are required for NLR-mediated plant immunity. *Nature* **2021**, *592*, 105–109. [CrossRef] [PubMed]
29. Ferguson, S.M.; De Camilli, P. Dynamin, a membrane-remodelling GTPase. *Nat. Rev. Mol. Cell Biol.* **2012**, *13*, 75–88. [CrossRef] [PubMed]
30. Antonny, B.; Burd, C.; De Camilli, P.; Chen, E.; Daumke, O.; Faelber, K.; Ford, M.; Frolov, V.A.; Frost, A.; Hinshaw, J.E.; et al. Membrane fission by dynamin: What we know and what we need to know. *EMBO J.* **2016**, *35*, 2270–2284. [CrossRef] [PubMed]
31. Mathur, J.; Shaikh, A.; Mathur, N. Peroxisome Mitochondria Inter-relations in Plants. In *Proteomics of Peroxisomes*; Springer: Singapore, 2018; Volume 89, pp. 417–433. [CrossRef]
32. Rose, R.J. Contribution of Massive Mitochondrial Fusion and Subsequent Fission in the Plant Life Cycle to the Integrity of the Mitochondrion and Its Genome. *Int. J. Mol. Sci.* **2021**, *22*, 5429. [CrossRef] [PubMed]
33. Fujimoto, M.; Arimura, S.I.; Ueda, T.; Takanashi, H.; Hayashi, Y.; Nakano, A.; Tsutsumi, N. Arabidopsis dynamin-related proteins DRP2B and DRP1A participate together in clathrin-coated vesicle formation during endocytosis. *Proc. Natl. Acad. Sci. USA* **2010**, *107*, 6094–6099. [CrossRef] [PubMed]

34. Ekanayake, G.; Smith, J.M.; Jones, K.B.; Stiers, H.M.; Robinson, S.J.; LaMontagne, E.D.; Kostos, P.H.; Cornish, P.V.; Bednarek, S.Y.; Heese, A. DYNAMIN-RELATED PROTEIN DRP1A functions with DRP2B in plant growth, flg22-immune responses, and endocytosis. *Plant Physiol.* **2021**, *185*, 1986–2002. [CrossRef] [PubMed]
35. Smith, J.M.; Leslie, M.E.; Robinson, S.J.; Korasick, D.A.; Zhang, T.; Backues, S.K.; Cornish, P.V.; Koo, A.J.; Bednarek, S.Y.; Heese, A. Loss of Arabidopsis thaliana Dynamin-Related Protein 2B Reveals Separation of Innate Immune Signaling Pathways. *PLoS Pathog.* **2014**, *10*, e1004578. [CrossRef] [PubMed]
36. Chaparro-Garcia, A.; Schwizer, S.; Sklenar, J.; Yoshida, K.; Petre, B.; Bos, J.I.B.; Schornack, S.; Jones, A.M.E.; Bozkurt, T.O.; Kamoun, S. Phytophthora infestans RXLR-WY Effector AVR3a Associates with Dynamin-Related Protein 2 Required for Endocytosis of the Plant Pattern Recognition Receptor FLS2. *PLoS ONE* **2015**, *10*, e0137071. [CrossRef]
37. Wu, G.; Cui, X.; Chen, H.; Renaud, J.B.; Yu, K.; Chen, X.; Wang, A. Dynamin-like proteins of endocytosis in plants are co-opted by potyviruses to enhance virus infection. *J. Virol.* **2018**, *92*, e01320-18. [CrossRef]
38. Tang, D.; Ade, J.; Frye, C.A.; Innes, R.W. A mutation in the GTP hydrolysis site of Arabidopsis dynamin-related protein 1E confers enhanced cell death in response to powdery mildew infection. *Plant J.* **2006**, *47*, 75–84. [CrossRef]
39. Leslie, M.E.; Rogers, S.W.; Heese, A. Increased callose deposition in plants lacking *DYNAMIN-RELATED PROTEIN 2B* is dependent upon *POWDERY MILDEW RESISTANT 4*. *Plant Signal. Behav.* **2016**, *11*, e1244594. [CrossRef]
40. Fujimoto, M.; Arimura, S.I.; Nakazono, M.; Tsutsumi, N. Arabidopsis dynamin-related protein DRP2B is co-localized with DRP1A on the leading edge of the forming cell plate. *Plant Cell Rep.* **2008**, *27*, 1581–1586. [CrossRef] [PubMed]
41. Finn, R.D.; Clements, J.; Eddy, S.R. HMMER web server: Interactive sequence similarity searching. *Nucleic Acids Res.* **2011**, *39*, W29–W37. [CrossRef] [PubMed]
42. Hruz, T.; Laule, O.; Szabo, G.; Wessendorp, F.; Bleuler, S.; Oertle, L.; Widmayer, P.; Gruissem, W.; Zimmermann, P. Genevestigator v3: A reference expression database for the meta-analysis of transcriptomes. *Adv. Bioinform.* **2008**, *2008*, 420747. [CrossRef] [PubMed]
43. Yue, J.; Xu, W.; Ban, R.; Huang, S.; Miao, M.; Tang, X.; Liu, G.; Liu, Y. PTIR: Predicted Tomato Interactome Resource. *Sci. Rep.* **2016**, *6*, 25047. [CrossRef]
44. Mi, H.; Muruganujan, A.; Casagrande, J.T.; Thomas, P.D. Large-scale gene function analysis with the PANTHER classification system. *Nat. Protoc.* **2013**, *8*, 1551–1566. [CrossRef] [PubMed]
45. Gupta, R.; Leibman-Markus, M.; Marash, I.; Kovetz, N.; Rav-David, D.; Elad, Y.; Bar, M. Root zone warming represses foliar diseases in tomato by inducing systemic immunity. *Plant Cell Environ.* **2021**, *44*, 2277–2289. [CrossRef] [PubMed]
46. Mehari, Z.H.; Elad, Y.; Rav-David, D.; Graber, E.R.; Meller Harel, Y. Induced systemic resistance in tomato (Solanum lycopersicum) against Botrytis cinerea by biochar amendment involves jasmonic acid signaling. *Plant Soil* **2015**, *395*, 31–44. [CrossRef]
47. Leibman-Markus, M.; Schuster, S.; Avni, A. LeEIX2 Interactors' Analysis and EIX-Mediated Responses Measurement. *Methods Mol. Biol.* **2017**, *1578*, 167–172. [CrossRef] [PubMed]
48. Schindelin, J.; Arganda-Carreras, I.; Frise, E.; Kaynig, V.; Longair, M.; Pietzsch, T.; Preibisch, S.; Rueden, C.; Saalfeld, S.; Schmid, B.; et al. Fiji: An open-source platform for biological-image analysis. *Nat. Methods* **2012**, *9*, 676–682. [CrossRef] [PubMed]
49. Huang, J.; Fujimoto, M.; Fujiwara, M.; Fukao, Y.; Arimura, S.I.; Tsutsumi, N. Arabidopsis dynamin-related proteins, DRP2A and DRP2B, function coordinately in post-Golgi trafficking. *Biochem. Biophys. Res. Commun.* **2015**, *456*, 238–244. [CrossRef] [PubMed]
50. Konopka, C.A.; Backues, S.K.; Bednareka, S.Y. Dynamics of Arabidopsis dynamin-related protein 1C and a clathrin light chain at the plasma membrane. *Plant Cell* **2008**, *20*, 1363–1380. [CrossRef]
51. Li, R.; Liu, P.; Wan, Y.; Chen, T.; Wang, Q.; Mettbach, U.; Baluska, F.; Samaj, J.; Fang, X.; Lucas, W.J.; et al. A Membrane Microdomain-Associated Protein, Arabidopsis Flot1, Is Involved in a Clathrin-Independent Endocytic Pathway and Is Required for Seedling Development. *Plant Cell* **2012**, *24*, 2105–2122. [CrossRef]
52. Zelazny, E.; Santambrogio, M.; Pourcher, M.; Chambrier, P.; Berne-Dedieu, A.; Fobis-Loisy, I.; Miege, C.; Jaillais, Y.; Gaude, T.; Miège, C.; et al. Mechanisms governing the endosomal membrane recruitment of the core retromer in Arabidopsis. *J. Biol. Chem.* **2013**, *288*, 8815–8825. [CrossRef]
53. Cui, Y.; Zhao, Q.; Gao, C.; Ding, Y.; Zeng, Y.; Ueda, T.; Nakano, A.; Jiang, L. Activation of the Rab7 GTPase by the MON1-CCZ1 Complex Is Essential for PVC-to-Vacuole Trafficking and Plant Growth in Arabidopsis. *Plant Cell* **2014**, *26*, 2080–2097. [CrossRef] [PubMed]
54. Geldner, N.; Dénervaud-Tendon, V.; Hyman, D.L.; Mayer, U.; Stierhof, Y.-D.D.; Chory, J.; Denervaud-Tendon, V.; Hyman, D.L.; Mayer, U.; Stierhof, Y.-D.D.; et al. Rapid, combinatorial analysis of membrane compartments in intact plants with a multicolor marker set. *Plant J.* **2009**, *59*, 169–178. [CrossRef]
55. Asaoka, R.; Uemura, T.; Ito, J.; Fujimoto, M.; Ito, E.; Ueda, T.; Nakano, A. Arabidopsis RABA1 GTPases are involved in transport between the trans-Golgi network and the plasma membrane, and are required for salinity stress tolerance. *Plant J.* **2013**, *73*, 240–249. [CrossRef] [PubMed]
56. Saint-Jore, C.M.; Evins, J.; Batoko, H.; Brandizzi, F.; Moore, I.; Hawes, C. Redistribution of membrane proteins between the Golgi apparatus and endoplasmic reticulum in plants is reversible and not dependent on cytoskeletal networks. *Plant J.* **2002**, *29*, 661–678. [CrossRef] [PubMed]
57. Pizarro, L.; Leibman-Markus, M.; Schuster, S.; Bar, M.; Meltz, T.; Avni, A. Tomato prenylated RAB acceptor protein 1 modulates trafficking and degradation of the pattern recognition receptor LeEIX2, affecting the innate immune response. *Front. Plant Sci.* **2018**, *9*, 257. [CrossRef] [PubMed]

58. Gupta, R.; Pizarro, L.; Leibman-Markus, M.; Marash, I.; Bar, M. Cytokinin response induces immunity and fungal pathogen resistance, and modulates trafficking of the PRR LeEIX2 in tomato. *Mol. Plant Pathol.* **2020**, *21*, 1287–1306. [CrossRef]
59. Spallek, T.; Beck, M.; Ben Khaled, S.; Salomon, S.; Bourdais, G.; Schellmann, S.; Robatzek, S. ESCRT-I Mediates FLS2 Endosomal Sorting and Plant Immunity. *PLoS Genet.* **2013**, *9*, e1004035. [CrossRef] [PubMed]
60. Aung, K.; Hu, J. The Arabidopsis peroxisome division mutant *pdd2* is defective in the *DYNAMIN-RELATED PROTEIN3A* (*DRP3A*) gene. *Plant Signal. Behav.* **2009**, *4*, 542–544. [CrossRef]
61. Aung, K.; Hu, J. Differential roles of *Arabidopsis* Dynamin-Related Proteins DRP3A, DRP3B, and DRP5B in Organelle Division. *J. Integr. Plant Biol.* **2012**, *54*, 921–931. [CrossRef]
62. Wang, F.; Liu, P.; Zhang, Q.; Zhu, J.; Chen, T.; Arimura, S.I.; Tsutsumi, N.; Lin, J. Phosphorylation and ubiquitination of dynamin-related proteins (AtDRP3A/3B) synergically regulate mitochondrial proliferation during mitosis. *Plant J.* **2012**, *72*, 43–56. [CrossRef] [PubMed]

MDPI AG
Grosspeteranlage 5
4052 Basel
Switzerland
Tel.: +41 61 683 77 34

Membranes Editorial Office
E-mail: membranes@mdpi.com
www.mdpi.com/journal/membranes